High-Performance Big Data Computing

High-Performance Big Data Computing

Dhabaleswar K. Panda, Xiaoyi Lu, and Dipti Shankar

The MIT Press
Cambridge, Massachusetts
London, England

The MIT Press would like to thank the anonymous peer reviewers who provided comments on drafts of this book. The generous work of academic experts is essential for establishing the authority and quality of our publications. We acknowledge with gratitude the contributions of these otherwise uncredited readers.

This book was set in Times New Roman by Westchester Publishing Services. Printed and bound in the United States of America.

Library of Congress Cataloging-in-Publication Data

Names: Panda, Dhabaleswar K., author. | Lu, Xiaoyi (Professor of computer science),
    author. | Shankar, Dipti, author.
Title: High-performance big data computing / Dhabaleswar K. Panda, Xiaoyi Lu, and Dipti Shankar.
Description: Cambridge, Massachusetts : The MIT Press, [2022] | Series: Scientific and engineering
    computation | Includes bibliographical references and index.
Identifiers: LCCN 2021038754 | ISBN 9780262046855 (hardcover)
Subjects: LCSH: High performance computing. | Big data.
Classification: LCC QA76.88 .P36 2022 | DDC 005.7—dc23/eng/20211020
LC record available at https://lccn.loc.gov/2021038754

10 9 8 7 6 5 4 3 2 1

Contents

Acknowledgments

We are grateful to our students and collaborators, Adithya Bhat, Rajarshi Biswas, Shashank Gugnani, Yujie Hui, Nusrat Islam, Haseeb Javed, Arjun Kashyap, Kunal Kulkarni, Tianxi Li, Yuke Li, Hao Qi, Md. Wasi-ur-Rahman, Haiyang Shi, and Jie Zhang, for their joint scientific work over the past ten years. We sincerely thank Shashank Gugnani, Haseeb Javed, Arjun Kashyap, Yuke Li, Hao Qi, and Haiyang Shi for their contributions to this collection or for proofreading several versions of this manuscript. Special thanks to Marie Lee, Kate Elwell, and Elizabeth Swayze from The MIT Press for their significant help in publishing this book. In addition, we are indebted to the National Science Foundation (NSF) for multiple grants (e.g., IIS-1447804, OAC-1636846, CCF-1822987, OAC-2007991, OAC-2112606, and CCF-2132049). This book would not have been possible without this support.

Finally, we dedicate this book to our loving families (P. S. Panda, S. M. Panda, Debashree Pati, Abha Panda, Zonghe Lu, Haiying Yu, Sherry Peng, Ada Lu, Alivia Lu, Alan Lu, Dr. R. Shivashankar, G. S. Usharani, and Manju G. Siddappa) for their love and understanding during the long process of writing this book over the past five years.

Dhabaleswar K. (DK) Panda, Xiaoyi Lu, and Dipti Shankar
March 19, 2022

1 Introduction

Human society is in a data explosion era, where data are growing exponentially. This era has been called the "big data era" with the 5Vs characteristics, which are volume, velocity, variety, veracity, and value. To tamp the challenges associated with five big Vs, a new field—high-performance big data computing—is emerging, which aims to bring high-performance computing (HPC), big data processing, and deep learning into a "convergent trajectory." This book aims to provide an in-depth overview of this field and the associated technical challenges, approaches, and solutions. This chapter provides a high-level overview of research topics and challenges in this field and an outline of the overall book.

1.1 Overview

During the last decade, big data has changed the way people understand and harness the power of data, both in the business and research domains. Big data has become one of the most important elements in business analytics. Big data, HPC, and deep learning/ machine learning (DL/ML) are converging to meet large-scale data processing challenges. Running high-performance data analytics workloads in the HPC and cloud computing environments are gaining popularity. According to the recent Hyperion research report (Norton et al., 2020), High-performance data analytics workloads have seen robust growth in the last few years, both in budget allocations as well as organizational focus. This trend is poised to grow over the next decade. The field of big data is being expanded into "huge data" (Wang et al.).

In this context, challenging issues are emerging along the following four major directions: (1) understanding big data characteristics and trends; (2) understanding the interplay among big data, HPC, and deep learning/machine learning; (3) understanding the trends of HPC technologies (processing, networking, and storage) to accelerate big data processing; and (4) understanding the benefits of accelerating big data processing.

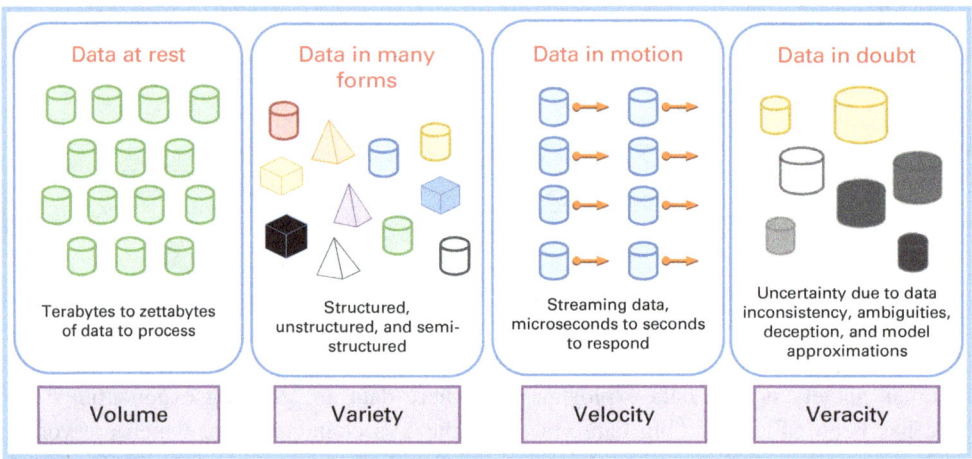

Figure 1.1
Four V characteristics of big data.

1.2   Big Data Characteristics and Trends

Traditionally, big data problems and solutions have been characterized with the 3Vs (volume, velocity, and variety). In recent years, a fourth V (veracity) has been added. These characteristics are illustrated in figure 1.1. *Volume* reflects the amount of data at rest to be processed. *Velocity* refers to data in motion. *Variety* refers to the vast variety in the type of data that must be processed. *Veracity* refers to data in doubt.

Data are being generated in all different forms and shapes from many different businesses, organizations, and entities. Figure 1.2 illustrates the amount of data being generated every minute of the day by various entities in 2020. For example, five hundred hours of videos are being uploaded to YouTube every minute. Around 41.67 million messages are being posted by WhatsApp users every minute. Amazon is shipping 6,659 packages per minute. This leads to a big challenge in designing appropriate systems to process big data analytics.

Efficient processing of big data with the 4Vs has many significant challenges with current-generation technologies, especially with the constantly growing data. Large volumes of data typically result in out-of-core data processing and movement, as well as significant input/output (I/O) bottlenecks. On the other hand, data in motion, popularly known as "big velocity," requires real-time data processing capability. This places high-performance expectations on the underlying computing resources, networks, and storage systems for computation, communication, and I/O. The third V (variety) has resulted in the development of several data processing frameworks by the big data community; for example, Hadoop, Spark, Flink, Storm, Kafka.

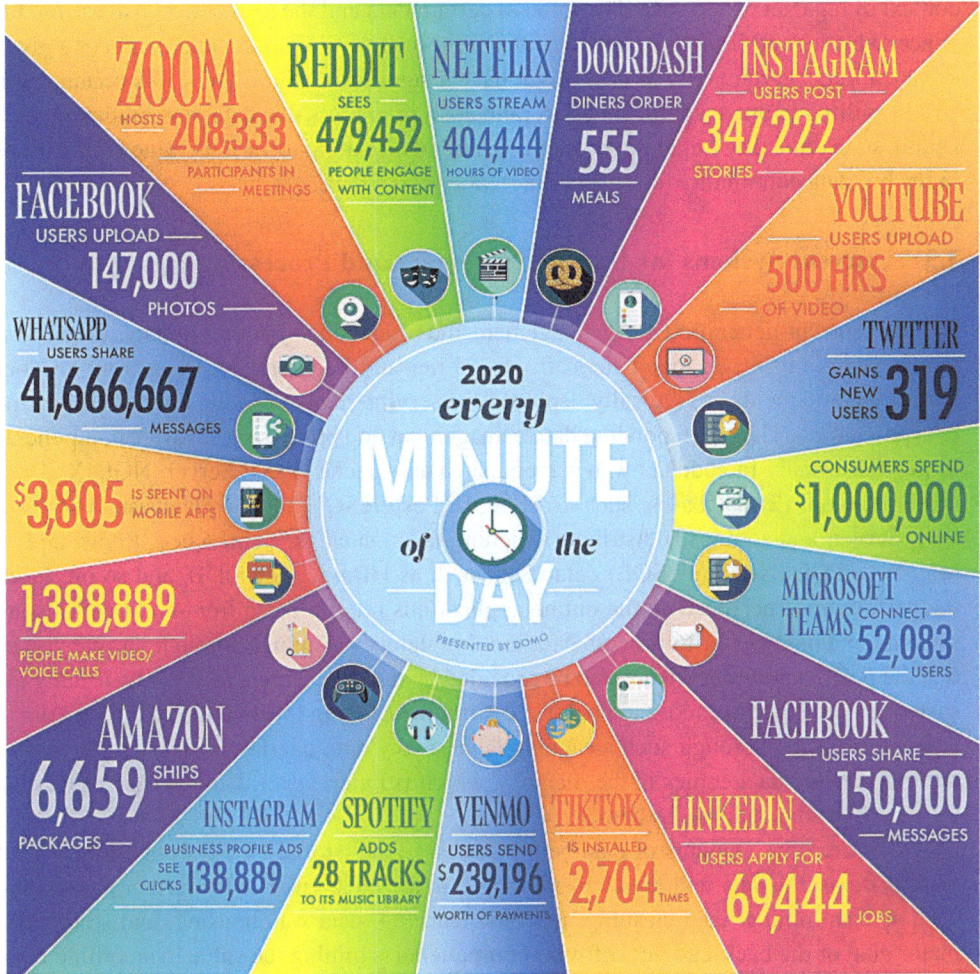

Figure 1.2
Data Never Sleeps 8.0. Source: Courtesy of (Domo).

However, it is unlikely that a single standardized specification or implementation will be converged upon in the near future, making it hard to move forward with highly optimized frameworks since each of them is designed differently. Current optimizations proposed in the literature and community have mostly been done in a case-by-case manner. Therefore, to address the challenges presented by the 4Vs for big data processing, there is a critical need to design next-generation big data software stacks capable of processing data in a high-performance and scalable manner that can optimally leverage the underlying network, computation, and storage capabilities. In this context, a fifth V (value) has been added in the context of big data processing. The value of certain data and the associated business intelligence to be derived from this data can differ from organization to organization. For a given organization, a certain type of data might have a significant value. Thus, this organization will be willing to put a significant effort (and cost) to process this data. For another organization, such criticality might not exist. Such a trend is leading to various value propositions in the big data community to process different types of data.

1.3   Current Systems for Data Management and Processing

Broadly, current-generation data management and processing systems on modern data centers are working in two major tiers: front-end tier and back-end tier. The front-end tier software components are typically deployed for serving data accessing queries and online data processing. The corresponding data management and processing software components in this tier usually include (1) web servers, such as Apache HTTP Server, NGINX, Tomcat, and so on; (2) databases, such as MySQL, PostgreSQL, Oracle Database, Microsoft SQL Server, and so on; (3) distributed caching layer, such as Memcached, Redis, and so on; (4) NoSQL (Not Only SQL) databases, such as HBase, MongoDB, and so on. From the performance perspective, the online applications require these front-end tier software systems in a data center to process data in a low-latency and high-throughput manner, so that the positive user experience can be provided by the data center. This is why usually the system administrators choose to deploy a distributed caching layer on top of the traditional database system. Through such a caching layer, many data queries can be directly served with the cached data copies in the memory, which is much faster than the case of loading the data from the database system.

With increasing amounts of data being processed and stored by the front-end tier software systems, the data will gradually be moved to the back-end tier for further processing, such as data mining, data cleaning, machine learning, data warehousing, and so on. The major goal of the back-end tier software components is mining the value in an offline fashion from the huge amount of data through data analytics and machine learning or deep learning jobs. The corresponding data management and processing software components in this tier usually include (1) distributed storage systems, such as Hadoop Distributed File System (HDFS), Ceph, Swift, and so on; (2) data analytics middleware, such as Hadoop, Spark, Flink, and so on; (3) machine learning or deep learning frameworks, such as TensorFlow, PyTorch, and so on; (4) different kinds of data analytics and machine learning tools or libraries, such as MLlib, Keras, and so on. From the performance perspective, high-throughput and horizontal scalability are the most important pursued properties for these back-end tier software systems in a data center.

In this book, we will discuss the programming models and software architectures of some example systems in both front-end tier and back-end tier. More details can be found in chapters 2 and 3.

1.4   Technological Trends

In the last few years, big data analytics and management software stacks have been significantly enhanced for performance and scalability. Among various factors, hardware evolution is one of the key factors driving the evolution of big data analytics and management systems. The last few years have witnessed a rapid increase in the number of processor cores and an equally impressive increase in memory capacity and network bandwidth on modern cluster-based systems in both HPC centers and data centers. This growth has been fueled by the current trends in multi-/many-core architectures, emerging heterogeneous memory technologies (e.g., DRAM, nonvolatile memory [NVM] (Qureshi et al., 2009, Kültürsay et al., 2013), or persistent memory [PMEM], high-bandwidth memory, NVM Express solid state drive [NVMe-SSD]), and high-speed interconnects such as InfiniBand, Omni-Path, RDMA (i.e., remote direct memory access) over converged enhanced Ethernet (RoCE), and so on.

These multi-/many-core architectures, heterogeneous memory, and high-speed interconnects are currently gaining momentum for designing next-generation HPC and cloud computing environments. These novel hardware architectures with higher performance and advanced features open up many opportunities to redesign the big data analytics and management software stacks to achieve unprecedented performance and scalability.

Thus, hardware-conscious or architecture-aware designs for big data analytics and management software stacks have been a fruitful research area. We have seen many exciting research results and promising performance improvements brought by architecture-aware optimizations, from emerging memory technologies (such as NVM/PMEM), high-speed interconnects (such as RDMA-enabled networks) on easing the I/O and communication bottlenecks, to multi-core/many-core architecture-based parallel processing for big data analytics and management software stacks.

In the HPC community, advanced technologies have been widely adopted to solve the challenges of a huge amount of scientific data to be processed and stored. Modern HPC systems and the associated middleware (such as message passing interface, or MPI, burst buffer, and parallel file systems) have been exploiting the advances in HPC technologies (multi-/many-core architectures, RDMA-enabled networking, NVRAMs and, SSDs) during the last decades. However, current-generation out-of-box big data analytics and management software stacks (e.g., Hadoop, Spark, Flink, Memcached) have not fully embraced such technologies. For instance, recent studies (Rahman et al., 2014, Lu et al., 2014, Islam et al., 2016b, Shankar, Lu, Islam, et al., 2016, Y. Wang et al., 2015, Lim et al., 2014, Huang et al., 2014, Arulraj et al., 2015) have shed light on the possible performance improvements for different big data middleware by taking advantage of RDMA over InfiniBand network, byte-addressability, and persistency of NVM. In this book, we will discuss more details about technological trends in modern HPC and data center clusters in chapter 4.

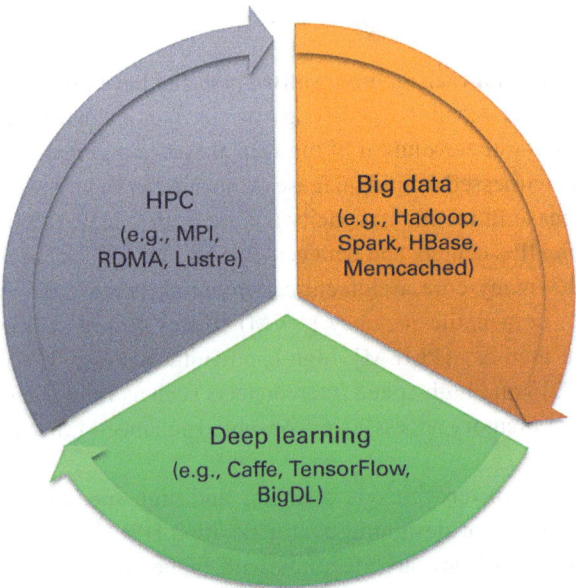

Figure 1.3
Convergence of HPC, big data, and deep learning.

1.5   Convergence in HPC, Big Data, and Deep Learning

In recent years, the community is seeing an important convergence among three big fields—HPC, big data, and deep learning—as shown in figure 1.3. As the HPC environment keeps providing more and more advanced capabilities, the big data community is able to take advantage of those capabilities recently. In the meantime, we also see the deep learning community is able to leverage the technological advances from both HPC and big data fields to form up its two critical pillars: unprecedented computing capabilities and a huge amount of data for model training. This convergence cycle is continuing over the years. We believe that this trend will benefit all three fields, and we will see increasingly better solutions being proposed and developed from these communities to achieve higher performance and scalability for end applications.

The convergence of HPC, big data, and deep learning is becoming the next game-changing business opportunity. This trend has led to many important research and development activities in the fields to bring HPC, big data processing, and deep learning into a convergent trajectory. As demonstrated in figure 1.4, from a user's perspective, we have to answer many critical questions and challenges to make this convergence happen. Some example questions may include the following:

- What are the major bottlenecks in current big data processing and deep learning middleware (e.g., Hadoop, Spark, TensorFlow, PyTorch)?
- Can these bottlenecks be alleviated with new designs by taking advantage of HPC technologies?
- Can RDMA-enabled high-performance interconnects, which are commonly deployed on HPC systems, benefit big data processing and deep learning systems and applications?
- Can HPC Clusters with high-performance storage systems (e.g., PMEM, NVMe-SSD, parallel file system) benefit big data and deep learning applications?
- How much performance benefits can be achieved through enhanced designs or codesigns?
- How to design benchmarks for evaluating the performance of big data and deep learning middleware on HPC clusters?

There are definitely more questions that can be added to this list. To help answer these questions, this book aims to provide an in-depth and systematic overview of the latest research findings in major and emerging topics for "HPC + big data + deep learning over HPC clusters and clouds."

As a starting point of exploring these research opportunities, we can try to deploy and run current-generation big data and deep learning jobs (e.g., Hadoop jobs, Spark jobs, TensorFlow jobs) on existing HPC infrastructures, as shown in figure 1.5. Through workload characterization and performance analysis, we can examine the potential bottlenecks for efficiency and scalability in this execution model and stack.

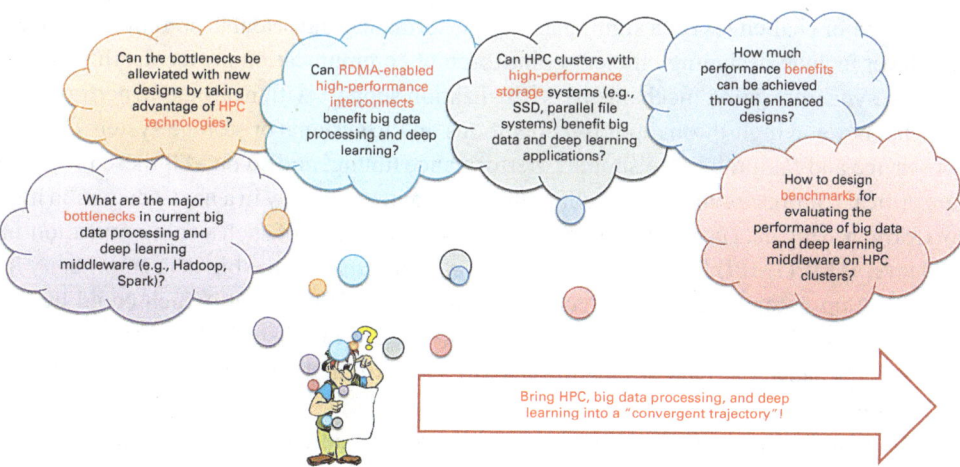

Figure 1.4
Challenges in bringing HPC, big data processing, and deep learning into a convergent trajectory.

Figure 1.5
Can we efficiently run big data and deep learning jobs on existing HPC infrastructure?

Figure 1.6 provides a high-level overview of the associated challenges of achieving high-performance big data computing on HPC and cloud computing systems. The bottom layer in this figure shows different kinds of advanced technologies provided by HPC and cloud computing infrastructures, such as high-speed networking technologies, high-performance and commodity computing system architectures, and advanced storage technologies. The upper layer in this figure shows the technology consumers, such as data-intensive applications, benchmarks, and workloads. In the middle, we have three important layers that help to deliver the near-peak performance from the hardware layer to the application layer. These three layers are the communication and I/O library layer, programming model layer, and big data processing and management middleware layer. Each of these layers needs to be designed efficiently to expose the maximum performance and flexibility to the upper layer components.

The major challenges of designing a high-performance and scalable communication and I/O layer include designing efficient point-to-point communication protocols, thread models and synchronization mechanisms, virtualization support with near-native performance, low-latency and high-throughput I/O operations on file systems or storage systems, quality of service and fault tolerance support, performance tuning, and so on. All these properties are critical features for the desired communication and I/O library in a next-generation high-performance big data computing stack. A successful example of such a communication and I/O layer is MPI (MPI) for the HPC community. Unfortunately, the big data community has not come up with a standardized communication and I/O layer yet, which could be seen as a "pre-MPI" stage (R. Lu et al., 2014, Lu et al., 2011). Historical lessons tell us that high-performance big data computing needs a standard and efficient communication and I/O infrastructure.

Traditionally, big data processing and management middleware, such as Hadoop, Spark, HBase, Memcached, and so on, are designed on top of conventional communication and I/O protocols, such as TCP/IP, remote procedure call (RPC), file system calls, and so on. These protocols are built with operating system-centric concepts and interfaces, such as Sockets,

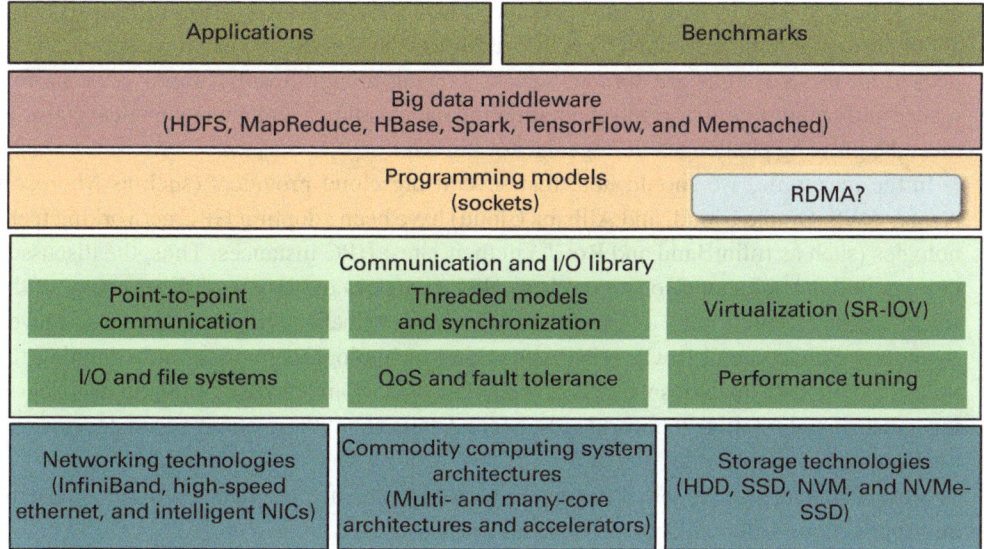

Figure 1.6
Challenges of high-performance big data computing. HDD, hard disk drive; NICs, network interface cards; QoS, quality of service; SR-IOV, single root input/output virtualization.

Portable Operating System Interface (POSIX), etc. These programming models typically have higher overhead due to the context switches and buffer copies between user-space and kernel-space (Rahman et al., 2014, Lu et al., 2014, Islam et al., 2016b, Shankar, Lu, Islam, et al., 2016). With more and more advanced technologies provided by the underlying hardware layer, new programming models and interfaces are becoming available and they can provide pure user-space and zero-copy communication and I/O protocols for the applications. For instance, RDMA is one such promising communication model, and it has been widely used in the HPC community for more than twenty years. In addition, PMEM and NVMe-SSD–based I/O programming models are emerging in the storage community as well, which have been demonstrated with high-performance benefits for data-intensive applications compared to the traditional POSIX-based I/O approaches (Klimovic et al., 2017, Cao et al., 2018, Xia et al., 2017, Islam et al., 2016b). These new programming models are not only significantly improving the performance and scalability of big data processing and management middleware, but they also open up a lot of new codesign opportunities for the upper layer systems and applications.

Rather than exploit these commodity hardware platforms and technologies (e.g., RDMA, NVMe, PMEM) in the research community, many high-tech companies (such as Google and Amazon) make their own proprietary chips, motherboards, networks, and so on. Their networking, I/O, and software stacks are presumably optimized to exploit the unique capabilities provided by their hardware devices. Due to the unavailability of technical details about those proprietary designs, we will not discuss them in this book. However, the major goal of all these designs are similar, which is trying to significantly improve the performance and scalability of current-generation big data analytics and management systems to meet the growing challenges of huge data or big data.

In the meantime, we should note that several big cloud providers (such as Microsoft Azure, AWS, Oracle Cloud, and Alibaba Cloud) have been adopting HPC networking technologies (such as InfiniBand and RoCE) in their latest HPC instances. Thus, the discussed designs in this book can also run on these HPC instances on the cloud. Even many of the social site data centers such as Facebook, Microsoft, Alibaba, and so on, have also moved to adopt InfiniBand and RoCE HPC networking technologies. In addition to running the traditional big data analytics workload, these data centers are currently running deep learning and artificial intelligence workloads. More details about these cloud-based designs will be discussed in chapter 11.

More technical challenges of designing high-performance big data computing systems and applications will be discussed in detail in chapter 5.

1.6   Outline of the Book

Based on the preceding discussions on research challenges of achieving high-performance big data computing, this book has been organized in the following five parts with twelve chapters in total as shown in figure 1.7.

• Chapters 1–4 describe the basic introductory concepts and background knowledge necessary for a good understanding of HPC, big data, deep learning, and so on. Chapter 1 has presented a global view of the field of high-performance big data computing. Chapter 2 describes popular data processing frameworks and programming models being employed for analyzing big data in HPC and data center environments. Chapter 3 presents an overview of the storage system layer being employed for accelerating data-intensive workloads in HPC and data center environments. Chapter 4 outlines the hardware capabilities and trends that are available and emerging to harness data processing in HPC and data center environments.

• Chapter 5 presents a more detailed discussion on the technical challenges and opportunities in accelerating big data computing. In particular, we summarize six concrete challenges from the perspectives of computation, communication, memory/storage, codesign, workload characterization and benchmarking, and system deployment and management.

• Chapter 6 presents a survey on the different benchmarks and workloads being designed and developed to evaluate current and emerging big data middleware systems.

• Chapters 7–9 include a detailed presentation on designs and accelerations for big data computing systems and applications with high-performance networking, computing, and

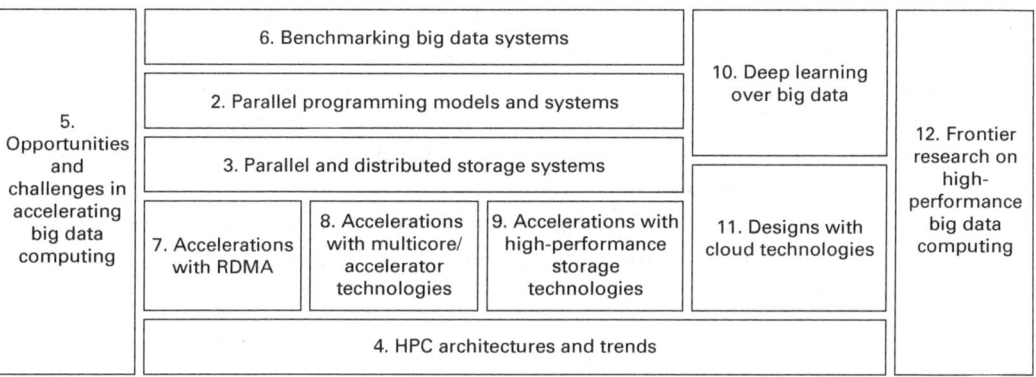

Figure 1.7
Outline of the book.

storage technologies. Chapter 7 presents the challenges and choices involved in designing RDMA-aware communication subsystems and protocols for big data middleware. Chapter 8 presents the designs on accelerating different components of the big data ecosystem on current and emerging heterogeneous computing architectures. Chapter 9 presents different architectural and design choices involved in building a file system or a data store using high-performance storage technologies for big data middleware. Comprehensive surveys of various state-of-the-art designs and accelerations on data processing and storage systems for big data are provided in these chapters.

• Chapters 10–12 provide in-depth discussions in some advanced research topics in the field of high-performance big data computing. Chapter 10 discusses the challenges of designing high-performance deep learning over big data (DLoBD) stacks. Some representative DLoBD stacks are analyzed in this chapter. Chapter 11 presents an overview of HPC cloud technologies and discusses several representative research studies in the community about how to satisfy the demands of efficiently running HPC and big data applications in the cloud. Chapter 12 discusses the on-going and possibly high-reward future research avenues for achieving high-performance big data computing.

1.7   Summary

In this chapter, we presented an overview of big data characteristics and trends. We also classified the current systems for data management and processing into two broad tiers: front-end tier and back-end tier. The corresponding design and optimization challenges of each tier are discussed as well. Then, we further overviewed technological trends in HPC and data center environments. Based on these discussions, we emphasized the opportunities and challenges of the emerging convergence in HPC, big data, and deep learning. We have organized the book into twelve chapters, which can help the readers to get better under-standings of each of these research topics. The major goal of this book is to provide detailed discussions on many important research and development activities in the fields to explain how big data and deep learning middleware/software can be redesigned and accelerated to take advantage of HPC technologies so that HPC, big data, and deep learning workloads can run on the same HPC system.

With our book, we hope that the communities from different fields can explore more collaboration opportunities to work toward bringing HPC, big data processing, and deep learning into a convergent trajectory.

2 Parallel Programming Models and Systems

This chapter presents an overview of popular data processing frameworks being employed for analyzing big data in HPC and data center environments. It discusses and lists examples of different frameworks designed to cater to 5Vs of big data. The chapter also provides application examples to demonstrate the programmability of the different big data processing paradigms.

2.1 Overview

High-performance big data processing has been driving business transformations and scientific research such as medical sciences to grow on a massive scale. To enable this, data processing frameworks that can crunch large amounts of data in a reasonable amount of time to provide valuable insights have evolved. To be more precise, big data frameworks need to be able to compute massive amounts of data from varied resources while managing data that can be imprecisely structured. Big data also needs systems that can process and analyze streams of unbounded or continuous data in a real-time manner. Thus, to cover each of these subtle and unique requirements, various big data processing frameworks have been proposed over recent years. We classify these into five broad categories, as shown in figure 2.1.

Each of the five different types of data processing frameworks in figure 2.1 differ in the way they process data. Batch processing frameworks such as Hadoop MapReduce (Dean and Ghemawat, 2008) and Spark (Zaharia et al., 2016) are widely used offline data analytical frameworks that are designed to process on large immutable datasets over distributed file systems. On the other hand, relational databases capable of query processing on structured data (e.g., MySQL (MySQL, 2020)) and unstructured data (e.g., HBase (Apache Software Foundation, 2021f)) have been providing valuable insights into data over the last decade. Solutions beyond MapReduce-like analytics have been proposed to cater to efficiently processing large-scale graphs, which are typically iterative workloads. With the emerging importance of artificial intelligence (AI)-driven systems, dedicated and optimized

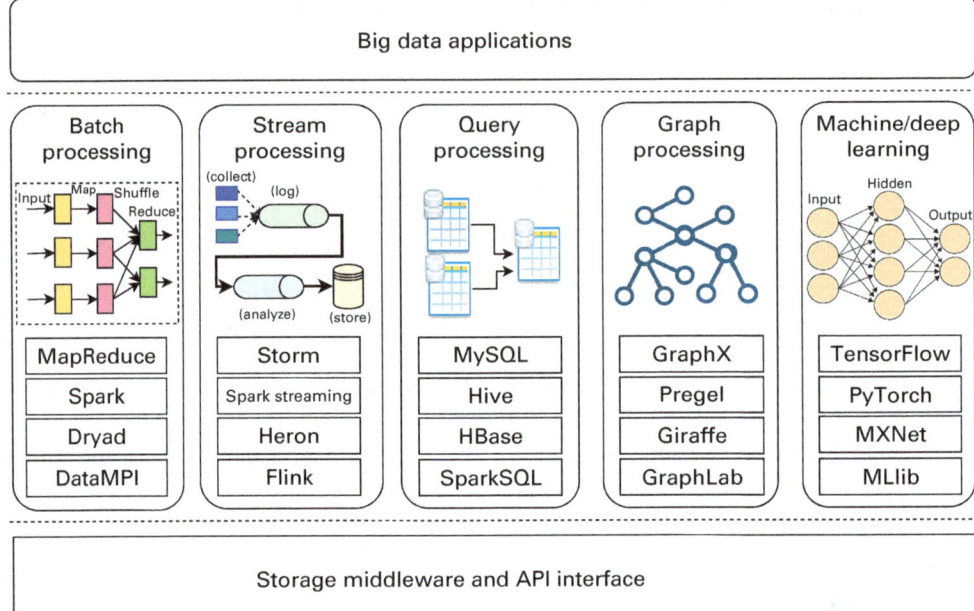

Figure 2.1
Programming models for distributed data processing.

machine learning and deep learning frameworks such as TensorFlow (Google, 2021b), Caffe2 (Markham and Jia, 2017), and BigDL (Wang et al., 2018) have evolved. With the evolution of high-performance accelerators and networks, various hardware vendors such as Intel, NVIDIA, and Mellanox have also been investing in "vertical integration" of their technologies with big data frameworks and making them available as vendor software stacks. These stacks are fine-tuned for high performance on vendor-specific hardware.

In this chapter, we will explore the fundamentals of these five big data processing paradigms and discuss widely used frameworks following each of these different approaches. Along the same lines, we also discuss popular solutions that exploit hardware/ software vertical integration.

2.2 Batch Processing Frameworks

Batch processing has had the longest history among the big data programming model paradigms. It involves computing over a large, static dataset and returning, possibly involving multiple pipeline rounds, the results when the computations are complete. Hadoop MapReduce and Spark are the two most popular batch processing frameworks available today.

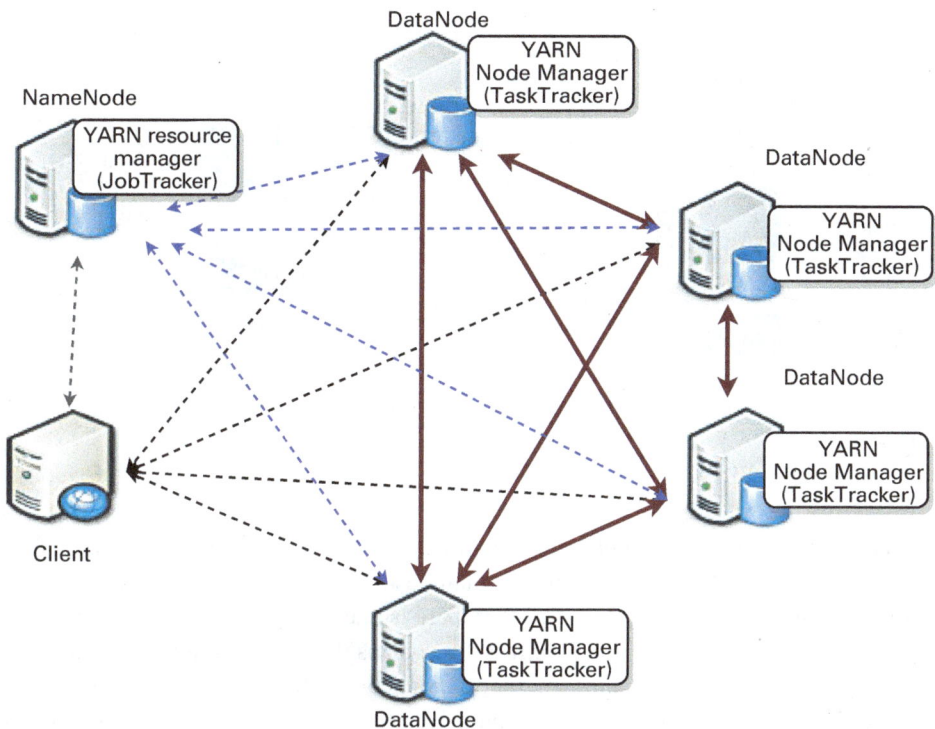

Figure 2.2
Overview of data processing with MapReduce.

2.2.1 Apache Hadoop MapReduce

Hadoop (Apache Software Foundation, 2021e) is a popular open-source implementation of the MapReduce programming model by Dean and Ghemawat (2008). The HDFS by Shvachko et al. (2010) is the primary data source and storage middleware for a Hadoop cluster. A Hadoop cluster consists of two types of nodes: NameNode and DataNode. The NameNode manages the file system namespace and the DataNodes store the actual data. The NameNode has a JobTracker process and all the DataNodes can run one or more TaskTracker processes. These processes, together, act as a master-slave architecture for a MapReduce job. A MapReduce job mainly consists of three basic stages: map, shuffle/merge/sort, and reduce. Figure 2.2 shows the flow of these different stages.

JobTracker is the service within Hadoop that is responsible for distributing the individual tasks of a job onto specific nodes in the cluster. Therefore, a single JobTracker coordinates with several TaskTrackers to enable the successful completion of a MapReduce job. Each TaskTracker launches one or more MapTasks, corresponding to the number of *splits* of

Input
HDFS

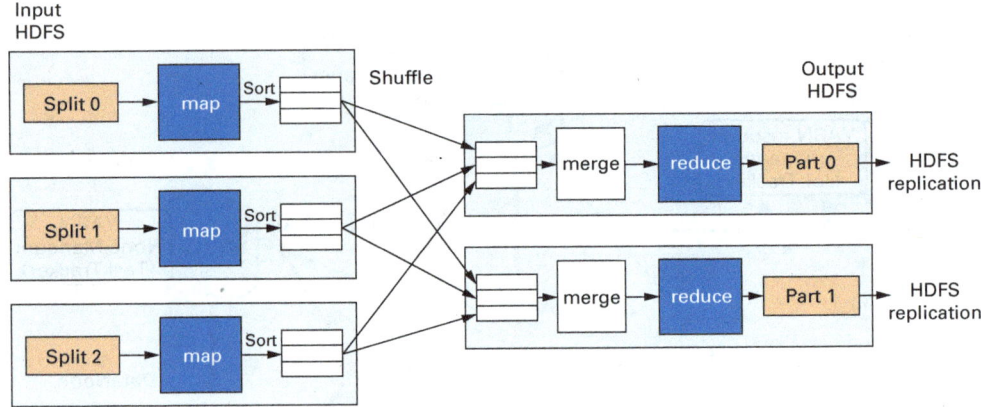

Figure 2.3
Overview of data processing with Hadoop MapReduce.

data as determined by the input datasets in the distributed file system. The map() function converts the original records into intermediate results and stores them on the local file system. Each of these files is sorted such that there is one data partition per ReduceTask. The JobTracker then launches the ReduceTasks as soon as the "map" outputs are available from the MapTasks. TaskTracker can spawn several concurrent MapTasks or ReduceTasks. Each ReduceTask starts fetching the map outputs from the MapTasks that have been completed. This stage is the shuffle/merge period, where the data from various map output locations are sent and received via hypertext transfer protocol requests and responses. While receiving these data from various locations, a merge-sort algorithm is run to combine these data blocks, which will then be presented to the "reduce" operation. Then, each ReduceTask loads and processes the merged outputs using the user-defined reduce() function. The final result is then stored in HDFS. This data flow is illustrated in figure 2.3.

In a typical Hadoop cluster (version 1.x), the master, or JobTracker, is responsible for accepting data processing jobs from clients and scheduling them while performing global resource management. Hadoop 2.x improves the scalability limitation by introducing a separation between per-node and overall resource management. To achieve this, Hadoop 2.x introduced YARN (Yet Another Resource Negotiator) (Vavilapalli et al., 2013), as a step in the direction of designing MapReduce 2.0. YARN decouples the resource management and scheduling capabilities of the JobTracker in Hadoop 1.x, by introducing two new components. The global *Resource Manager* is responsible for assigning resources to all the jobs running on the Hadoop cluster, while the *Node Manager* is similar to the Task-Tracker in the earliest version of Hadoop MapReduce (version 1.x) with one Node Manager running per node. An *Application Master* is started for every new job in the cluster. The Application Master coordinates with the Resource Manager and Node Managers to execute the corresponding tasks.

On the other hand, the recent Hadoop 3.x framework introduces major changes to the storage schemes in HDFS (discussed in chapter 3.2.2.1). Hadoop 3.x also adds minor changes to improve the scalability and reliability of the timeline server that is responsible for maintaining current and historic information regarding applications in YARN.

2.2.1.1 WordCount example with MapReduce

Toward illustrating the MapReduce programming model from the application design perspective, we present a simple WordCount example. The WordCount example counts the number of unique words in a document dataset consisting of text files. The code for the WordCount application is presented in listing 2.1. As we can see, MapReduce requires us to define a Mapper class that defines the map() function to compute on individual HDFS data blocks, and a user-defined reduce() function implemented by extending the Reduce class to merge the results computed by the map tasks.

The dataset stored in HDFS is divided among the map tasks with locality-aware scheduling in mind. As seen from the code, the map() function that enables the map task creates a (word, count) tuple for every word that it encounters in the HDFS data blocks assigned to it. The tuples are shuffled to the reduce tasks, where they are sorted and merged such that they are grouped by key. Once shuffled and sorted, the reduce() function is invoked to counter the number of occurrences of each word. The output containing the words and their corresponding counters are stored in an output file written to HDFS.

Listing 2.1
Hadoop WordCount application.

```java
import java.io.IOException;
import java.util.StringTokenizer;

import org.apache.hadoop.conf.Configuration;
import org.apache.hadoop.fs.Path;
import org.apache.hadoop.io.IntWritable;
import org.apache.hadoop.io.Text;
import org.apache.hadoop.mapreduce.Job;
import org.apache.hadoop.mapreduce.Mapper;
import org.apache.hadoop.mapreduce.Reducer;
import org.apache.hadoop.mapreduce.lib.input.FileInputFormat;
import org.apache.hadoop.mapreduce.lib.output.FileOutputFormat;

public class WordCount {

  public static class TokenizerMapper
       extends Mapper{

    private final static IntWritable one = new IntWritable(1);
    private Text word = new Text();
```

```java
    public void map(Object key, Text value, Context context
                    ) throws IOException, InterruptedException {
      StringTokenizer itr = new StringTokenizer(value.toString());
      while (itr.hasMoreTokens()) {
        word.set(itr.nextToken());
        context.write(word, one);
      }
    }
  }

  public static class IntSumReducer
        extends Reducer {
    private IntWritable result = new IntWritable();

    public void reduce ( Text key
          , Iterable values
                        , Context context
                        ) throws IOException, InterruptedException {
      int sum = 0;
      for (IntWritableval : values) {
        sum += val.get();
      }
      result.set(sum);
      context.write(key, result);
    }
  }

  public static void main(String[] args) throws Exception {
    Configuration conf = new Configuration();
    Job job = Job.getInstance(conf, "word count");
    job.setJarByClass(WordCount.class);
    job.setMapperClass(TokenizerMapper.class);
    job.setCombinerClass(IntSumReducer.class);
    job.setReducerClass(IntSumReducer.class);
    job.setOutputKeyClass(Text.class);
    job.setOutputValueClass(IntWritable.class);
    FileInputFormat.addInputPath(job, new Path(args[0]));
    FileOutputFormat.setOutputPath(job, new Path(args[1]));
    System.exit(job.waitForCompletion(true) ? 0 : 1);
  }
}
```

2.2.2 Apache Spark

Spark (Apache Software Foundation, 2021k) by Zaharia et al. (2016) is an open-source data analytics cluster computing framework originally developed in the AMPLab at the University of California at Berkeley. It was designed for specific types of workloads in cluster computing namely—iterative workloads such as machine learning algorithms that reuse a working set of data across parallel operations and interactive data mining. To optimize for these types of workloads, Spark employs the concept of in-memory cluster computing, where datasets can be cached in memory to reduce their access latency. Spark's architecture is based on the concept of resilient distributed datasets (RDDs). In general, RDDs are a fault-tolerant collection of objects distributed across a set of nodes that can be processed in parallel. These collections are created by reading data from a storage middleware or are obtained as the output of another Spark job. They are resilient as they can be reconstructed if any portion of the dataset is lost by recomputing the transformed distributed dataset.

Each Spark application decomposes its work into multiple RDDs and runs a process for each of these processable data units. Thus, a Spark job comprises an independent set of processes coordinated by a SparkContext object, which is created by the user's application, and called the *driver program*. For easy compatibility, the SparkContext has been designed to work either as a standalone or with existing popular cluster managers such as Apache Mesos and Hadoop YARN. In stand-alone mode, the environment consists of one Spark master and several Spark worker processes, as shown in figure 2.4. Spark also utilizes ZooKeeper to enable high availability. Once the cluster manager allocates resources across applications, Spark acquires executors on the worker nodes, which are responsible for running computations and storing data for the application. The application code is then sent to the executors. Finally, SparkContext sends tasks for the executors to run. At the end of the execution, tasks will be executed at the workers and results will be returned to the driver program. As this chapter mainly focuses on network communication aspects of Spark, we only consider the stand-alone cluster manager.

2.2.2.1 Dependencies in Spark

The tasks run by the executor are made up of two types of operations, both of which are supported by the RDDs: an *action*, which performs a computation on a dataset and returns a value to the *driver*, and a *transformation*, which creates a new dataset from an existing dataset. The transformation operation specifies the processing-dependent directed acyclic graph (DAG) among RDDs.

As illustrated in figure 2.5, these dependencies come in two forms: *narrow* dependencies (e.g., map, filter), where each partition of the parent RDD is used by at most one partition of the child RDD and *wide* dependencies (e.g., GroupByKey, join), where multiple child partitions may be dependent on the same partition of the parent RDD. Wide dependencies involve data shuffling across the network. Thus, wide dependencies are communication-intensive and a potential performance bottleneck for most Spark applications.

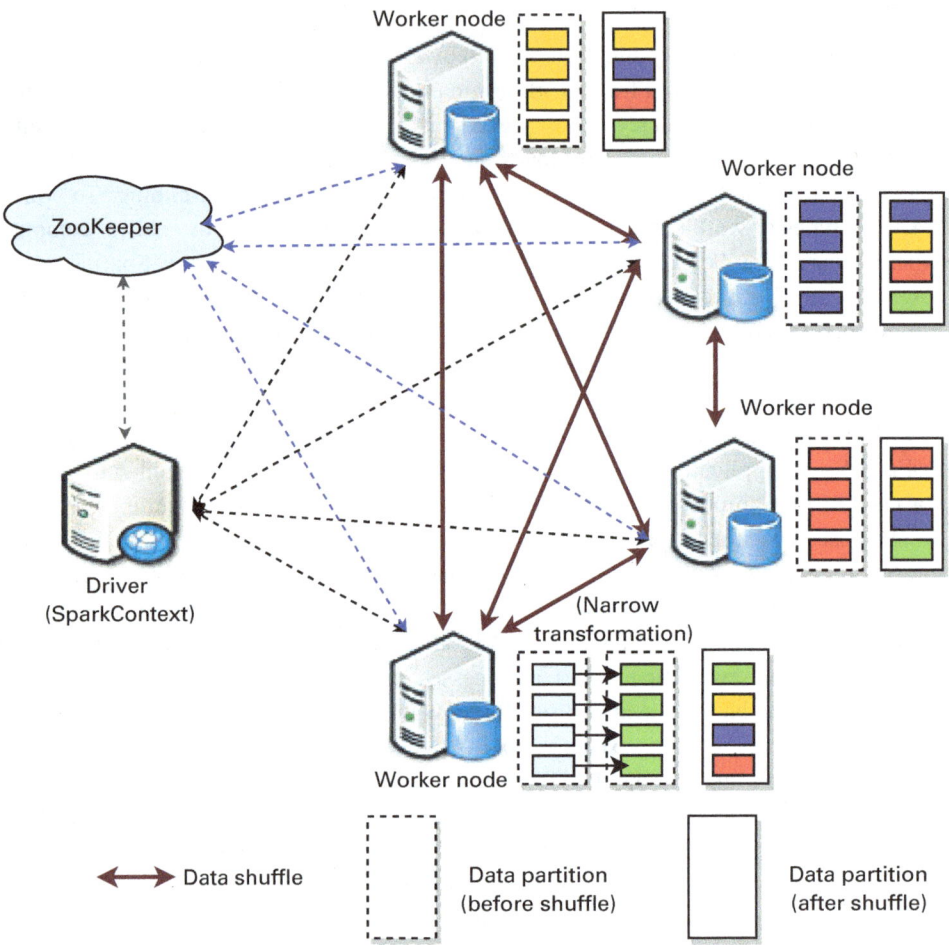

Figure 2.4
Overview of Spark architecture.

Although Spark has similarities to Hadoop, it represents a new cluster computing framework with several subtle differences. Because RDDs can enable in-memory
computing and can be flexibly stored in persistent storage, they enable designing better
pipelines for iterative workloads that, unlike Hadoop MapReduce, do not need to read input
data from an external storage system on every iteration.

2.2.2.2 WordCount example with Spark

Toward illustrating the MapReduce-like programming model of Spark, we present a simple
WordCount example presented with Hadoop MapReduce. The code for the application is
presented in listing 2.2.

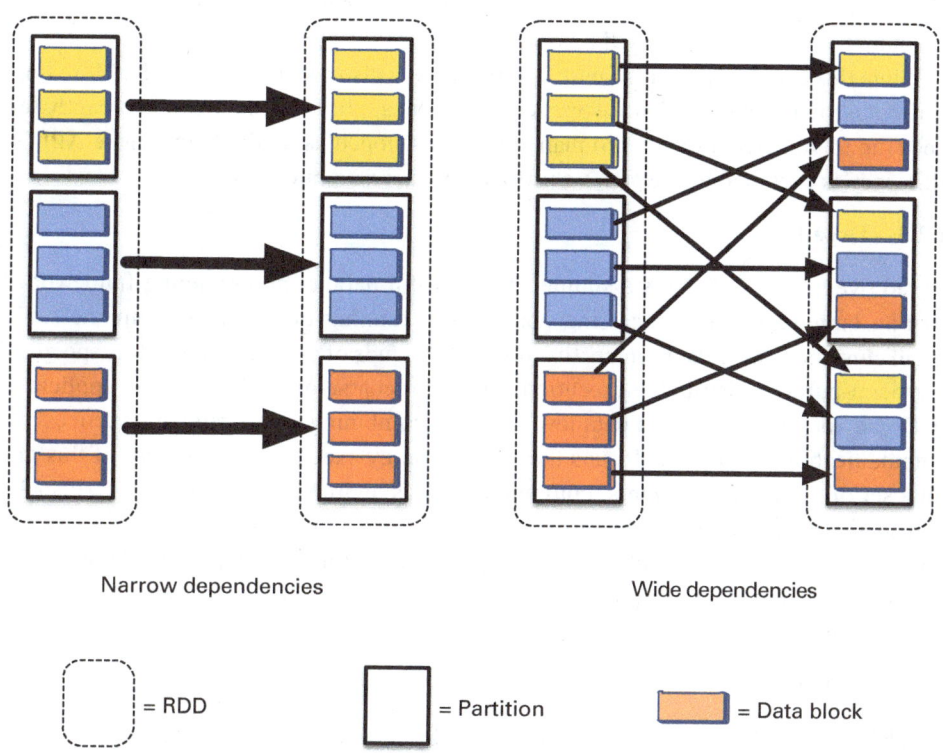

Narrow dependencies   Wide dependencies

☐ = RDD   ☐ = Partition   ☐ = Data block

Figure 2.5
Spark dependencies.

Listing 2.2
Spark WordCount application.

```
JavaRDD<String> textFile = sc.textFile("hdfs:///user/wordfiles");
JavaPairRDD<String, Integer> counts = textFile
    .flatMap(s -> Arrays.asList(s.split(" ")).iterator())
    .mapToPair(word -> new Tuple2<>(word, 1))
    .reduceByKey((a, b) -> a + b);
counts.saveAsTextFile("hdfs://...");
```

As seen from the code, unlike Hadoop MapReduce, Spark requires only three lines of code to implement the same application. An input RDD is defined to convert input text files into a list of words, and a narrow map transformation is applied to the input RDD to define (word, count) tuples with count as one. A wide transformation reduceByKey is applied on the map-transformed RDD to count the number of occurrences per word. As in Hadoop, the output is written into an HDFS-resident text file.

While this example presents Spark's Scala APIs, it also provides Java APIs. More importantly, it is vital to observe that the Spark WordCount application requires significantly fewer lines (up to 90 percent reduction) than the one developed using the MapReduce API. This efficiency is attributable to Spark's ability to abstract data into RDDs.

2.2.3 Dryad

Dyrad (Microsoft, 2021b) is a simple and powerful parallel programming framework for writing large-scale data processing applications running on multicore computers, from small clusters to large data centers that house thousands of compute nodes. It is specifically a general-purpose distributed execution engine for coarse-grain data parallel applications. A Dryad application forms a dataflow graph by combining computational "vertices" with communication "channels." It runs the application by executing the vertices of this graph on a set of available computers, communicating as appropriate via files, TCP/IP sockets, and shared memory. Multiple vertices are scheduled on multiple cores or across multiple computers (or nodes) to enable concurrency. Toward optimally using available resources, a Dyrad application can discover the size and placement of data during runtime and modify the dataflow graph as the computation progresses.

2.2.4 DataMPI

DataMPI Team (2021) is an efficient and flexible communication library, which provides a key-value pair-based communication interface to extend MPI leverage to designed parallel scientific applications on HPC clusters to processing big data (Lu, Fan, et al., 2014). It attempts to bridge the two diverse fields of HPC and big data, which both require massive computational power for processing data. By utilizing the efficient communication technologies that leverage high-performance networking technologies, DataMPI can speed up the emerging data-intensive computing applications. DataMPI currently supports multiple modes for various big data applications, such as MapReduce, streaming, and iterative workloads, to enable single program, multiple data applications.

2.2.5 Vertically Integrated Solutions

As discussed in sections 2.2.1 and 2.2.2, MapReduce-based frameworks have a network-intensive shuffle phase that distributes the outputs of the map phase with the reduce phase. With the evolution of high-performance networks (discussed in depth in chapter 4) that enable low-latency data transfers by software giants such as Mellanox (now a part of NVIDIA (Mellanox, 2021)) have offered vendor software stacks such as Unstructured Data Accelerator (UDA) for optimizing the data shuffling across the network for the Hadoop MapReduce framework, based on Hadoop-A (Wang et al., 2011). Mellanox UDA (Mellanox, 2011) is a software plug-in that speeds up Hadoop networks and enhances the scaling of Hadoop clusters running data analytics–heavy applications. Hadoop clusters based on

Mellanox InfiniBand and ten-gigabit Ethernet RoCE adapter cards (discussed in depth in chapter 4.4) can now transfer data between servers using a novel data-moving protocol that combines RDMA with an optimized merge-sort algorithm. Along similar lines, Intel also released its HPC distribution of Apache Hadoop for optimized performance over parallel file systems such as Lustre (discussed in chapter 3.2.1).

For the more recent and widely used framework today, Apache Spark, Mellanox has released a vertically integrated solution known as Spark-RDMA (Mellanox, 2018) plug-in that accelerates the data fetching over the networking using its RDMA/RoCE technology to improve overall job execution time.

2.3 Stream Processing Frameworks

Streaming data processing is the processing paradigm in which applications will continuously process in-motion data. In other words, while the data are generated as streams, like a series of events, applications will receive and process them in a real-time fashion. There are many streaming data processing applications in modern data center environments, such as processing on financial trades, sensor events, user activities on websites, and so on. Efficiently processing on streaming data is a nontrivial task. Thus, many streaming data processing systems are proposed in the community. This section describes some popular examples of stream processing systems.

2.3.1 Apache Storm

Apache Storm (Apache Software Foundation, 2021l) is one of the first frameworks to address the problem of real-time big data processing. It is a distributed, fault-tolerant data processing system. Each store cluster consists of Nimbus node(s) and Supervisor nodes(s). The architecture of a Storm pipeline is shown in figure 2.6.

As shown in figure 2.6, each job submitted to Storm is described as a *topology*, representing various stream sources called *spouts* and transformations called *bolts* that are applied on the stream (Toshniwal et al., 2014). A user submits a topology to Nimbus which then allocates a subset of the topology to each Supervisor node in the cluster. Storm achieves fault tolerance using ZooKeeper. Both Nimbus and Supervisors in a Storm topology are stateless, so to coordinate between them and to handle failures, ZooKeeper is used. Every node in the cluster writes its state to ZooKeeper. In the case of a node failure, it can be restarted by obtaining its state from ZooKeeper and resuming its role in the topology. Storm offers at least once message delivery semantics, which implies that messages in the topology may be delivered more than once and it is up to the application to handle duplicates.

2.3.2 Apache Spark Streaming

Apache Spark Streaming (Apache Software Foundation, 2021t) is a stream processing layer on top of the Spark data processing engine. The basic unit of processing in Spark Streaming is the Discretized Stream (DStream) (Zaharia et al., 2013), which is an abstraction for

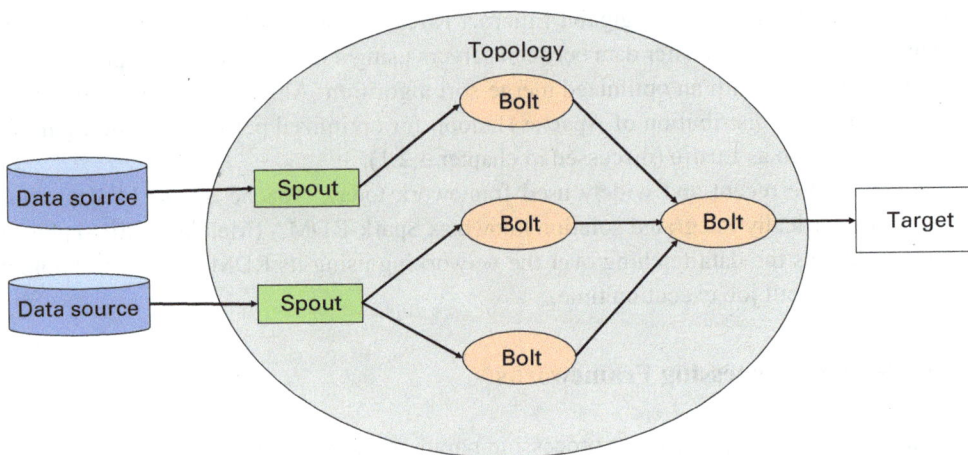

Figure 2.6
Overview of streaming processing with Apache Storm.

a collection of RDDs to be processed together. Unlike Storm, however, Spark follows a microbatch policy for stream processing whereby streaming data coming into the systems are first collected in small batches called DStreams before being processed by the system. This microbatch policy inherently adds significant latency to the data pipeline because, even though messages arrive in the system in real time, they are not processed until the batch duration is reached. However, Spark Streaming offers exactly one delivery semantics; therefore, the application does not have to worry about duplicates. Also having the same underlying core Spark engine means that streaming logic can easily be incorporated with other Spark modules such as batch, graph, or machine learning.

2.3.3 Apache Heron

Apache Heron (Apache Software Foundation, 2021g) is a streaming framework used in production at Twitter (Kulkarni et al., 2015) and is heavily inspired by Storm. Heron jobs, called topologies, also operate on a DAG model consisting of sprouts and bolts. It offers full compatibility with Storm while claiming to be two to five times more efficient. Heron executes spouts and bolts in isolation for the purpose of debuggability. It applies strict restrictions on resource consumption such that resources used by a topology should not exceed the amount available during execution. If there is an attempt to consume more resources, the container will be throttled, leading to a slowdown of the topology. When a topology slows down due to some slow containers, a back pressure mechanism is used to allow the topology to readjust its speed accordingly to minimize data loss. A client writes topologies using Heron API that are sent to a scheduler. The scheduler provides the required number of resources and spawns containers on different nodes. The master container handles the topology and sends its location based on its host-port pair to an ephemeral ZooKeeper that handles the coordination among other containers. All slave containers run stream managers, metric managers, and instances of spouts/bolts. The stream manager controls the flow of data among the whole topology. A Heron instance on these containers runs the processing logic of spouts or bolts. The gateway thread in the instances communicates with the stream manager to send and receive messages. It passes them on to the task-execution thread that applies the processing logic.

2.3.4   Apache Flink

Apache Flink (Apache Software Foundation, 2021d) provides a uniform architecture for both batch and real-time data processing by treating them both as streams (Carbone et al., 2015). The only difference being a batch processing job would be a bounded data stream, whereas a real-time data processing job would be an unbounded data stream. Figure 2.7 shows the architecture of Apache Flink.

Flink has an underlying layer that provides optimizations for various join, shuffle, and partitioning operations that make for faster data processing. Flink, unlike Spark, also supports iterations in the data pipeline natively, meaning that data produced from an iteration can just be fed back to the pipeline for subsequent iterations. This feature makes it highly suitable for iterative and machine learning workloads. As shown in figure 2.7, a program written in Flink is submitted to the JobManager, which acts as master of the entire Flink cluster. The JobManager decomposes the job into smaller components and schedules them on various TaskManagers in the cluster. These TaskManagers are responsible for executing the task and sending the results back to the JobManager while communicating with other TaskManagers in the process.

2.4   Query Processing Frameworks

Support for immoderate querying of large volumes of data is becoming increasingly important in several software domains. Query processing frameworks represent a subclass of big data frameworks geared toward providing data analysis capabilities on huge amounts of both structured and unstructured data. Their primary purpose is to abstract away the finicky details of data storage and retrieval and expose a simplistic and expressive query language to analyze data. While batch processing places more importance on throughput and data volume, query processing places emphasis on both latency and throughput. Conventional database systems were designed for structured data and exposed a well-defined query language to interact with the database. In contrast, the growing trend is to have systems that can support querying and storage of arbitrary data. Typical frameworks have an architecture with a group of workers—usually distributed across many servers—to handle client queries. Synchronization is handled by a (usually centralized) coordination service. Data storage and distribution depend on the data organization model; examples of which include

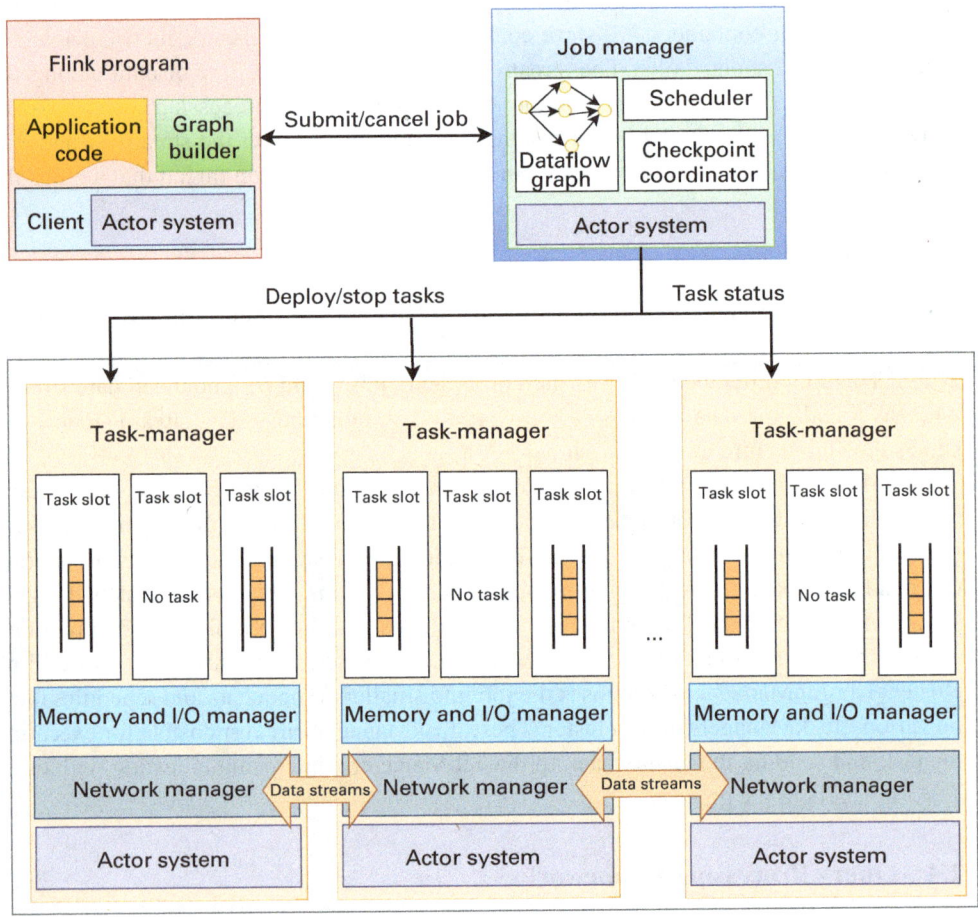

Figure 2.7
Overview of streaming processing with Apache Flink.

distributed tables or simply distributed binary objects. Some popular examples of query processing frameworks are described in the following sections.

2.4.1   MySQL

MySQL (Oracle, 2021) is a popular open-source database solution from Oracle. It exposes the traditional Structured Query Language (SQL) interface to query and retrieve data. MySQL is an implementation of a relational database management system, which presents data to users in tabular form (or relations). High stability, good software support, and high availability are the primary reasons for the popularity of MySQL. However, due to its highly structured relational architecture, MySQL is not able to efficiently handle unstructured data. Furthermore, the scalability of MySQL has often been called into question. For these reasons, users have been increasingly adopting emerging NoSQL and NewSQL frameworks.

2.4.2 Hive

Apache Hive (Apache Software Foundation, 2021h) is a data warehousing solution that provides an SQL-like interface to data stored in Apache Hadoop. Internally, Hive uses a query compiler to convert the query into an execution plan. By analyzing the query, the compiler converts it to a DAG, which is then used to define MapReduce tasks to query data. By storing data in the scalable HDFS and leveraging the MapReduce paradigm, Hive inherits the scalability of the underlying infrastructure. With added features, such as compression, indexing, and multiple storage formats, Hive has quickly become the favored solution for data warehousing.

2.4.3 HBase

Apache HBase (Apache Software Foundation, 2021f) is an open-source distributed nonrelational database based on Google's Bigtable. HBase stores and operates on data stored in memory, relying on HDFS to provide fault tolerance. Data are stored as column-oriented key-value pairs. While HBase itself does not provide a sophisticated query interface, projects such as Apache Phoenix as well as several academic works have implemented an SQL-like layer for HBase. These works typically build primary and secondary indices for each data table to speed up query processing and leverage the MapReduce paradigm to execute the query plan. HBase has become popular for storing vast amounts of data. Until recently, it was used by Facebook to implement its messaging platform.

2.4.4 Spark SQL

Spark SQL (Apache Software Foundation, 2021r) is a module for Apache Spark brings SQL support for analyzing data stored in RDDs. It allows developers to intermix SQL queries with traditional Spark analytics creating powerful pipelines, all within one application. Spark SQL converts queries to internal MapReduce-based code using a code generator and cost-based optimizer, resulting in fast and scalable query processing. Furthermore, it provides full mid-query fault tolerance and can even import existing data from Hive or Parquet tables.

2.4.5 Vertically Integrated Solutions

With the rising computing and memory capabilities of offload-based accelerators such as graphics processor units (GPUs) (discussed in detail in chapter 4), the GPU analytics community and big data community have worked together on solutions such as BlazingSQL (BlazingSQL, 2021). BlazingSQL is a GPU-accelerated SQL engine built entirely on top of NVIDIA's RAPIDS AI (NVIDIA, 2021n). Based on the columnar representation of data adopted by Apache Arrow (Apache Software Foundation, 2021b), the GPU

DataFrame (cuDF) (RAPIDS, 2021) was created, which can hold data in one contiguous block for optimal processing on the GPUs. With these cuDFs, the RAPIDS AI's Complete Unified Device Architecture (CUDA)-based cuDF library is leveraged within BlazingSQL to perform data operations from simple sorting to analytical operations such as aggregations and joins.

2.5   Graph Processing Frameworks

From fraud protection to predicting emerging trends, social graphs are the center to deriving comprehensible relationships in data. Processing these graphs requires massive processing power and customized frameworks. Some popular examples of graph processing frameworks are described in the following sections.

2.5.1   GraphX

GraphX (Gonzalez et al., 2014) is a specialized embedded graph processing framework built on top of Apache Spark. It is designed to transparently leverage the inherent fault-tolerance capabilities in Spark that achieve feature and performance parity with other specialized graph frameworks. Graph operations are implemented using basic dataflow operations such as join, map, and group-by. Existing applications can be implemented on top of GraphX using its API based on the Pregel graph operator.

2.5.2   Pregel

Pregel (Malewicz et al., 2010) is a system for large-scale graph processing. Its underlying model expresses programs as a series of iterations with vertices exchanging messages and mutating graph topology within an iteration. Pregel exposes a simplistic C++ API where applications can define their custom subroutines to compute and mutate. In this manner, arbitrary graph algorithms can be implemented on arbitrary graph representations with ease.

2.5.3   Giraph

Apache Giraph (Apache Software Foundation, 2021q) is an iterative system for highly scalable graph processing. For instance, Facebook is currently using it to analyze social graphs formed by users and their connections (Ching et al., 2015). Giraph originated as the open-source rival to Pregel, the graph processing architecture developed at Google. Both systems are inspired by the bulk synchronous parallel distributed computation model (Valiant, 1990) pioneered by Leslie Valiant. Giraph adds several features beyond the basic Pregel model, including master computation, sharded aggregators, edge-oriented input, and out-of-core computation, to name a few. With a growing community of users worldwide and a fast development cycle, Giraph is an obvious choice for unshackling the potential of structured datasets at a massive scale.

2.5.4 GraphLab

GraphLab (Low et al., 2012, 2014) is a framework for implementing efficient parallel machine learning algorithms. It builds on the MapReduce abstraction by compactly expressing asynchronous iterative algorithms with sparse computational dependencies while ensuring data consistency. It provides a data model to capture both data and computation dependencies, a concurrent access model providing sequential consistency, and an aggregation framework to manage the global state. The GraphLab data model is designed to overcome the shortcomings of existing models (MapReduce, DAG, and systolic) in expressing iterative algorithms. Like Giraph, GraphLab is being developed as an open-source project written in C++. Several case studies have successfully demonstrated that it can outperform other abstractions by orders of magnitude.

2.5.5 GraphBLAS

Orthogonal to iterative processing techniques, several works in the field of graph data processing are focused on leveraging matrix-based linear programming approaches to effectively separate the concerns of hardware/software developers from the application logic. GraphBLAS (Buluç et al., 2017a,b) is one such effort that mathematically defines a set of core matrix-based graph operations that can be used to implement a broad class of graph algorithms in various programming environments. It employs custom representations, from various sizes of integers to bit-oriented storage. Projects such as SuiteSparse (Davis, 2019), IBM GraphBLAS (IBM, 2018), GraphBLAST (C. Yang et al., 2019), and so on, implement the GraphBLAS standard. Similarly, data stores such as Redis also integrated GraphBLAS into their systems through modules such as RedisGraph (Redis, 2021a).

2.5.6 Vertically Integrated Solutions

With the widespread popularity of GPUs (discussed in detail in chapter 4), the GPU analytics community has also invested in graph analytics due to the increasing need for dealing with large-scale graphs in a timely manner. A major advancement in this direction is the vertically integrated library solution based on NVIDIA's RAPIDS AI (NVIDIA, 2021n). The RAPIDS cuGraph library (NVIDIA, 2021b) is a collection of GPU-accelerated graph algorithms. It runs on GPU DataFrames (RAPIDS, 2021) at the Python layer for efficient data passing.

2.6 Machine Learning and Deep Learning Frameworks

With the ubiquity of massive computational power, AI has achieved tremendous momentum in recent years. The applications of AI are wide-ranging from self-driving cars to assisting medical research to identify early stages of cancer. AI applies machine learning, deep learning, and other techniques to solve real-world problems. Machine learning is a subfield of AI that focuses on providing systems the ability to learn and improve without being explicitly programmed. Deep learning is a subset of machine learning that has gotten a lot of attention due to its inference accuracy.

Many machine learning and deep learning frameworks and tools have been proposed in the community, such as Caffe (Jia et al., 2014), Facebook Caffe2 (Markham and Jia, 2017), Microsoft Cognitive Toolkit (Seide and Agarwal, 2016), Intel BigDL (Wang et al., 2018), Google TensorFlow (Abadi, Agarwal, et al., 2016), MLlib for Apache Spark (Meng et al., 2016), PyTorch (Paszke et al., 2019), MXNet (Chen et al., 2015), and many others. In this section, we discuss some of the popular and impactful frameworks that were developed.

2.6.1   Google TensorFlow

Google's TensorFlow (Google, 2021b) is one of the most popular frameworks for performing distributed machine learning and deep learning. It has been gaining a lot of momentum recently in big data, deep learning, and high-performance computing communities.

TensorFlow (Abadi et al., 2016) framework was developed by the Google Brain Team in November 2015. TensorFlow leverages dataflow graphs to do the distributed deep neural network training. The dataflow graph consists of nodes that represent mathematical operations and the edges that signify the multidimensional data arrays (i.e., tensors) communicated among the nodes. The execution model of distributed TensorFlow can be attributed to four distinct components: client, master, a set of workers, and several parameter servers. Figure 2.8 illustrates the interaction among these components. The computational graph is built by a user-written client TensorFlow program. The client then creates a session to the master and sends the graph definitions as a protocol buffer. Afterward, the master delegates and coordinates the execution (after pruning and optimizing) of the subgraphs to a set of distributed worker and parameter server processes. Each of these processes can leverage heterogeneous environments seamlessly. For example, a combination of CPU, GPU, and/or tensor processing unit (TPU) (Google, 2021a) can be leveraged to finish their tasks.

The parameter servers are responsible for updating and storing the model parameters, while the workers send optimization updates of the model to and get the updated model from the parameter servers. The parameter exchanging process (or tensor transmission) is the main communication phase, and the default open-source TensorFlow can support different communication channels such as gRPC, gRPC+Verbs, and gRPC+MPI to handle it, as shown in figure 2.9.

To achieve parallelism, distributed TensorFlow supports both data parallel training and model parallel training (Abadi et al., 2016). In data parallel training, TensorFlow parallelizes the computation of the gradient for a minibatch across minibatch elements. This technique replicates the TensorFlow graph (which does the majority of the computation) across different nodes and each of the replicas operates on a different set of data. The gradients are combined after each iteration of computation by these replicated models. Then,

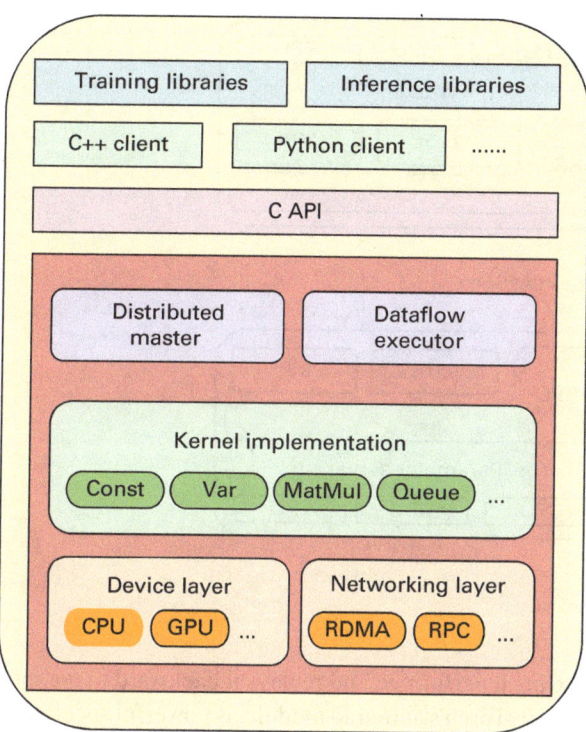

Figure 2.8
Overview of TensorFlow stack.

the parameters update can be applied synchronously as well as asynchronously. Thus this will have the same effect as running the sequential graph computation with the accumulated minibatch size but much faster. On the other hand, in model parallel training, for the same batch of data, different portions of the graph computation are distributed onto different nodes.

2.6.2 Facebook's PyTorch

PyTorch (Paszke et al., 2019) is an open-source machine learning library for Python, based on Torch (maintained by Ronan and Soumith (2021); developed by Collobert et al. (2002)), used for applications such as natural language processing. Developed by Facebook's AI research group, PyTorch employs tensor computations, which like NumPy (NumPy, 2021), are multidimensional arrays that can be computed and processed on GPUs. Unlike most deep learning frameworks, NumPy does not follow a symbolic approach but focuses on extensibility and low overhead through defining purely imperative programs. This technique of differentiation utilized by PyTorch is referred to as automatic differentiation. A

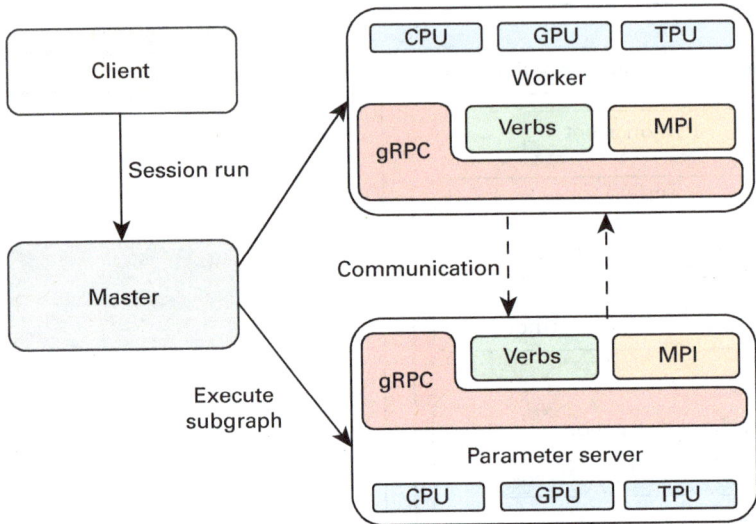

Figure 2.9
Overview of distributed TensorFlow environment.

recorder records what operations have been performed and replays it backward to compute the gradients. This technique, known as PyTorch's autograd module, is powerful specifically when building neural networks to save time on one epoch by calculating differentiation of the parameters at the forward pass itself. This autograd system makes it simple to define computational graphs and take gradients. To wrap the raw and low-level autograd capabilities, PyTorch defines the torch.nn module to help application designers define complex neural networks with ease.

2.6.3   Apache MXNet

Apache MXNet (Chen et al., 2015) is a deep learning framework that offers efficiency and convenience for defining deep neural networks. This open-source framework represents neural networks as a mixture of declarative symbolic expression and imperative tensor computation and provides flexible APIs to construct/optimize custom networks. It is generally employed for implementing data parallelization, that is, any CPU or GPU devices should contain all operations of the computation graph. MXNet uses NDArray as the primary tool for storing and transforming data. For evaluation, MXNet presents binded symbolic expression as a computational graph and transforms its graph to optimize for efficiency. For distributed training, it employs a parameter server-based framework, by employing a key-value store to synchronize data among devices.

2.6.4 Spark MLlib

Spark MLlib (Meng et al., 2016) is Apache Spark's machine learning library that supports common machine learning algorithms, such as regression, classification, clustering, and collaborative filtering over the Spark framework. It also provides utilities and tools to support these algorithms, including data handling, persistence, pipelining, and so on. MLlib supports local vectors and matrices on a single machine and employs Spark RDDs to define distributed matrices. With RDD-based APIs, local vectors and matrices are presented as simple data models that serve as public interfaces. The underlying linear algebra operations are provided by the Breeze linear algebra library.

2.6.5 Vertically Integrated Solutions

Vendor software stacks play a massive role in the machine learning community today, specifically due to the need for accelerators like GPUs for performing machine and deep learning at scale. Among these, the most popular vertically integrated solution is cuML, which is based on the popular NVIDIA RAPIDS AI library collection. cuML (NVIDIA, 2021c) is a set of libraries that incorporate machine learning algorithms and mathematical primitive functions using APIs that are consistent with those used by other RAPIDS projects. These GPU-based implementations can complete massive datasets ten to fifty times faster than their CPU counterparts can. Data scientists, analysts, and software developers use cuML to run typical tabular machine learning (ML) tasks on GPUs without having to learn CUDA programming. cuML's Python API is user-friendly as it closely resembles that of scikit-learn.

With the introduction of Google's accelerators, that is, TPUs (discussed in chapter 4.2.4), TensorFlow can also be considered a vertically integrated solution for machine learning.

2.7 Interactive Big Data Tools

Because big data processing is not always straightforward, there have been several efforts toward integrating easy-to-use human components into the data processing pipeline. This involves using interactive visualization tools, also popularly known as notebooks, to understand and analyze massive datasets. These tools enable scientists and business analysts to effectively find patterns in raw data and employ the results to enabling larger-scale data processing via other big data frameworks. For instance, data scientists perform feature extraction or training on a small subset of the dataset with interactive visualization tools for big data toward choosing and refining their machine learning algorithms for processing their data at scale. In this section, we discuss a few of the plethora of interactive open-source and hosted tools available.

2.7.1 Open-Source Tools

Jupyter (Project Jupyter, 2021) is the most popular notebook utilized by data scientists for interactive computing. It enables data analysts and researchers to integrate software code, computational output, text-based explanations, and graphical representations into a single document. It supports programming languages such as Python, R, and Julia and can be combined with graphing libraries such as Plotly (Plotly, 2021) and data manipulation tools such as Pandas (NumFOCUS, 2021). Similarly, Apache Zeppelin (Apache Software Foundation, 2021m) is a web-based notebook that combines data exploration, visualization, sharing, and collaboration features to SQL, Scala, and so on and can interface with big data frameworks such as Hive and SparkSQL. Additionally, JavaScript libraries such as D3.js (Bostock, 2020) also make big data digestible.

2.7.2 Hosted and Commercial Visualization Tools

With interactive and collaborative computing becoming a staple, several organizations have invested in hosting tools for data analytics. These include Deepnote (Deepnote, 2021), Google Colab (Bisong, 2019), and Databricks Collaborative Notebooks (Databricks, 2021), which support end-to-end life cycle development of machine learning–models. In addition to this, machine learning–specific tools such as TensorBoard (Tensorflow, 2021) are also available, which is designed as a managed service to track and collaborate on TensorFlow experiments.

Orthogonally, specifically for business analytics, popular visualization tools employed are Tableau visualization software (Chabot, 2009, Apache Software Foundation, 2021n), PowerBI (Powell, 2017), and so on.

2.8 Monitoring and Diagnostics Tools

Complementary to big data processing frameworks, it is vital to have monitoring and diagnostics tools that enable debugging and monitoring the performance of these frameworks. While each of the frameworks discussed in this chapter provide some form of built-in diagnostic tools, in this section, we discuss a few of the plethora of popular external tools that can be used with platforms such as Hadoop, and Spark for managing and monitoring the distributed clusters that are found in both literature and large-scale use.

2.8.1 Starfish

Due to its highly configurable nature, frameworks such as Hadoop heavily rely on the users to set the right range of configuration parameters to optimize the overall system performance. Toward systematically supporting Hadoop performance tuning, academic projects such as Starfish (Herodotou et al., 2011) present a platform that combines cost-based database query optimization, static and dynamic program analysis, dynamic data sampling and run-time profiling, and statistical machine learning to provide an in-depth understanding of the performance implications of various cluster configurations. The MapReduce programs can leverage this without any additional customization and minimal overhead of the job execution.

2.8.2 Dynatrace

Dynatrace (Dynatrace, 2021b) is a generic software platform based on AI to monitor and optimize application performance and development, infrastructure, and user experience. It analyzes both present and historical data to provide in-depth cluster output analysis. By analyzing dependencies inside the application stack, it helps pinpoint problematic components. Specifically for big data, Dynatrace automatically identifies the Hadoop (Dynatrace, 2021c) and Spark (Dynatrace, 2021a) components and displays main stats and a timeline map for each.

2.8.3 Sparklint

With performance being key to data processing on frameworks such as Spark, it is vital to monitor and tune Spark jobs for efficiency. Open-source projects, such as Sparklint (Groupon, 2017) support monitoring and tuning by providing advanced metrics and better visualization about the Spark job's application resource utilization to help identify inefficiencies. It analyzes individual Spark workers for overspecified or unbalanced resource usage, inaccurate data partitioning, and suboptimal worker locality using the Spark metrics API and a custom event listener. It can be quickly added to every Spark job and presents data for review through a web interface, as well as making suggestions for common performance issues, with just changes to the cluster configuration.

2.8.4 Datadog

Specifically, for the cloud, software-as-a-service–based application management services such as Datadog (Datadog, 2021) are used to track servers, databases, tools, and resources in real time. It provides capabilities to perform end-to-end monitoring for the Hadoop framework, including Hadoop MapReduce and HDFS (Mouzakitis, 2016), Hive (Gottschling, 2019), Flink (Tai, 2020), and so on.

2.9 Summary

Big data processing spans distinct paradigms, including batch processing, stream processing, query processing, graph processing, and machine/deep learning. These data processing models have led to the development of various kinds of analytics and processing systems, from the earliest batch processing systems such as Google Bigtable to modern general-purpose frameworks such as Apache Spark and AI-centric frameworks such as TensorFlow.

These big data processing systems and frameworks have been becoming the important building blocks of today's large-scale data analytical infrastructure deployed across different fields, from business analytics to medical discoveries.

Usually, these big data processing systems and frameworks have been well optimized for commodity clusters. However, in the recent past, there has been an inclination toward employing modern HPC clusters for deploying these frameworks. These HPC clusters are equipped with advanced hardware architectures, such as high-speed interconnects, GPUs, field programmable gate arrays, NVRAMs, NVMe-SSDs, and so on, that offer computing power to match the growing velocity of big data. This has opened up a lot of new research opportunities and challenges to the data analytics and HPC communities to explore more design alternatives that can help maximize time-to-value for large datasets. This chapter aims to help the readers to get a brief understanding of traditional big data processing systems and their general usage scenarios. With this as the basis, the following chapters present in-depth discussions on the challenges of employing these frameworks in HPC environments, the need for codesigning these frameworks to fully take advantage of modern hardware architectures, and the corresponding approaches proposed to ensure high-performance data analytics.

3 Parallel and Distributed Storage Systems

This chapter presents an overview of the storage middleware layer being employed for accelerating data-intensive workloads in HPC and data center environments. It discusses and lists examples of different types of storage middleware designed to cater to varied big data workloads. The chapter also provides an overview of different application interfaces that are being leveraged to store and manage large volumes of data for big data analytics.

3.1 Overview

The massive growth of data in today's world has necessitated the design of big data middleware (e.g., Hadoop MapReduce and Apache Spark) that is capable of delivering high performance. From an architectural point of view, storing and managing massive volumes of data are major challenges to enabling this. Because big data demands the storage of a massive amount of data ranging to the order of exabytes or petabytes in size, several middleware solutions have been designed to enable storage infrastructure to scale out on multiple servers. In addition to accommodating massive volumes of big data, there is a need for storage middleware that can cater to the veracity and variety of data and the corresponding application requirements. The evolution of storage middleware has not only been driven by the varying application needs, but also by the emergence of newer storage hardware devices. For instance, the transition to NVMe and Peripheral Component Interconnect Express (PCIe) SSDs has contributed to improvements in random and sequential performance compared to traditional hard disk drives (HDDs).

We categorize the different big data storage middleware being employed in the cloud and HPC cluster environments into four broad categories: file storage, block storage, object storage, and memory-centric storage (figure 3.1). File and block are file system–based methods of storage access. Both these methods have been traditionally used to enable applications to access data on block-based storage devices such as SSDs and HDDs. The key difference between the two is how the application is enabled to interface with the data blocks on the storage device; the file system enables access via file-based APIs such as POSIX,

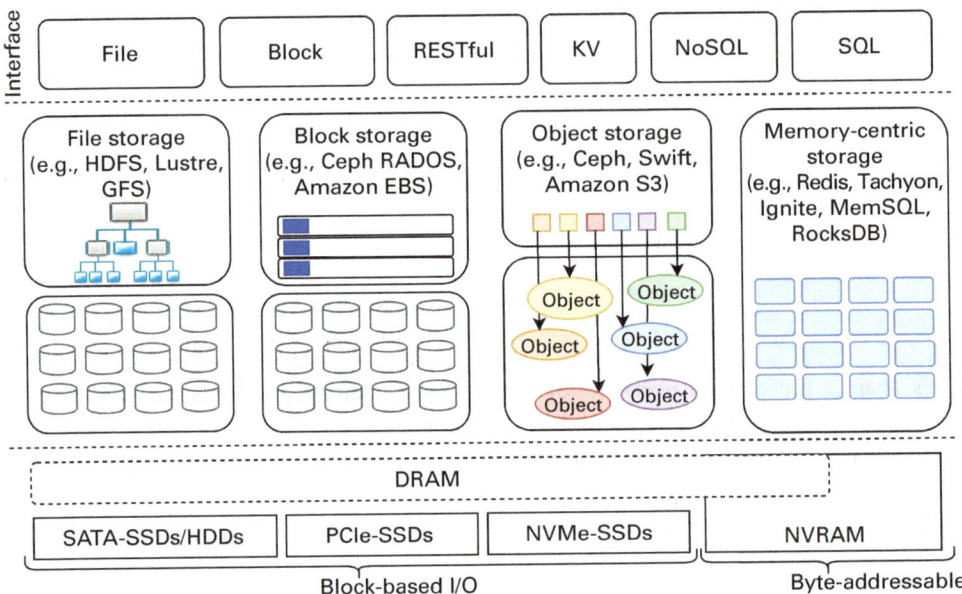

Figure 3.1
Various types of parallel and distributed storage systems. PCIe, Peripheral Component Interconnect Express; RADOS, Reliable, Autonomic Distributed Object Store; SATA, Serial AT Attachment.

that is, Portable Operating System Interface (The Open Group, 2011), and block storage allows direct access to raw data blocks. With the big data boom, it was observed that a majority of the data being produced is "unstructured," immutable, and scales many petabytes across geographically distributed clusters. This required designing newer approaches that were more suitable for dealing with data storage as distinct units, along with customizable metadata for each of these distinct units rather than files or blocks. Scalable solutions such as object storage that can deal with exponential growth in data have been proposed. On the other hand, as DRAM chips are getting cheaper every year, there has been a lot of focus on "in-memory processing" that focuses on leveraging "active memory" as much as possible. To complement this, both scientific research and data center analytics have turned to memory-centric storage systems, including in-memory key-value (KV) stores, in-memory databases, and file systems. Figure 3.1 also illustrates the different application interfaces (e.g., file, block, REST/RESTful services (Fielding, 2000), KV, SQL (ISO/IEC, 2016), NoSQL (NoSQL Database.org, 2021)) that big data workloads employ to access these distinct storage solutions.

In this chapter, we will explore the fundamentals of these four storage middleware systems and discuss real-world examples with corresponding interfaces being used to develop large-scale big data applications.

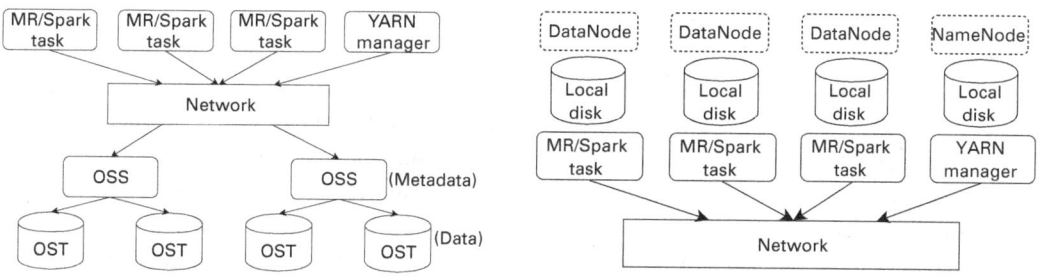

(a) Parallel file system (e.g., Lustre) (b) Distributed file system (e.g., HDFS)

Figure 3.2
File system architecture. MR, MapReduce; OSS, Object Storage Server; OST, Object Storage Target.

3.2 File Storage

Big data demands the storage of a massive amount of data. This makes it necessary to have advanced storage infrastructure: a high-performance file system that can handle the outcome of this big data growth by scaling out on multiple servers.

3.2.1 Parallel File System

Parallel file systems are a category of file systems in which data blocks are striped, in parallel, across multiple storage devices on a separate cluster of storage nodes. It is the primary data storage tier on HPC clusters. A parallel file system enables each running application to access data to and from the storage tier over a common networked connection. The design approach allows all nodes to access the same file simultaneously, thus delivering concurrent read and write capabilities.

Several parallel file systems have been proposed and implemented recently and are in use today. Most of them are composed of two specialized servers: the *data server* and the *metadata server*. A data server consists of a cluster of storage devices, such as object storage devices or network-attached disks. A metadata server is responsible for metadata, which is information such as data sizes, permissions, and locations among multiple data servers.

Figure 3.2a presents a representative architecture of parallel file systems. A key design concept of this architecture is the decoupling of metadata and data paths. Clients talk to metadata server for all namespace operations, such as open(), and communicate with the data server for all file I/O operations, such as read() and write(). Because the data tier is isolated, there is no "locality" when it comes to access the storage layer. In this context, we describe two major parallel file system designs.

3.2.1.1 Lustre

Lustre (OpenSFS and EOFS, 2021), introduced by (Braam and Zahir, 2002), is a high-performance parallel file system that provides a coherent, global POSIX-compliant namespace for very large-scale HPC platforms. It can support hundreds of petabytes of data storage and hundreds of gigabytes per second in simultaneous, aggregate throughput. It does not distribute metadata to keep its management simple. A technique called *data striping* divides each file into portions of a fixed size that are named *stripes*. The stripes are stored onto multiple data servers in a round-robin manner. This design choice enables parallel file systems to deliver optimal throughput to upper-layer applications. The stripe size in different parallel file systems depends on the system's target applications. For instance, Lustre has stripe sizes ranging between 64 KB and 1 MB. Lustre proposes a stripe granularity locking design, which does not allow the same stripe to be accessed by multiple clients concurrently, to ensure data consistency. It employs a version-based recovery mechanism, This tracks dependencies per object and allows changes to be applied only if its preversion matches the current object version. This enables Lustre recovery to continue even if multiple clients fail at the same time the server does.

3.2.1.2 IBM GPFS

IBM General Parallel File System (GPFS) (IBM, 2021), introduced by Schmuck and Haskin (2002), is a cluster file system that can support concurrent access to a single file system (shared-disk system) or set of file systems from multiple nodes (shared-nothing system). Similar to Lustre, this scale-out solution provides a global namespace, a shared file system access across GPFS-based clusters, and simultaneous file access from multiple compute nodes. However, unlike Lustre, it offers high recoverability and data availability through replication.

3.2.1.3 POSIX file system APIs

POSIX (The Open Group, 2011) is a family of standards specified by the IEEE Computer Society of the Institute of Electrical and Electronics Engineers for maintaining compatibility between variants of Linux and other operating systems. It defines the API, that enables us to access the SSDs, HDDs, or any other block-based storage devices. It provides a rich set of standard APIs (such as `open()`, `read()`, `write()`, `ioctl()`) to access files through the storage peripheral device driver. Typical file systems such as IBM GPFS and Lustre employ these file system APIs to access the underlying storage system.

A simple example of copying data from one file to another with the C POSIX APIs to the file system is illustrated in the code in listing 3.1. As seen in the example, a file descriptor is created with the mode of choice, and data can be read to or written from any position within the file. The file descriptor can be repositioned to any offset within the file (Linux, 2021b).

Listing 3.1
POSIX file system API example.

```
int copy_file(char* input_file, char* output_file, long copy_offset)
{
    /* Create input file descriptor */
    input_fd = open (input_file, O_RDONLY);
    if (input_fd == -1) {
        perror ("open");
        return 2;
    }

    /* Create output file descriptor */
    output_fd = open(output_file, O_WRONLY | O_CREAT, 0644);
    if(output_fd == -1){
        perror("open");
        return 3;
    }

    /* Re-position within the output file */
    file_posn = lseek(output_fd, copy_offset, SEEK_CUR);

    /* Copy process */
    while((ret_in = read (input_fd, &buffer, BUF_SIZE)) > 0){
        ret_out = write (output_fd, &buffer, (ssize_t) ret_in);
        if(ret_out != ret_in){
            /* Write error */
            perror("write");
            return 4;
        }
    }

    /* Close file descriptors */
    close (input_fd);
    close (output_fd);

    return 0;
}
```

3.2.2   Distributed File System

A distributed file system typically operates in an environment where the data may be spread across multiples nodes across a high-performance network. They have been designed to provide efficient, reliable access to data using large clusters of commodity servers. Today, they play a vital role in enabling data-intensive applications to run seamlessly for offline data analytics. In addition to offline processing jobs, these distributed storage systems facilitate data warehousing middleware such as Apache Hive (Apache Software Foundation, 2021h) to manage and process large datasets using SQL. We discuss two prominent examples of distributed file systems that are being leveraged today.

3.2.2.1 Hadoop Distributed File System

The HDFS (Shvachko et al., 2010) is a distributed file system that is employed as the primary storage for a majority of big data computing frameworks. Figure 3.2b illustrates the basic architecture of HDFS.

Architecture: An HDFS cluster consists of two types of nodes: NameNode and DataNode. The NameNode is the metadata manager for the distributed storage clusters that are responsible for maintaining the file system namespace and directory tree structures and is backed by a secondary NameNode. Files are divided into blocks of size 64 MB (configurable) and stored or accessed directly from the DataNodes. To enable fault tolerance, HDFS usually replicates each block to three or more DataNodes. In addition to this, recent Hadoop releases (3.x onward) have introduced storage-efficient resilience via erasure coding-aware distributed fault-tolerance (Apache Software Foundation, 2020a).

A client contacts the NameNode while performing any file system operations. For writes, the client receives block IDs and a list of corresponding DataNode addresses from the Name-Node. Each block is split into smaller packets and sent to the first DataNode in the pipeline, and the first DataNode replicates each packet to subsequent DataNodes. HDFS, however, only allows append-only write operations, in other words, existing files are immutable. Any updates require writing new files to replace the existing ones. Similarly, the client reads data from the nearest DataNode. HDFS enables big data computing frameworks such as Hadoop MapReduce and Spark to exploit "data locality" by facilitating moving computation close to the data instead of moving large data to the computation. For instance, a typical Hadoop cluster is deployed such that the task managers (e.g., YARN) scheduling the computation to the processing cores run locally to the DataNodes. Thus, "map" tasks can be scheduled such that the data blocks assigned are located at the DataNode running on the Hadoop cluster node.

HDFS APIs: HDFS APIs are currently not POSIX-compliant as APIs offers only a subset of the behavior of a POSIX file system. However, the LocalFileSystem interface in Hadoop provides access to the underlying file system of the platform it is running on and can be leveraged to run over POSIX-complaint file systems such as Lustre (details discussed in chapters 7 and 9) that can be extremely beneficial in noncommodity clusters. A simple example that copies a file from the local file system to the HDFS cluster is illustrated in listing 3.2. As shown in the example, an input stream using BufferedInputStream is created to read the file from the local file system, and an output stream is created to the file location in HDFS.

Listing 3.2
HDFS API usage example.

```java
public void addFile(String source, String dest, Configuration conf)
  throws IOException
{

  FileSystem fileSystem = FileSystem.get(conf);

  /* Check if the file already exists */
  Path path = new Path(dest);
  if (fileSystem.exists(path)) {
   System.out.println("File " + dest + " already exists");
   return;
  }

  /* Create a new file and write data to it */
  FSDataOutputStream out = fileSystem.create(path);
  InputStream in = new BufferedInputStream(
        new FileInputStream(new File(source)));

  byte[] b = new byte[1024];
  int numBytes = 0;
  while ((numBytes = in.read(b)) > 0) {
   out.write(b, 0, numBytes);
  }

  /* Close all the file descriptors */
  in.close();
  out.close();
  fileSystem.close();
}
```

3.2.2.2   Google File System and Colossus

Google File System (GFS), introduced by Ghemawat et al. (2003), is the predecessor to HDFS that is built on an architecture similar to figure 3.2b. It was implemented to meet Google's rapidly growing data processing needs. It relies on a large number of commodity servers to define a distributed cluster that can span up to thousands of machines. A GFS cluster consists of a single master and multiple chunk servers. Similar to the HDFS Name-Node, the Master manages the namespace, file-to-block mapping, and the current location of the blocks. These are backed by Shadow Master nodes.

Unlike HDFS, GFS can write at a specific offset in an existing file in addition to atomic data appending operations. However, concurrent writes to the same region are not serializable. They can contain data fragments from multiple clients such that the state of a file on GFS while consistent can be nondeterministic. In addition to this, during write operations, a GFS client simultaneously pushes the data to all the replicas across multiple chunk servers to enable efficient network bandwidth utilization. This is a contrast to HDFS's fault-tolerance mechanism, which delegates the replication to the primary server (or DataNode).

In 2010, Google introduced the next generation of GFS called Google Colossus (Fikes, 2010). Unlike GFS, which was designed specifically for batch-based processing, Colossus is built to cater to real-time data. Along with the novel web indexing system, Caffeine (Google, 2010) that replaces MapReduce to build real-time updatable search indexes, Google leverages this new infrastructure to power its web services and cloud storage.

3.3   Object Storage

Object storage systems represent an architecture of data storage that store and manage data in the form of *objects*. In stark contrast to file- and block-based storage systems, object stores encapsulate data, metadata, and a globally unique identifier within the object. They are heavily used by big cloud service providers such as Amazon, Google, and Microsoft to store vast amounts of unstructured data. The growing popularity of simple cloud services has unintentionally increased the importance of object storage systems in recent times.

The primary purpose of these systems is to provide simple application interfaces, decoupling of hardware and software, and data management capabilities such as replication and erasure coding. The semantics of objects vary widely based on implementation. However, most implementations provide access to objects using the representational state transfer. This allows objects to be accessed over hypertext transfer protocol/hypertext transfer protocol secure (HTTP/HTTPS) connections using simple GET/PUT methods. Object stores are great at abstracting out lower layers of storage, including hardware. It is possible to utilize multiple disks across several (possibly geo-distributed) servers. From an application viewpoint, the physical location of data does not matter. Finally, object stores also provide fault-tolerance in the form of replication and erasure coding at the object granularity. These features make object stores a great base for building modern data center applications requiring high scalability and availability.

With their infinitely scalable flat structure, object stores are increasingly being employed for tasks that have traditionally required file systems, such as data analytics, media streaming, archiving, web server, and so on. Hence it has emerged as a low-cost alternative to distributed file systems especially as big data analytical jobs have migrated to the cloud.

Some popular object stores and their architectures are described in the following sections.

3.3.1   Amazon S3

Amazon S3 (Amazon-Web-Services, 2019) is a production object store developed by Amazon. Durability and availability are the key selling points of S3, with Amazon claiming 99.999999999% durability of data. Data are stored as objects inside *buckets* that serve as

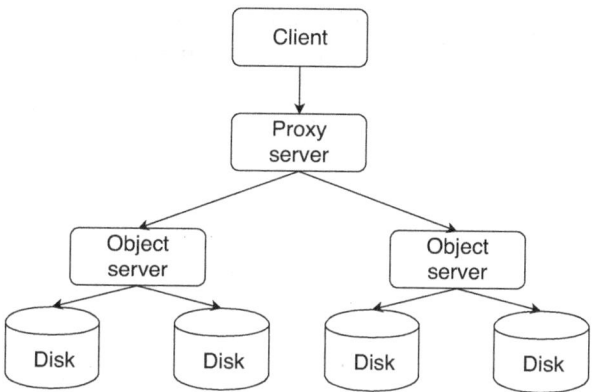

Figure 3.3
OpenStack Swift: architecture overview.

the object store equivalent of directories. The architecture of S3 has not been made publicly available but is believed to be similar to Amazon's DynamoDB.

3.3.2 OpenStack Swift

OpenStack Swift (OpenStack, 2021b) is an open-source object store developed as part of the OpenStack software suite. Swift provides similar API and semantics as S3 does. However, the internal architecture is likely different. Swift stores all data in object servers while using proxy servers as a gateway for all requests. Data are distributed across object servers using a consistent hashing mechanism. Swift provides eventual consistency to end applications as a way to maintain high availability despite the high failure rate in cloud environments. Figure 3.3 shows the overall architecture of Swift.

3.3.3 Ceph

Ceph (Ceph, 2021) is another popular object store that is also a part of the OpenStack software suite. Ceph is architecturally similar to both S3 and Swift. The primary difference between Ceph and other object stores is in its data consistency model and API. Ceph provides strong consistency as opposed to eventual consistency provided by other solutions. Furthermore, Ceph provides several convenient interfaces to access data, including the POSIX file system, block device, and Internet Small Computer Systems Interface (iSCSI).

3.3.4 Object Storage API Example

A client can create, modify, and get objects and metadata by using the object storage APIs, running custom commands on store-specific command-line interfaces (e.g., AWS CLI for S3 (Amazon, 2021e)), and native REST API (Fielding, 2000). To illustrate managing and accessing objects in an object store, we employ Amazon S3 as an example. Listing 3.3 presents a sample application that uses the boto3 (Amazon, 2021c) Python APIs to store (put_object()) and retrieve (get_object()) an object from Amazon S3.

Listing 3.3
Amazon S3 Python API example.

```
import boto3
from botocore.exceptions import ClientError

KEY = os.urandom(32)
BUCKET = 'your-bucket-name'

s3_client = boto3.client('s3')
s3_client.create_bucket(Bucket=BUCKET)

# Put object into object store
s3.put_object(Body='HappyFace.jpg',
    Bucket=BUCKET,
    Key='HappyFace.jpg',
    ServerSideEncryption='AES256',
    StorageClass='STANDARD_IA')

# Getting the object
response = client.get_object(
    Bucket=BUCKET,
    Key='HappyFace.jpg',
    )

print(response)
```

3.4 Block Storage

Block storage is a storage middleware layer typically leveraged in cloud environments. Files are split into evenly sized blocks of data, each with its own address. It is more abstract than the file storage system as it does not provide any additional metadata information to provide more context regarding how or for what the data blocks are used. Block storage is employed to accelerate a large number of enterprise workloads, especially in storage-area network (SAN) arrays that directly access data blocks using the SCSI protocol. For instance, several blocks (e.g., in a SAN system) build a file. A block consists of an address and the SAN application gets the block if it makes an SCSI-Request command to this address. The storage application decides where the data blocks are stored inside the system and on what specific disk or storage medium, how the blocks are combined in the end, and how to access them. Blocks in a SAN do not have metadata that are related to the storage system or application. In other words, blocks are data segments without description, association, and without an owner to the storage solution. Everything is handled and controlled by the SAN software. As a consequence, SAN and block storage are often used for performance-hungry applications such as databases or transactions because the data can be accessed, modified, and saved. Some popular object stores and their architectures are described in the following sections.

3.4.1 Amazon EBS

Amazon EBS (Elastic Block Store) (Amazon, 2021a) is a storage service that provides persistent block storage volumes for Amazon EC2 (Elastic Compute Cloud) instances. EBS transparently handles data storage for applications, giving the illusion of limitless storage. By separating underlying storage media from the user, EBS enables fast scale-up or scale-down. Common use cases for EBS include big data analytics engines, database systems, and stream processing applications.

3.4.2 Cinder

Cinder (OpenStack, 2021a) is a block storage service for OpenStack that is primarily utilized by the nova component. Like EBS, Cinder virtualizes the management of block storage devices, allowing users to consume resources without knowledge of where and on what device data are deployed. The basic resources offered by Cinder include Volumes, Snapshots, and Backups, which fulfill different user requirements. Cinder differs from EBS mostly in terms of resource sharing. Unlike EBS, with Cinder, a block device can only be attached to one instance at a time.

3.4.3 Blizzard

Blizzard (Mickens et al., 2014) is a recent cloud block storage solution proposed by Microsoft Research. Blizzard is POSIX-compliant, which implies that the exposed block storage can be utilized by cloud-oblivious applications. Blizzard uses a novel striping scheme to realize disk parallelism and allows out-of-order write commits to provide low latency while guaranteeing crash consistency. The resulting storage service shows significant improvement over EBS for several SQL workloads.

3.4.4 Block Storage API Example

In block storage, each storage volume acts as an individual hard drive, and data are saved to the storage media in fixed-sized chunks called blocks. Point-in-time copies of a volume referred to as a "snapshot," can be used to store and retrieve data from the block storage. Typically, REST APIs are used to manage block storage. An HTTPS POST method is used to store a block, and a GET method is employed to retrieve a block snapshot. Additionally, direct APIs are also provided for various block storage systems.

Similar to object stores, we illustrate accessing data in a block store using Amazon EBS as an example. Listing 3.4 presents a sample application that uses the `boto3` (Amazon, 2021c) Python APIs to store (put) and retrieve (get) a snapshot block from Amazon EBS.

Listing 3.4
Amazon EBS Python API example.

```python
import boto3
from botocore.exceptions import ClientError

client = boto3.client('ebs')

response = client.put_snapshot_block(
    SnapshotId='string',
    BlockIndex=123,
    BlockData=b'bytes'|file,
    DataLength=123,
    Progress=123,
    Checksum='string',
    ChecksumAlgorithm='SHA256'
)

response = client.get_snapshot_block(
    SnapshotId='string',
    BlockIndex=123,
    BlockToken='string'
)
```

3.5 Memory-Centric Storage

Inspired by the performance and scalability offered by in-memory computing, memory-centric storage or in-memory storage has been a significant step toward alleviating the disk I/O bottleneck. Early memory-centric designs involved combining the power of distributed in-memory storage with the durability of traditional databases and file systems, without requiring the need to redesign existing storage solutions. A distributed in-memory caching solution such as Memcached (Dormando, 2021) is a prominent example of this. Over the years, memory-centric storage systems have been designed from scratch to make the best use of available DRAM together with the disk durability in one system, while meeting the consistency needs of different applications (e.g., Redis (Redis Labs, 2021d), SingleStore [formerly MemSQL (SingleStore Inc., 2021)], and Apache Cassandra (Apache Software Foundation, 2021c)).

3.5.1 Memory-Centric Databases

Traditional SQL databases, which are table-based databases relying on managing datasets based on a predefined structure or schema, have long been used to provide data services for online transactional processing (OLTP) and online analytical processing workloads (e.g., MySQL, PostgreSQL). These reliable and relational databases have evolved to create a generation of systems, known as NoSQL databases, that can handle the massive amounts of unstructured data, with the scalability, reliability, and availability requirements that modern big data applications demand. On the other hand, with the growth of in-memory data processing, memory-centric databases are not bottlenecked by the block-based system's page structures and buffer management, and the memory-centric databases have evolved to directly access and manipulate data in byte-addressable memory. High-performance database management systems such as HyPer (Kemper, 2021) and MemSQL (SingleStore Inc., 2021) have been demonstrating performance on par with NoSQL systems while meeting the atomicity, consistency, isolation, and durability needs of relational database management systems.

These SQL and NoSQL databases are built on complex transactional and analytical processing systems by leveraging the underlying storage devices, and there are several books dedicated to discussing them. Here, we discuss two prominent memory-centric design examples.

3.5.1.1 Cassandra

Cassandra (Apache Software Foundation, 2021c) essentially employs a hybrid key-value store model to define a columnar (i.e., tabular) NoSQL data store. Data are partitioned into rows with tunable consistency, and these rows are organized into tables. Cassandra's storage engine enables defining column families with compound primary keys. The first column in a compound key is used to identify the table partition (i.e., serves as the partition key). All the rows sharing a given partition key are sorted by the remaining components of the primary key to facilitating auto-clustering within a partition. Cassandra employs a sparse column engine to complement its schema less system. Because space is only used by columns present in a row, defining a table with thousands of columns does not cause suboptimal storage otherwise inherent in sparse data.

Cassandra employs a distributed shared-nothing database that runs on a homogeneous cluster, as shown in figure 3.4. It has been designed from the ground up to handle large volumes of data and to facilitate providing high volumes of write and read throughput. With its shared-nothing architecture, a Cassandra cluster has no masters, slaves, or elected leaders. This peer-to-peer and decentralized data management design allows it to be highly available while eliminating the possibility of a single point of failure.

A Cassandra client application example that reads and writes data from the Cassandra cluster is illustrated in listing 3.5. Initially, a connection is established with one (or more) of the Cassandra server nodes that act as a driver. The driver discovers other nodes in the cluster automatically. Once connected, a "keyspace" is created if necessary or an existing

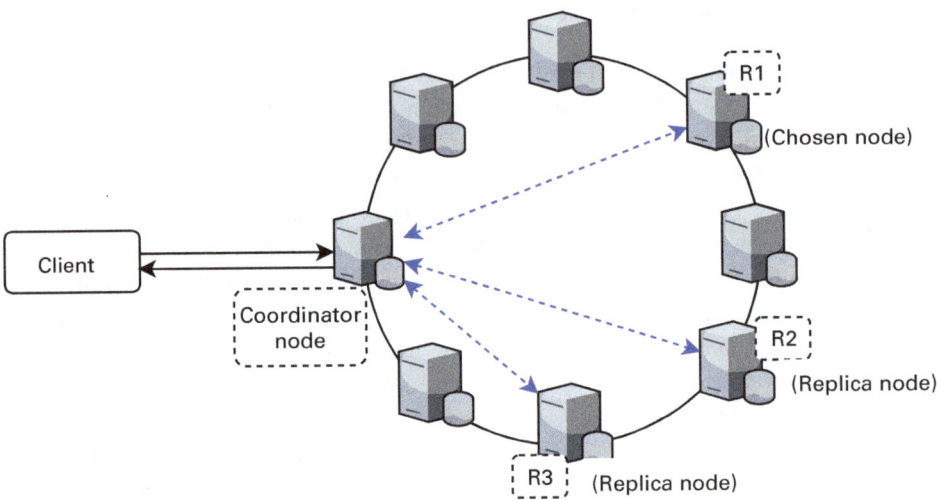

Figure 3.4
Apache Cassandra: architecture overview.

one can be used. Keyspaces serve as an outermost object that determines how data are replicated onto the nodes and consists of column families (i.e., a table), indexes, and so on. A table can be created in the designated keyspace, and data are inserted using an SQL-like INSERT command. Similarly, data can be read from the Cassandra cluster using SQL-type SELECT queries.

Listing 3.5
Cassandra Java client API example.

```java
import com.datastax.driver.core.Cluster;
import com.datastax.driver.core.ResultSet;
import com.datastax.driver.core.Session;

public class Create_Table {
 public static void main(String args[]){
 /* create Cluster and Session objects */
 Cluster cluster = Cluster.builder().addContactPoint("server1").
     build();
 Session session = cluster.connect();

 /* create keyspace with replication factor 1*/
 String ks_query = "CREATE KEYSPACE my_keyspace WITH replication
     " + "= {'class':'SimpleStrategy', 'replication_factor':1}; "
     ;
 session.execute(ks_query);
```

```
/* create table in 'my_keyspace' */
session.execute("USE my_keyspace");
String create_table_query = "CREATE TABLE my_table(id int
    PRIMARY KEY, " + "my_data text );";
session.execute(create_table_query);

/* insert data */
String insert_query = "INSERT INTO id (my_data)" + " VALUES(1, '
    some_data_here');";
session.execute(insert_query);

/* read data from table */
String select_query = "SELECT * from my_table";
ResultSet result = session.execute(query);
System.out.println(result.all());
  }
}
```

3.5.1.2 Apache Ignite

Apache Ignite (Apache Software Foundation, 2021p) is a memory-centric storage system that includes an in-memory data grid, in-memory database capabilities, and support for streaming analytics, real-time processing, and emerging machine learning workloads. While it can be integrated with existing database systems, such as Cassandra, it also provides a durable memory architecture. This configurable durable memory component treats DRAM not just as a caching layer, but also as a comprehensive and functional storage layer. If the durability is not set, then it can function as a distributed in-memory database or an in-memory data grid, depending on what type of interface is required: SQL or NoSQL APIs. If the durable memory is employed, then it can function as a distributed, horizontally scalable database that guarantees full data consistency and is tolerant of any cluster failures. Memory-centric designs like these are enabling utilizing a single storage middleware to cater to different big data online transactional and offline analytical applications.

Ignite supports both key-value cache and SQL for modeling and accessing data. A Java-based Ignite client example that stores and retrieves data from a key-value cache in Ignite is illustrated in listing 3.6. The client connects to the Ignite server cluster via `setAddresses()` at initialization. A pointer to the remote cache can be created with a designated name via `getOrCreateCache`. An Ignite cache can be partitioned, replicated, or local, and the key-to-node mapping is determined via the affinity function (Apache Software Foundation, 2021a). The default scheme employed is rendezvous hashing (Thaler and Ravishankar, 1996). More details on the SQL model in Ignite can be found in Apache Software Foundation (2021u).

Listing 3.6
Ignite Java client API example.

```java
import org.apache.ignite.client.ClientException;
import org.apache.ignite.client.IgniteClient;
import org.apache.ignite.IgniteCache;
import org.apache.ignite.IgniteException;
import org.apache.ignite.configuration.ClientConfiguration;

public class IgniteExample {
 public static void main(String[] args) {
 try (IgniteClient client = Ignition.startClient(new
      ClientConfiguration().setAddresses("server1:1234"))
 {
  final String CACHE_NAME = "put-get-example";
  IgniteCache<String, String> cache = client.getOrCreateCache(
     CACHE_NAME)

  cache.put("key", "value");

  cachedVal = cache.get(key);
  assertEquals(cachedVal, "value");
  System.out.format("Loaded [%s] from the cache.\n", cachedVal);
 }
 catch (ClientException e) {
  System.err.println(e.getMessage());
 }
 catch (IgniteException e) {
  System.err.println(e.getMessage());
 }
 catch (Exception e) {
  System.err.format("Unexpected failure: %s\n", e);
 }
 }
 }
}
```

3.5.2   Key-Value Stores

A key-value store is a storage paradigm designed to manage data as a collection of key-value pairs, where a *key* is used as a unique identifier to store or retrieve a data object referred to as the *value*. Both keys and values are abstract data objects, in other words, they can be constituted by anything ranging from simple strings to complex data structures. A typical key-value store deployment enables aggregating low-latency memories (e.g., DRAM) and fast persistent storage (e.g., SSD) across multiple distributed nodes. In this big data era, the key-value store serves as the backbone of many large-scale data-intensive applications.

A typical key-value store deployment enables aggregating low-latency memories (e.g., DRAM) and fast persistent storage (e.g., SSD) across multiple distributed nodes. A

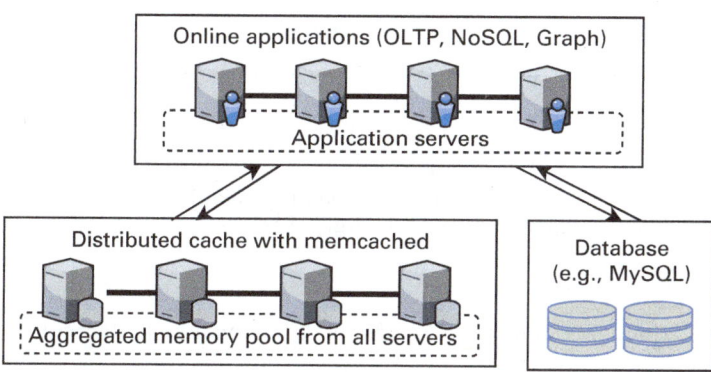

Figure 3.5
Memcached (distributed caching over DRAM).

key-value store typically supports APIs for storing and retrieving data from the distributed key-value server cluster as follows: (1) `Put(K, V)`: A Put (also known as Set) operation inserts or updates a key-value pair (K, V), and (2) `Get(K)`: A Get operation retrieves value data V that is corresponding to the key K. APIs are also available to enable bulk inserts (Multi-Put) and bulk retrievals (Multi-Get).

Key-value stores are leveraged to accelerate data-intensive applications spanning various big data application domains, including, online data processing, offline analytical, graph processing, and machine learning workloads. To give some perspective, we discuss real-world scenarios wherein key-value stores are being leveraged and an overview of three prominent examples of key-value stores.

3.5.2.1 Memcached

Memcached (Dormando, 2021) is a general-purpose distributed memory caching system. It is often used to speed up dynamic database-driven websites by caching data and objects in DRAM aggregated across distributed nodes to alleviate the load on the back-end database. Memcached implements a distributed hash table that employs a consistent hashing mechanism to locate the designated server for every key-value pair.

Figure 3.5 gives an overview of the Memcached architecture. As seen from the figure, Memcached augments the performance of existing and complex database systems by serving as an external high-performance layer. This caching layer is typically maintained by the application servers, thus enabling legacy database backends to continue to function, while harnessing higher performance.

A client example with the C++ Libmemcached library (Aker, 2011) that stores and retrieves a key-value pair from a Memcached cluster is illustrated in listing 3.7. The client connects to the Memcached cluster by specifying server list while creating the `memcached_st` object or via APIs such as `memcached_server_add`. Once connected, key-value pairs can be stored in the distributed cluster using `memcached_set()`

Listing 3.7
Libmemcached client API example.

```c
int main(int argc, char **argv)
{
 const char *config_string =
    "—SERVER=host10.example.com —SERVER=host11.example.com"
 memcached_st *memc= memcached(config_string, strlen(
     config_string));

 if (!memc) {
  fprintf(stderr, "Couldn't add server: %s\n", memcached_strerror
     (memc, rc));
  exit(1);
 }

 char *key= "foo";
 char *value= "value";

 memcached_return_t rc = memcached_set ( memc
              , key
              , strlen(key)
              , value
              , strlen(value)
              , (time_t)0
              , (uint32_t)0
              );

 if (rc != MEMCACHED_SUCCESS) {
  fprintf(stderr, "Couldn't store key: %s\n", memcached_strerror(
     memc, rc));
 }

 retrieved_value = memcached_get(memc, key, strlen(key), &
     value_length, &flags, &rc);
 if (rc == MEMCACHED_SUCCESS) {
  fprintf(stderr, "Key retrieved successfully\n");
  printf("The key '%s' returned value '%s'.\n", key,
     retrieved_value);
  free(retrieved_value);
 }
 else {
  fprintf(stderr, "Couldn't retrieve key: %s\n",
     memcached_strerror(memc, rc));
 }

 memcached_free(memc);
}
```

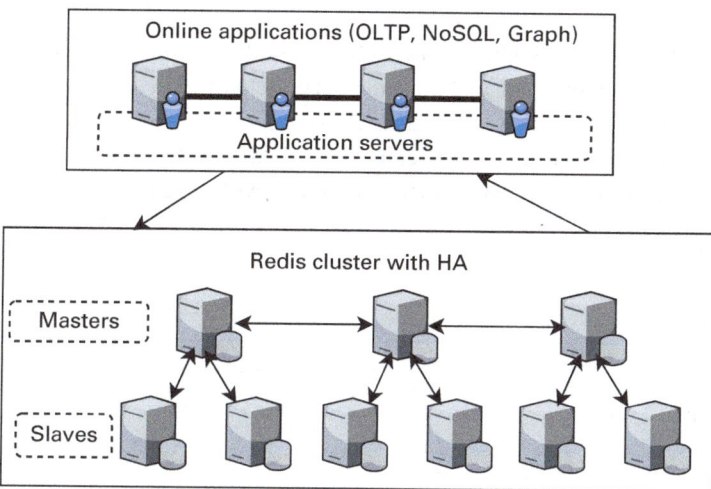

Figure 3.6
Redis (distributed in-memory data store). HA, High Availability.

and fetched using `memcached_get()`. More details on these APIs and their variations
can be found in Aker (2011).

3.5.2.2 Redis

While designs like Memcached are highly beneficial to accelerate the response times by
alleviating the pressure on the backend servers, the caching layer is volatile and has limited
applicability as a stand-alone solution. However, Memcached design laid the basis and
led to the growth of modern in-memory key-value stores that can function as stand-alone
entities with the performance of DRAM and all the capabilities of durable data stores. A
prime example of this is Redis (Redis Labs, 2021d), a general-purpose in-memory data
structure store. It provides a memory-centric, persistence-capable, and highly available key-
value store design with weaker consistency than SQL databases. Unlike Memcached, Redis
follows a master-slave architecture to build a fault-tolerant system. Figure 3.6 gives an
overview of the Redis architecture.

As Redis has a flexible in-memory data model built into its architecture, it has usage
scenarios beyond being a distributed cache. It is being leveraged as a data store for metadata
management, a message broker, or as the primary NoSQL database for performance-critical
data. In-memory data stores such as Redis are being used to boost the performance of
offline analytics jobs for Hadoop and Spark. While a Redis server is essentially a single-
threaded server instance at its core, it provides a *sentinel mode* (Redis Labs, 2021c) for
high availability via redundancy and a *cluster mode* (Redis Labs, 2021b) for automatically
managing data across multiple Redis server instances.

A client example with the Java-based library Jedis (Leibiusky, 2021) that stores and retrieves a key-value pair from a Redis cluster (i.e., cluster mode) is illustrated in listing 3.8. One or more Redis instances that make up the server cluster are specified as a `HostAndPort` set during initialization of the `JedisCluster` object. Once connected, key-value pairs can be stored in the distributed cluster using `JedisCluster.set()` and fetched using `JedisCluster.get()`. More details on these APIs and their variations can be found in Leibiusky (2021).

Listing 3.8
Java Redis (Jedis) API example.

```java
import redis.clients.jedis.HostAndPort;
import redis.clients.jedis.JedisCluster;

public class JedisClusterExample {
  public static void main(String[] args) {

    /* create Java Redis instance to connect to
       remote Redis cluster */
    Set<HostAndPort> connectionPoints =
            new HashSet<HostAndPort>();
    connectionPoints.add(new HostAndPort("server1", 7379));
    connectionPoints.add(new HostAndPort("server1", 7380));
    connectionPoints.add(new HostAndPort("server2", 7379));

    JedisCluster cluster = new JedisCluster(connectionPoints);

    /* write key-value pair into table in Redis */
    cluster.set("key", "value");

    /* read value for a key from Redis table */
    String value_from_remote_store = cluster.get("key");
    assertEquals(value_from_remote_store, "value");

    cluster.close();
  }
}
```

3.5.2.3 RAMCloud

RAMCloud (Ousterhout et al., 2015) is an in-memory storage system that provides low-latency access to large-scale datasets, but stores all data only in DRAM. To support petabyte-scale workloads, it aggregates DRAM from thousands of servers into a single coherent key-value store. It ensures the durability of DRAM-based data by keeping backup copies on secondary storage while maintaining only one copy of data online. It employs a memory-efficient log-structured mechanism to manage data in the DRAM and the secondary storage backups, for enabling high performance. RAMCloud also employs a crash recovery mechanism that exploits the capabilities of the entire cluster to work concurrently toward enabling recovery performance to scale with the cluster size.

Each storage server in the RAMCloud cluster contains two modules: (1) a master that manages objects in its DRAM and (2) a backup that stores segment replicas on persistent storage. A central coordinator manages the cluster of masters and backups. A RAMCloud client application accesses data using RPCs. An example with the Java-based client that stores and retrieves a key-value pair from a RAMCloud cluster is illustrated in listing 3.9. The client connects to the RAMCloud cluster via the coordinator. Once connected, a table can be created and key-value pairs can be stored using RAMCloud.write(). Data can be fetched as RAMCloudObject and value retrieved using RAMCloudObject.get_value() API call. More details on these APIs and their variations can be found in PlatformLab (2021).

Listing 3.9
RAMCloud Java API example.

```java
import edu.stanford.ramcloud.RAMCloud;
import edu.stanford.ramcloud.RAMCloudObject;

public class RAMCloudExample {
 public static void main(String[] args) {

  /* create RAMCloud instance and connect to coordinator */
  String COORDINATOR_LOCATOR = "tcp:host=4.3.2.1,port=12345";
  RAMCloud ramcloud = new RAMCloud(COORDINATOR_LOCATOR);

  /* create table */
  long tableId = ramcloud.createTable(tableName);

  /* write key-value pair into table in RAMCloud */
  long version = ramcloud.write(tableId, "key", "value");

  /* read value for a key from RAMCloud table */
  RAMCloudObject obj = ramcloud.read(tableId, "key");
  String value_from_remote_store = obj.getValue();

  assertEquals(value_from_remote_store, "value");
  assertEquals(version, obj.getVersion());
 }
```

3.5.3 In-Memory File Systems

Scaling distributed file systems such as HDFS are bottlenecked by the performance of the slow persistent storage systems that they rely on for resilience and durability. Slow writes have been shown to significantly hurt the performance of the job, especially for iterative jobs where the current job relies on the output of the previous job. While one approach to alleviate this was to move to in-memory processing frameworks such as Apache Spark, a new type of file systems, known as in-memory file systems, was introduced. In-memory file systems are distributed file systems that enable reliable data sharing at memory speed across cluster computing frameworks via the file system interface. They enable high throughput for writes and reads by aggregating byte-addressable memories across distributed nodes, without compromising fault tolerance.

MemHDFS (Apache Software Foundation, 2020b) is a variation on HDFS, where all data are stored in DRAM memory, and the necessary fault-tolerance is availed via in-memory replication. While this alleviated the I/O performance bottleneck, maintaining several copies in memory is expensive in terms of resource usage. Along this direction, designs such as Tachyon (currently known as Alluxio) (Li, Ghodsi, et al., 2014) that circumvent the limitations of replication have been introduced. These advanced in-memory virtual file system designs leverage the concept known as "lineage," which relies on reexecuting the operations to recreate any lost data for recovering from any potential failures.

Recent innovations in-memory hardware designs have led to PMEM technologies that promise DRAM-like performance, persistence, and high density. These can be directly attached to the processors, similar to DRAM, and offer the opportunity to rethink current persistent storage system designs. The promise of very low-latency and durable storage systems has led to several research works that are designing shared and distributed file systems over PMEM, including, Linux PMFS (Intel, 2015a), NVFS (NVM- and RDMA-aware HDFS) introduced by Islam et al. (2016b), and NOVA (Xu and Swanson, 2016).

3.6   Monitoring and Diagnostics Tools

With data being the center of all things in data analytics, it is essential to monitor the health of the data storage layers. Most data stores and file systems, several of which are discussed in this chapter, provide built-in commands and graphical user interfaces that enable querying server metrics for analyzing their performance.

Open-source tools such as Ganglia (Massie, 2018), Prometheus (Prometheus, 2021), Riemann (Kingsbury et al., 2021), among others, are available on many data analytics systems. For the cloud, proprietary monitoring tools such as Datadog (Datadog, 2021) can be used to track servers and databases in real time. It provides capabilities to monitor HDFS (Mouzakitis, 2016), Cassandra (Datadog, 2015), Redis (Redis Labs, 2021d), and so on. Additionally, other proprietary tools such as RapidLoop OpsDash (RapidLoop, 2021) and SolarWinds AppOptics (Harzog, 2019) enable monitoring for commercial deployments of storage systems such as MySQL, MongoDB, Memcached, and others.

3.7 Summary

This chapter presents a landscape overview of four different types of storage middleware systems: file storage, block storage, object storage, and memory-centric storage. All four kinds of storage systems have been widely used in many HPC, data center, and cloud environments. They have been playing crucial roles in daily data analytics and processing applications all over the world. To give a better understanding of these systems, this chapter also discussed some representative real-world examples with corresponding interfaces being used to develop large-scale big data applications.

With the affordability and performance offered by non-volatile storage devices like PMEM and NVMe-SSDs, there is a need to accelerate the adoption of their new I/O programming primitives. We envision that this has and will open up many research opportunities along the high-performance storage direction. This chapter gives us a high-level overview of high-performance parallel and distributed storage systems. Early and ongoing works on advanced designs for high-performance parallel and distributed data stores that leverage this new evolution of storage devices are further discussed in chapter 9.

4 HPC Architectures and Trends

This chapter outlines the hardware capabilities that are available to harness data processing in the HPC environment. It presents an overview of the emerging trends in state-of-the-HPC architectures, concerning network, storage, and computation. It also discusses an overview of the software interface that enables application designers to exploit these HPC resources.

4.1 Overview

Most modern HPC clusters follow a simple architecture, as illustrated in figure 4.1. The server machines, more commonly referred to as "nodes" that play different roles, are interconnected by high-performance network interconnects. The design and deployment of the individual HPC nodes comprise three vital hardware-related aspects: computation, network, and storage. With this basis, the HPC cluster spans four main components: compute nodes, storage nodes, I/O nodes, and high-performance nodes.

Compute node: A compute node (CN) is comparable to a stand-alone server machine and serves as a set of computational units on which data are processed. These nodes are equipped with memory (typically a set of DRAM) that is accessible by all its CPUs. Today's CNs include multicore CPUs with support for AVX2/AVX-512 (i.e., 512-bit Advanced Vector Extensions for single instruction multiple data [SIMD] instructions on the x86 architecture) instruction sets, assisted by offload-based accelerators like CUDA-enabled NVIDIA GPUs (NVIDIA, 2021m) and AMD GPUs (AMD, 2021a).

Each of these nodes is also equipped with node local storage, including (1) high-bandwidth NVM storage, such as NVMe-SSDs (NVMe, NVM Express, 2021), and (2) emerging PMEM with near-DRAM latencies, such as 3DXPoint (Handy, 2015).

Storage node: The storage node (SN) layer or the SAN fabric constitutes a large global and shared parallel file system (PFS), such as Lustre (Braam and Zahir, 2002). It is internally made up of one or mode metadata server nodes that store namespace metadata for all SNs and several OSS nodes that store the actual data. The CNs (PFS clients) access the data in this shared storage layer through its unified namespace.

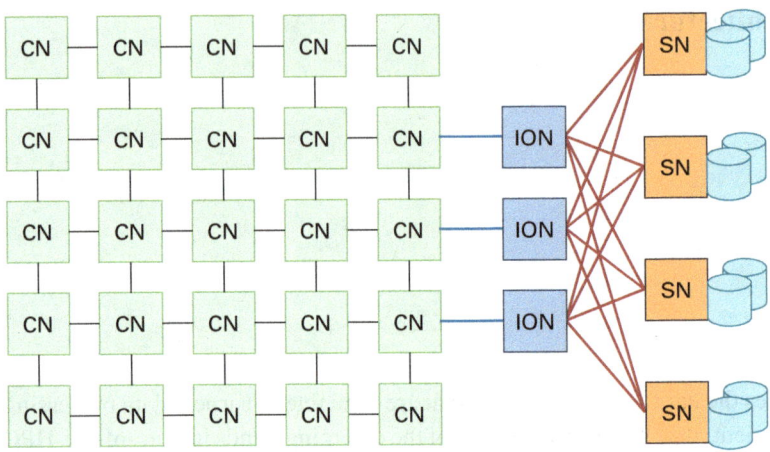

Figure 4.1
Overview of a typical HPC system architecture.

I/O nodes: I/O nodes (IONs) are specialized nodes that bridge the internal interconnect of the computing system and the SAN fabric of the PFS. Fast NVM-based arrays built with PCIe-/NVMe-SSDs are installed in IONs and operate in a transparent manner for accessing data via the SNs. On some systems, this layer also incorporates a burst buffer layer.

High-performance network: High-performance network interconnects enable the use of novel low-latency communication protocols such as InfiniBand (IBTA, 2021a), RoCE (Infiniband Trade Association, 2010), iWARP (Internet Wide Area RDMA Protocol) (Dalessandro et al., 2005), Amazon Elastic Fabric Adapter (EFA) (Amazon, 2021d), and so on. These interconnects connect the storage and compute capabilities of individual nodes, as illustrated in figure 4.1.

On a production HPC system, the CNs can be allocated dynamically to deploy short-running and long-running data-intensive computational applications, through a job scheduler or job batching system such as Slurm (SchedMD, 2020b). To extract high performance from the advanced network, storage, and computation technologies, extensive native library support and APIs have been proposed in the HPC community in the past and currently. We discuss each of these three aspects of the HPC architecture, along with a detailed overview of available native libraries, in the following sections.

4.2   Computing Capabilities

As today's applications and their working datasets grow in both complexity and size, modern CPU processor designs are heading toward enabling workload-optimized functionality, faster data movement, and more efficient data storage. The HPC system consists of CNs that leverage one or more of the following four major types of computing technologies to accelerate data-intensive applications: (1) multi-core CPUs, (2) GPUs, (3) field programmable gate arrays (FPGAs), and (4) application-specific integrated circuits (ASICs).

4.2.1 Multicore CPU Architectures

CPU processor cores form the heart of the HPC cluster. Today's multicore CPUs are equipped with dozens of cores and often sit side by side with accelerators such as GPUs.

4.2.1.1 Hardware trends

There are currently four dominant players in the HPC and data center that are enabling fast and efficient CPU processor designs.

Intel Xeon: Intel is the most dominant player with its Xeon family (Intel, 2021g). Intel Skylake (Doweck et al., 2017) supports up to eight-way multiprocessing, up to twenty-eight cores per socket, and incorporates AVX-512 x86 extensions. They support SSE (Streaming SIMD Extension; 128-bit), AVX2 (256-bit), and AVX-512 (512-bit) vector instructions. Intel Cascade Lake (Arafa et al., 2019) packs in forty-eight cores built from a pair of twenty-four–core chips (two sockets), connected via using Intel Ultra Path Interconnect. It introduces a dozen Double Data Rate 4 (DDR4) memory channels to support Intel Optane PMEM dual in-line memory modules (DIMMs) in a hierarchical storage configuration. Current versions of Cascade Lake use PCIe 3.0, but PCIe 4.0 is on the roadmap. Cascade Lake also introduces extensions to its AVX-512 vector processing module that accelerate deep learning calculations. These AVX-512 Vector Neural Network Instructions (VNNI), also being referred to as DL Boost, are extended to support the brain floating-point format. The recent Intel Cooper Lake microarchitecture introduced the third-generation Intel Xeon scalable processors (14-nm microarchitecture) as the successor to Cascade Lake, which was launched in June 2020 (Intel, 2019d). Cooper Lake supports up to fifty-six cores, Intel Optane PMEM, DDR4-3200 memory, and higher deep learning workload performance. Cooper Lake is claimed to be the first x86 processor to equip built-in deep learning training accelerations through the newly introduced `bfloat16` support. The upcoming 10nm++ Sapphire Rapids Xeon processors (Intel, 2020c) are projected to include the next-generation DDR5 RAM with support for the PCI-e 5.0 standard (insideHPC, 2021) and Compute Express Link running up to 32.0 GT/s (gigatransfers per second) to deliver the low-latency and high-bandwidth I/O solutions, each with fifty-six cores and possibly up to seventy-two cores per node (Morgan, 2019).

IBM Power: IBM Power family of CPUs is powered by the Power instruction set architecture (ISA) developed by the OpenPOWER Foundation, led by IBM. The POWER9 CPUs (Sadasivam et al., 2017) are superscalar, multithreading, symmetric multiprocessors based on the Power ISA. They deliver up to forty to forty-four cores, backed by 2 TB of DRAM. They also introduce a state-of-the-art I/O subsystem technology, including next-generation NVIDIA NVLink, PCIe Gen-4, and OpenCAPI (Coherent Accelerator Processor Interface). The upcoming POWER10 processors will be manufactured using a 7-nm process and support multiple important enhancements to its predecessor, POWER9, including more cores, better microarchitecture designs, more memory controllers, and more advanced I/O subsystem supports for OpenCAPI 4.0 and NVLink3 (IBM, 2020).

AMD EPYC: AMD EPYC (Lepak et al., 2017) is a family of x86 64-bit processors designed by AMD. This processor is based on the Zen microarchitecture and is manufactured on a 14-nm process. The current EPYC CPUs provision a dual-socket 32-core enterprise server microprocessor. Along this line, the next-generation CPUs with 7-nm processor Zen 2 microarchitecture (Singh et al., 2020), code-named "EPYC Rome" have also been introduced. They pack up to 64 cores per socket, 128 total cores (256 threads), and eight memory channels. The most recent AMD server chips, "EPYC Milan," are based on the Zen 3 microarchitecture (AMD, 2021e) with up to 64 cores and support eight-channel DDR4 SDRAM and 128 PCIe 4.0 lanes.

ARM: While tremendous progress has been made in improving CPUs' capabilities, it is also becoming vital to reduce the operational costs and environmental impact by considering "green" architectures. Specifically for data-intensive and scientific computing, one such energy-efficient architecture that is making a big impact today is Fujitsu's A64FX ((Fujitsu, 2021)) 64-bit ARM architecture processors (ARM, 2021b). An A64FX processor consists of forty-eight cores that employ the 512-bit ARMv8.2-A Scalable Vector Extension and four nonuniform memory access nodes (twelve cores/node). Each CPU is equipped with 32 GB of high-bandwidth memory (HBM) that provides a peak memory bandwidth of 1 TB/s connected via the Tofu interconnect. The A64FX processors are capable of achieving three tera floating-point operations per second (TFLOPS) of peak bandwidth while being ten times more power-efficient than an x86 processor. It is being used to power the Fugaku cluster ((Monroe, 2020)), which ranks first in the Top500 list and tenth in the Green500 list as of November 2020 (TOP500.org, 2020). Along a similar direction, ARM is also working on high-performance processors known as Neoverse V1 (ARM, 2021a), for machine learning and AI applications.

4.2.1.2   CPU-Optimized libraries

A vector or SIMD-enabled CPU processor can simultaneously execute an operation on multiple data operands, using a single instruction. One such technology, referred to as AVX, is available on Intel processors or compatible non-Intel processors that support SIMD. These vector lengths vary from 128 bits to 512 bits.

Specifically, with 512-bit extensions to AVX for x86 ISA, we have the opportunity to operate on an entire cache-line in a single instruction. The example in listing 4.1 helps to illustrate a simple program to add two arrays (or vectors) using AVX-512 vector instructions via Intel Intrinsics (Intel, 2021e). The _mm_512_* instructions enable the add operation to be performed simultaneously on sixteen 32-bit floating-point numbers.

Listing 4.1
Dot product example with AVX-512.

```c
float dot512(float *x1, float *x2, size_t len) {
  assert(len % 16 == 0);

  __m512 sum = _mm512_setzero_ps();

  if (len > 15) {
    size_t limit = len - 15;
    for (size_t i = 0; i < limit; i += 16) {
      __m512 v1 = _mm512_loadu_ps(x1 + i);
      __m512 v2 = _mm512_loadu_ps(x2 + i);
      sum = _mm512_add_ps(sum, _mm512_mul_ps(v1, v2));
    }
  }

  float buffer[16];
  _mm512_storeu_ps(buffer, sum);

  return buffer[0] + buffer[1] + buffer[2] + buffer[3] +
         buffer[4] + buffer[5] + buffer[6] + buffer[7] +
         buffer[8] + buffer[9] + buffer[10] + buffer[11] +
         buffer[12] + buffer[13] + buffer[14] + buffer[15];
}
```

Several libraries and tools are leveraging these SIMD CPU capabilities for different operations. Prominent examples include the following.

Intel ISA-L: This library provides tools to maximize storage throughput, security, and resilience, as well as minimize storage space in use. It provides a set of highly optimized functions for redundant array of independent disks, erasure code, cyclic redundancy code, cryptographic hash, encryption, and compression. These functionalities exploit built-in SIMD CPU instructions via Intel SSE, AVX2, and AVX-512 to accelerate the computing performance.

Intel MKL: MKL stands for math kernel library (Intel, 2021f). It provides a set of optimized math routines for science, engineering, and financial applications. Specifically, oneDNN (Intel, 2021h) (formerly known as Intel MKL DNN) supplies a set of mathematical programs for common components in deep neural networks (DNNs). MKL-DNN is leveraged by Caffe (Intel, 2019b), OpenVINO (OpenVINO, 2021), TensorFlow (Ould-Ahmed-Vall et al., 2017), PyTorch (Greeneitch et al., 2019), and other popular software frameworks. These functionalities include matrix multiplication, batch norm, normalization, and convolution. The library is optimized for deploying models across Intel's CPUs. For example, it enables AVX-512 VNNI support on Cascade Lake processors.

OpenMP SIMD directives: SIMD capabilities are also being integrated into parallel programming libraries such as OpenMP (Dagum and Menon, 1998) that are designed to leverage the symmetric multiprocessor model. A section of code that is to be executed in parallel is marked by a special directive (omp pragma). When the execution reaches a parallel section, this directive will cause slave threads to form and each executes the parallel section of the code independently. When a thread finishes, it joins the master and continues to execute the code following the parallel section.

OpenMP provides SIMD integration via the omp simd directive (OpenMP, 2018) that tells the compiler that it should consider vectorizing the parallelizable loop. OpenMP SIMD directives make OpenMP more accessible due to OpenMP's clear syntax and portability beyond Matrix Math Extensions (MMX) and AVX intrinsics. Listing 4.2 demonstrated how the dot product example in listing 4.1 can be simplified with the ease of use of libraries such as OpenMP.

Listing 4.2
Dot product example with OpenMP.

```
float dot_openmp_simd(float *x1, float *x2, size_t len) {
    float sum = 0;

    #pragma omp simd reduction(+:sum)
    for (int i=0; i<len; ++i) {
        sum += x1[i] * x2[i];
    }

    return sum;
}
```

4.2.2   GPU Architectures

General-purpose GPUs have become today's top choice of hardware for accelerating computational workloads on modern HPC clusters. Unlike CPUs, which are optimized for low latencies, GPUs provide a throughput-oriented architecture that provides about three times more concurrent threads than multicore CPUs do. A single GPU device consists of multiple processor clusters that contain multiple streaming multiprocessors (SMs). Each SM accommodates a layer-1 instruction cache layer with its associated cores. Typically, one SM uses a dedicated L1 cache and a shared L2, before fetching data from the global DRAM-like memory.

4.2.2.1 Hardware trends

Alongside the multicore CPUs, two dominant vendors are working toward introducing GPU architectures that deliver high performance for deep learning, HPC, and cloud data center systems.

NVIDIA GPUs: NVIDIA's Volta architecture (NVIDIA, 2017) and flagship high-end GPU, the Tesla V100, packs 21.1 billion transistors and almost 100 billion via connectors to enable 33 percent larger capacity than its Pascal GPU predecessor (NVIDIA, 2021k). The Volta GPUs powered by tensor cores designed specifically for deep learning deliver up to twelve times higher peak TFLOPS for training and six times higher peak TFLOPS for inference, with the highly tuned 16 GB HBM2 memory subsystem capable of delivering 900 GB/s peak memory bandwidth. The NVIDIA Tesla V100 accelerator is built with the Quadro GV100 GPU (NVIDIA, 2021l). NVIDIA released its Ampere architecture-based A100 accelerator in May 2020 (NVIDIA, 2021g). The A100 GPU supports 19.5 TFLOPS of 32-bit floating-point performance, 6,912 CUDA cores, and 40-GB GPU memory with 1.6 TB/s memory bandwidth. These advanced GPUs can be combined using NVIDIA NVLink interconnect technology (Foley and Danskin, 2017) to scale memory and performance, creating a massive computing solution in a single workstation/server.

Based on NVIDIA's latest GPUs, a purpose-built system optimized for deep learning with fully integrated hardware and software that can be deployed quickly and easily, referred to as NVIDIA DGX (NVIDIA, 2021j), has been introduced. The NVIDIA DGX-1 (NVIDIA, 2021h) was the first such system introduced, built on the Pascal and later versions on Volta GPUs. The next generation, NVIDIA DGX-2 (NVIDIA, 2021i) system, combines sixteen interconnected GPUs for both speed and scale. Powered by NVIDIA DGX software and the scalable architecture of NVIDIA NVSwitch, the DGX-2 platform promises a ten times greater deep learning performance to meet today's complex AI challenges. The A100 GPUs have also been deployed on the third-generation DGX server, including eight A100 GPUs. The DGX A100 workstation has 15 TB of PCIe Gen-4 NVMe storage, two sixty-four–core AMD Rome 7742 CPUs, 1-TB main memory, and Mellanox High Dynamic Range (HDR) (200 Gb/s) InfiniBand interconnect.

AMD GPUs: AMD Radeon Instinct GPUs (MI60 and MI50) (AMD, 2021b), released in 2018, introduced designs based on AMD's Vega architecture. These GPUs are optimized for deep learning operations, fast double-precision operations, and 16–32 GB of hyper fast HBM2 memory delivering 1-TB/s memory speeds. The cards also support PCIe Gen-4 and direct GPU-to-GPU links using AMD's Infinity Fabric, which is being leveraged by their Ryzen and EPYC CPU cores, promising to offer up to 200 GB/s of bandwidth among four GPUs maximally (which is approximately three times faster than PCIe Gen-4). AMD has recently released the Instinct MI100 GPU, which is built on the 7-nm+ processors and equipped with 32 GB of HBM2 memory, around 8,000 cores, and 100 TOPs (Tera operations per second) INT8 (Integer data type with eight bytes) computing performance (AMD, 2021c).

Intel GPUs: The Ponte Vecchio GPUs (Intel, 2019c) are Intel's first GPUs based on the
7-nm Xe-HPC microarchitecture. Typically used alongside the next-generation Intel Xeon
Scalable Sapphire Rapids processor, such as in the upcoming Aurora supercomputing cluster (Argonne National Lab, 2021), these GPUs are projected to provide over 1 exaFLOPS
64-bit floating-point performance. Each GPU packs over 100 billion transistors over forty-
seven compute tiles to aid two GPU dies and eight HBM2 memories (four for each GPU)
in a single package.

4.2.2.2   Software interface for GPUs

The software libraries used to program GPUs have traditionally been empowered via
vendor-specific libraries (e.g., NVIDIA's CUDA library (NVIDIA, 2021a)). On the other
hand, the recent emergence in the demand for GPU-accelerated computing has led to the
introduction of universal platforms, such as ROCm (AMD, 2021d).

CUDA: NVIDIA created a parallel computing architecture and platform for its GPUs
called CUDA (NVIDIA, 2021a), which provides application developers to express simple
processing operations in code. In GPU-accelerated applications, the sequential part of the
workload runs on the CPU, while the computation-intensive portion of the application runs
on thousands of GPU cores in parallel. The computational tasks offloaded from the CPU
to the GPU are referred to as GPU kernels. CUDA enables applications to be developed in
several popular languages, including C, C++, Fortran, Python, among others, and to express
parallelism through simple API extensions.

Listing 4.3 presents a CUDA program example for adding two vectors, along with the
different stages of a GPU computation. First, on the CPU side, a cudaMemcpy is called to
transfer array or vector data from the host to the GPU device. Next, the CUDA kernel to be
executed on the GPU side is invoked. When the kernel is launched, the number of threads
per block (`blockDim`) and the number of blocks per grid (`gridDim`) are specified (total
number of threads = [`blockDim * gridDim`]). Each GPU thread evaluates one copy
of the kernel and computes on one or more vector indices to add multiple array elements
in parallel. Once the GPU kernel completes and synchronizes with the CPU host, data are
copied back from the GPU device to the CPU host.

Listing 4.3
Vector sum with CUDA.

```
#define  SIZE  65536
#define  NUMBER_OF_BLOCKS  1
#define  THREADS_PER_BLOCK  65536

__global__ void vecsum_kernel(const float *v1, const float *v2,
    float *out, size_t num) {
    /* compute current thread id */
    int idx = blockIdx.x * blockDim.x + threadIdx.x;
```

```
/* each thread iterates one or more vector element */
while ( idx < num_elements ) {
  out[idx] = v1[idx] + v2[idx];
  idx += gridDim.x * blockDim.x;
}
}

void vecsum_offload(float *v1, float *v2, float *out) {
/* GPU allocated memory */
float *dev_in1, dev_in2, dev_out;
cudaMalloc(&dev_in1, SIZE);
cudaMalloc(&dev_in2, SIZE);
cudaMalloc(&dev_out, SIZE);

/* copy data to GPU */
cudaMemcpy(dev_in1, v1, SIZE, cudaMemcpyHostToDevice);
cudaMemcpy(dev_in2, v2, SIZE, cudaMemcpyHostToDevice);

/* kernel launch on GPU */
sum_vectors_gpu <<<NUMBER_OF_BLOCKS, THREADS_PER_BLOCK>>>
  (dev_in1, dev_in2, dev_out, SIZE);

/* retrieve results from GPU */
cudaMemcpy(vout, dev_out, SIZE, cudaMemcpyDeviceToHost);
}
```

ROCm: ROCm is an HPC-class platform for GPU-based computing that is language-independent. It brings a rich foundation to advanced computing by seamlessly integrating the CPU and GPU to solve computation-intensive workloads. This software enables the high-performance operation of AMD GPUs for computation-oriented tasks in the Linux operating system. ROCm provides direct support for OpenCL, Python, and several C++ variants. The Heterogeneous-Compute Interface for Portability tool in ROCm offers a vendor-neutral dialect of C++ that is ready to compile for either AMD or NVIDIA GPUs.

4.2.3 Specialized Hardware: FPGAs

An alternative to CPUs and GPUs, FPGAs are specialized hardware that serve as reconfigurable integrated circuits. FPGAs allow hardware implementations of algorithms at a circuit level. This approach offers excellent performance at very low power, but it also offers more flexibility by allowing the designer to change the underlying hardware to best support changing the software. Currently, these are used primarily in machine learning inferences, video algorithms, and thousands of small-volume specialized applications. The current major FPGA designers include Intel and Xilinx. While they promise performance, they need additional programming skills to use this specialized FPGA hardware. Some prominent examples include the following.

4.2.3.1 Intel FPGA

Stratix 10 SX FPGAs provide high-performance PCIe-based FPGA Programmable Acceleration Cards for data center and HPC environments. These high-bandwidth devices leverage the "acceleration stack" for Intel Xeon CPU with FPGAs and provide application developers with a robust platform to deploy FPGA-based accelerated workloads.

4.2.3.2 AMD Xilinx Virtex and Versal

Virtex is the flagship family of FPGA products developed by Xilinx (2021) (soon to be a part of AMD). The Virtex UltraScale FPGAs connect as many as 120 transceivers capable of data rates up to 30.5 Gb/s combined with huge on- and off-chip memory capabilities. As successors of its FPGAs, Xilinx has also introduced the Versal Adaptive Compute Acceleration Platform (ACAP) (Gupta, 2020). It provides vector and scalar processing elements tightly coupled to the next-generation programmable logic of the FPGAs. These are interconnected via a high-bandwidth network-on-chip.

4.2.4 Specialized Hardware: ASICs

The alternative to CPUs, GPUs, and FPGAs is to design custom-developed ASICs dedicated to performing fixed operations extremely fast because the entire chip's logic area can be dedicated to a set of narrow functions.

4.2.4.1 Google TPU

TPUs are ASICs that are custom-developed by Google. They are used for accelerating machine learning workloads. Specifically, it accelerates the performance of linear algebra computation, which is used heavily in these machine learning applications. TPUs minimize the time-to-accuracy while training large, complex neural network models.

4.2.4.2 Intel Nervana and Habana

Built solely to train deep learning models at lightning speed, the Intel Nervana Neural Network Processors (Intel, 2021d) pack a large amount of HBM memory and local SRAM much closer to where computing happens. This means that more of the model parameters can be stored on-die to save significant power for an increase in performance. The Intel Nervana Neural Network Processor features high-speed on- and off-chip interconnects enabling multiple processors to connect, acting as one efficient chip and scale to accommodate larger models for deeper insights. The Nervana processors will be succeeded by the Intel Habana Gaudi AI Training Processor (Habana, 2019), which is specifically designed for training deep learning models. It employs eight Tensor Processor Core 2.0 cores, on-chip SRAM, and four HBM2 devices that provide 32-GB capacity, and 1-TB/s bandwidth. It also natively integrates 10×100 Gb/s Ethernet ports of RoCE version2. Similarly, the Habana Goya AI Inference Processor (HL-1000 PCIe card) (Habana, 2019) provides throughput of about

15,000 images per second for a ResNet-50 (50-layer residual neural network) workload at a latency of around 1 ms.

4.2.4.3   Cerebras WSE

Technology specialists such as Cerebras (2021a) have introduced AI-specific chips to enable high-performance for machine and deep learning applications. The Cerebras Wafer Scalable Engine (WSE-2) (Cerebras, 2021b) is a silicon wafer packed with 2.6 trillion transistors and 850,000 AI-optimized, fully programmable cores to deliver world-leading AI compute density. The data movement between cores and memory happens entirely on-chip, which enables a high-bandwidth and lowest-latency communication fabric. Unlike traditional computing devices that work with a very small cache, the 7-nm–based WSE-2 utilizes 40 GB of SRAM spread across the chip to enable fast and memory-centric computing. Its performance is projected to be several thousand magnitudes greater than the latest GPUs with a thousand times more capacity on the device. Cerebras CS-2 system is built to accelerate AI applications using the WSE-2 chips. It integrates with open-source machine learning frameworks like TensorFlow and PyTorch.

4.2.4.4   SambaNova

Technology specialists such as SambaNova (2021b) provide AI/machine learning–centric computing solutions through their SambaNova Reconfigurable DataFlow Unit (RDU). The RDU (Emani et al., 2021) is a next-generation processor that provides native dataflow processing and programmable acceleration. It employs a tiled architecture that consists of an array of reconfigurable processing and memory units. These are interconnected by a high-speed, 3D on-chip switching fabric. This reconfigurable architecture enables a large number of data-parallel patterns to be efficiently programmed as a combination of computation, memory, and communication networks. The SambaNova Systems DataScale system is equipped with very large memory using several RDUs in the rack. With this architecture and their dedicated software stack, it can help accelerate the training of machine learning models that otherwise would not fit on the GPU memory. It is projected to leverage next-generation DDR5 memories to scale even further (SambaNova, 2021a).

4.2.4.5   Graphcore

Semiconductor technologists such as Graphcore (2021) have designed and launched dedicated AI processors. These ASICs, referred to as intelligence processor units (IPUs), are massive parallel processors that are designed to store a complete machine learning model, and to compute with it, within the processor. The latest second-generation IPUs referred to as the Colossus Mk2 GC200 IPU processors, are designed on a 7-nm chip with 59.4 billion transistors with 1,472 computational cores and 900 MB of on-chip memories (Toon, 2020). These GC200 processors have been employed to build the IPU-Machine M2000 (four GC200s per machine) and provide 1 PetaFlop of AI computing, supported by up to 450 GB exchange memory and 2.8-Tb/s IPU-Fabric for low-latency communication, in a single 1U (1.75 inches) blade, toward supporting the performance needs of most AI workloads (Moor Insights and Strategy, 2020).

4.2.4.6 Emu

Traditional CPU-GPU systems are typically bottlenecked by data movement to/from memory, which can adversely affect large-scale big data applications. Toward alleviating this, similar to Cerebras WSE, Emu (Dysart et al., 2016) proposes a new architecture that introduces "processing-in-memory," which combines memory and logic on the same die. The computation is moved around to where the memory resides by employing the novel concept of "migratory threads." The system consists of a set of nodes interconnected by RapidIO-based system interconnect, each subdivided into "nodelets." Every nodelet contributes its memory to the global address space to deploy a shared memory system based on the partitioned global address space model (Padua, 2011). It enables fine-grained access without the need for caches or buses. With this modular approach, big data applications can achieve high performance because there is no need to move datasets during the workflows.

4.3 Storage

Big data applications involve a diverse set of data-intensive workloads. For instance, while offline data analytics involves batch jobs with large datasets, emerging deep learning applications deal with a large number of small files. More importantly, both require storing and accessing these large datasets. Thus, high-performance storage that enables high-speed, concurrent access to data, is vital to high-performance data analytics.

4.3.1 Storage Hardware Trends

With the emergence of cost-efficient flash-based storage devices, such as PCIe-SSDs and NVMe-SSDs (NVMe, NVM Express, 2021), modern and emerging HPC nodes are equipped with SSD-based fast local storage. Emerging technologies such as phase-change memory (PCM) (Wong et al., 2010) and 3D XPoint (Micron, 2019), which provide byte-addressability and near-DRAM latencies, are also being widely studied as the next-generation storage technology to bridge the gap between DRAM and block-based NVMe-SSDs, as shown in figure 4.2.

4.3.1.1 PCIe-Based SSDs

PCIe-based SSDs represent a new class of SSDs that can be attached to the CPU via the PCIe bus. PCIe-SSDs have replaced SATA SSDs owing to the faster latency and bandwidth of the PCI fabric. These SSDs are backed by new storage technologies such as NAND (NOT AND logic gate) or 3D XPoint memory (Intel, 2015b). In general, most SSDs have multiple channels that can be operated in parallel. Even within a channel, pipelining and interleaving can be used to achieve high degrees of parallelism in the form of a plane-, chip-,

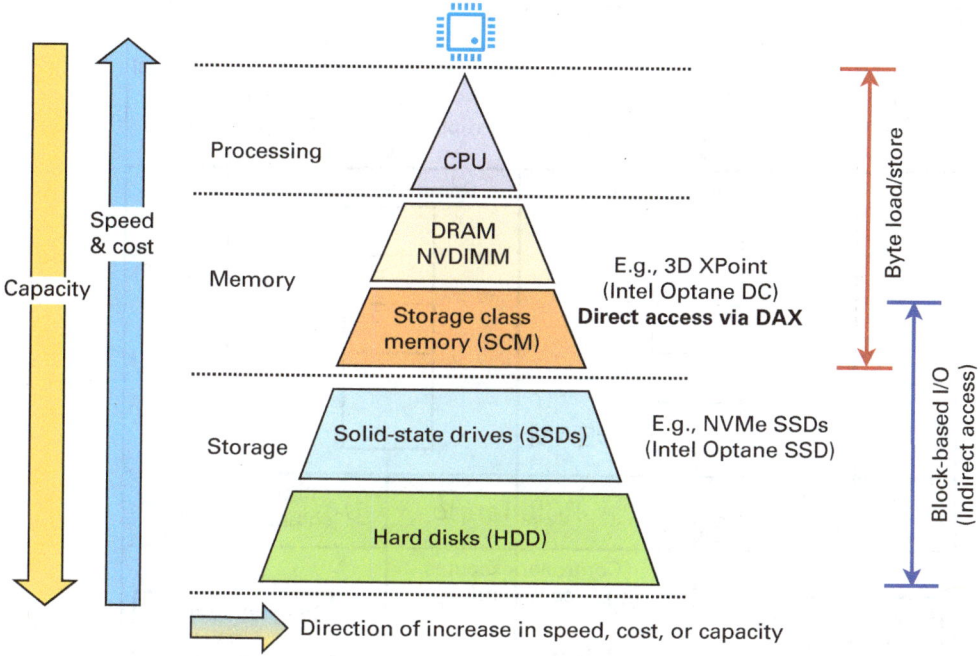

Figure 4.2
Storage device hierarchy.

and die-level parallelism. Thus, PCIe-SSDs offer the possibility of achieving highly parallel I/O. The I/O commands are placed in host memory or controller memory within the SSD. Data transfer is processed as direct memory access (DMA) requests over the PCI fabric. With the recent release of PCIe 6.0 (PCI-SIG, 2019), a new class of SSDs has emerged that can satisfy the demanding I/O requirements of modern data center applications.

4.3.1.2 NVMe

NVMe (NVMe, NVM Express, 2021) is a standard for interacting with PCIe-based SSDs. NVMe replaces the aging and unscalable Advanced Host Controller Interface (AHCI) standard (Intel, 2021b), which is a technical standard to specify the operation of SATA host controllers. NVMe has been built from the ground up for low-latency and high-throughput processing. Unlike the AHCI protocol, NVMe allows multiple hardware I/O queues to be created for submitting and completing requests. There can be up to sixty-four thousand queues and each queue can have up to sixty-four thousand pending requests. This significantly improves the possibility of parallel I/O given the huge number of flash channels in modern SSDs. Figure 4.3 presents an overview of the NVMe request processing pipeline. The host first writes the command in the DRAM-based submission queue and rings the controller doorbell to let it know that a command needs to be processed. The controller

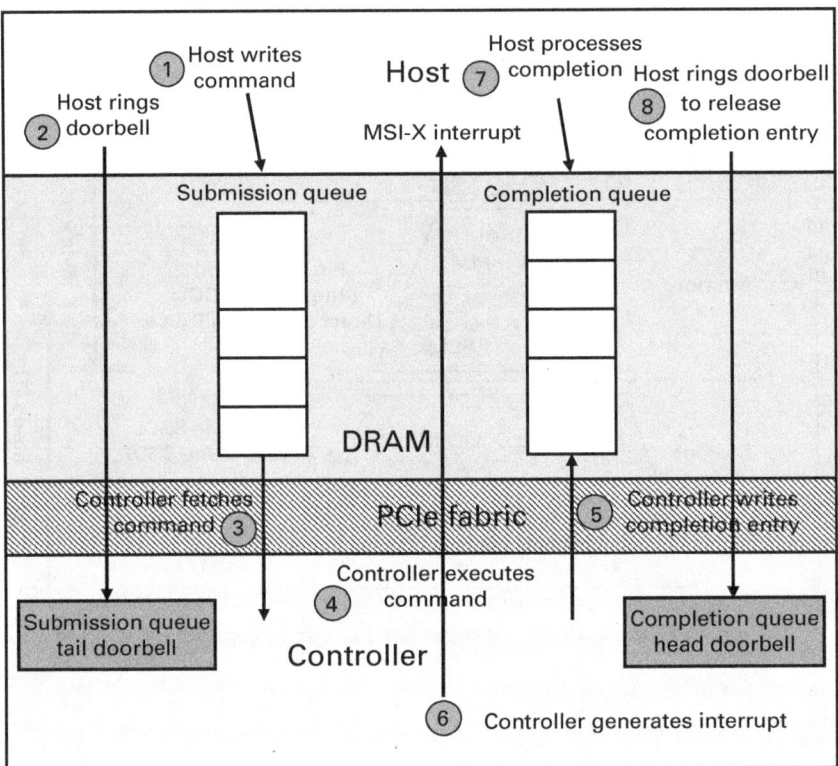

Figure 4.3
NVMe command processing.

fetches the command using a DMA read and executes it. After execution, the controller generates a completion entry and places it in the completion queue (CQ) using a DMA write. Then, it generates an enhanced message signaled interrupt (MSI-X) interrupt (which allows per-core signaling) to let the host know of request completion. Finally, the host rings the doorbell again to release the completion entry. This pipeline allows for some flexibility. For example, the submission and CQs can be placed in controller memory instead of host memory. Furthermore, the host can itself poll the CQ to check for request completion to reduce latency. The most recent version of the NVMe specification (version 4.1) (NVMe, NVM Express, 2021), released in 2019, has matured into a standard that is now incorporated into SSDs employed in modern data centers and HPC clusters.

NVMe devices are revolutionizing big data middleware via RDMA, that is, NVMe over Fabrics (NVMe-oF) (NVMe Express, 2016), and remote persistent memory protocols (Talpey, 2016, 2019, Li et al., 2020). Accelerations for big data processing and management middleware with these NVMe-based storage technologies will be discussed in chapter 9.

4.3.1.3 Persistent memory

PMEM is a new hardware technology combining the performance of DRAM with the non-volatility of SSDs. In essence, it serves as a PMEM for applications requiring performance with durability. The biggest advantage is probably its DRAM-like byte addressability that allows standard *load* and *store* instructions to be used for accessing data. Emerging memory technologies, such as PCM (Lee et al., 2010, 2009), ReRAM (Akinaga and Shima, 2010), spin-transfer torque memory RAM (Tulapurkar et al., 2005), and memristor (Strukov et al., 2008), are being considered for realizing PMEM. These features are fundamentally changing the design principles of many persistent storage systems, as presented in (Facebook, 2018, Kim et al., 2016, Wu and Reddy, Huang et al., 2014, Islam, 2016b, Engel and Mertens, 2005, Lee, 2015)

PMEM can be placed on the memory bus with DRAM packages as nonvolatile dual in-line memory modules. The nonvolatility guarantees that data are persistent immediately following write. Hence, PMEM write latency is expected to be worse than DRAM. For example, PCM is expected to have latency two to five times slower than that of DRAM (Lee et al.). Similarly, the emerging Intel Optane DC memory (Izraelevitz et al., 2019) is estimated to provide peak read and write bandwidths of 33.2 GB/s and 8.9 GB/s, respectively, with slower-than-DRAM write latencies of about 346 ns. New advances are continuing to be made, such as the Intel Optane Memory H20 announced during Intel's recent Memory and Storage 2020 event (Intel, 2020b). These current and emerging PMEM technologies are also introducing remote PMEM storage and access protocols (data access via RDMA (Talpey, 2016, 2019).

The persistent memory programming model (Rudoff, 2013) enables accessing to PMEM (or NVRAM) as memory-mapped files, through the Direct Access (DAX) feature (Rudoff, 2017). This allows applications to perform load/store directly into PMEM. Because this model does not use a page cache, the operating system or the user application needs to only flush the appropriate CPU cache-lines, to get the relevant data changes into the persistence domain. Durable PMEM writes involve the following two steps:

1. **Flush phase:** During this phase, all the cache-lines associated with the data to be persisted are written out to PMEM. For x86 architecture, this can be enabled using *clflush*, *clflushopt*, and *clwb* instructions.

2. **Drain phase:** During this phase, a persistent store barrier is invoked to ensure that the flushed data are globally visible. For x86 systems, memory instructions such as *sfence* can be employed as the memory barrier.

4.3.1.4 Burst-buffer and parallel file system

In current generation HPC systems, the CNs usually have had limited-capacity local storage, due to its traditional Beowulf architecture model (Sterling et al. 2003). In this architecture, as illustrated in section 4.1, the SNs form a dedicated and shared storage cluster that hosts a high-performance PFS (e.g., Lustre (Braam and Zahir, 2002), GPFS (IBM,

2021)) and is connected to the CNs via high-performance interconnects such as InfiniBand. HPC applications have relied on PFSs such as Lustre for their large-scale data storage and access needs. However, due to the increasing demands of handling big data applications on HPC systems, the traditional PFS layer could easily become a performance bottleneck in the I/O pipelines of big data applications. Thus, modern HPC clusters have turned toward hardware or software-based burst buffer technologies. The purpose of burst buffers is to absorb bulk data produced by big data applications, which is usually much faster than what the PFSs can provide. A burst buffer system is typically deployed between the application layer and the PFS layer as an intermediate high-performance storage layer. Currently, several vendors have developed and are continuing to release burst buffer systems, such as DDN's Infinite Memory Engine (IME) and Cray's DataWarp.

DDN IME: IME is a solution from DDN (2021) that leverages a native flash-only storage platform for I/O staging and caching the file and enables PFSs to scale. DDN IME introduces a new layer on top of the PFS that can improve the performance of network and storage through adaptive I/O and data coalescing for I/O-intensive applications. For emerging machine learning and deep learning applications, DDN IME can efficiently feed the diverse I/O requirements on CPU- and GPU-based heterogeneous environments.

Cray DataWarp: Cray DataWarp (Henseler et al., 2016) is an I/O accelerator technology that supports the bursty I/O patterns typical to data-intensive HPC applications, similar to DDN's IME. It is aimed at designing a flash-based storage appliance that can be leveraged as a global storage cache. The DataWarp I/O blades are essentially Cray XC40 compute blades equipped with local SSDs and are connected via the high-speed Cray Aries HPC interconnect. Every XC series blade type leverages the Aries interconnect.

4.3.2 I/O Libraries

To access each of the different storage devices we have discussed, different software interfaces are available. We describe a few of these in this section.

Intel SPDK: Intel Storage Performance Development KIt (SPDK) (Intel, 2017) is a user-space storage library designed for modern applications requiring high-performance data access. All necessary drivers are moved to user-space and polling is used instead of interrupts, thereby enabling high-performance access to data. Apart from legacy storage protocols (such as POSIX, iSCSI, and so on), SPDK offers support for the NVMe standard, including NVMeoF/NVMf. SPDK directly uses the NVMe command set to interact with PCIe-SSDs, completely bypassing the operating system's software stack. For processing NVMe requests using SPDK, users should create a queue pair (QP), which is essentially a set of submission and completion queues. Requests for each QP can be processed completely in parallel. Synchronization within a QP is left to the user's discretion. In general, each application I/O thread is recommended to create a separate QP, allowing for maximum possible parallel I/O and eliminating the need for synchronization. I/O operations are processed asynchronously. Applications need to explicitly process completions by asking the SPDK runtime to poll the CQs. This asynchronous mechanism allows for partial or even complete overlap of I/O and computing. The SPDK design solves most of the performance problems that plague the POSIX NVMe driver, as presented by Yang et al. (2017).

Listing 4.4 illustrates an instance of how to use SPDK to write a block of data into the NVMe device. This is a snapshot of a more detailed "hello world" example provided in the SPDK examples repository (spdk.io). On initializing and setting up connection to the NVMe device (briefed as `init_spdk()` [see spdk.io for more details]), an I/O QP is allocated. Once the I/O data buffer to be written into file is ready, the `spdk_nvme_ns_cmd_write()` is used to start a write function along with the callback function `rw_complete()` that is invoked on write completion. The write completion can be monitored via `spdk_nvme_qpair_process_completions()`.

Listing 4.4
SPDK write example.

```c
struct workreq_t {
 struct ns_entry *entry;
 char   *buf;
 atomic_flag  is_cmpl;
};

static void rw_complete(void *arg,
  const struct spdk_nvme_cpl *cmpl)
{
 struct workreq_t *wr = arg;
 wr->is_cmpl = spdk_nvme_cpl_is_error(cmpl) ? -1 : 1;
}

void write_example_spdk (...) {
/* init NVMe device */
init_spdk(..);

struct workreq_t wr;
wr.entry->qpair =
    spdk_nvme_ctrlr_alloc_io_qpair(wr.entry->ctrlr, 0);
wr.is_cmpl = 0;
wr.buf = rte_zmalloc(NULL, 0x1000, 0x1000);
sprintf(wr.buf, "Hello Big Data!");

rc = spdk_nvme_ns_cmd_write(entry->ns,
 wr.entry->qpair, wr.buf, 0, /* LBA start */,
 1 /* num. of LBAs */, rw_complete, &wr, 0);

/* wait till write completes */
while (!wr.is_cmpl) {
 spdk_nvme_qpair_process_completions(wr.entry->qpair, 0);
}
```

```
spdk_free(wr.buf);
spdk_nvme_ctrlr_free_io_qpair(wr.entry->qpair);
/* detach an free NVMe device */
cleanup(...);
}
```

Intel PMDK: Intel Persistent Memory Development Kit (PMDK) (Intel, 2019e) is a collection of libraries and tools for storing and maintaining data in PMEM devices. These libraries build on the DAX feature that allows applications to directly access PMEM as memory-mapped files.

The libpmem library provides low-level PMEM support. In particular, it provides support for the CPU for flushing changes to the PMEM device. A simple example with the libpmem to access PMEM is illustrated in listing 4.5. As the example shows, the interaction with the PMEM-resident data via DAX is similar to performing operations on memory. It can be persistently written to the PMEM device for durability by employing PMDK libpmem functions such as pmem_memcpy_persist(). Data can also be modified in the cached buffers and flushed to durability only when necessary using functions such as pmem_persist().

Listing 4.5
PMDK persistent store example.

```
char *pmemaddr;
size_t len;
int is_pmem;

/* create a pmem file and memory map it */
if ((pmemaddr = pmem_map_file("/pmem-fs/myfile",
    4096, PMEM_FILE_CREATE,
    0666, &len, &is_pmem)) == NULL || is_pmem == 0) {
  exit(1); /* error; exit with failure */
}

/* store a string and flush to the persistent memory */
pmem_memcpy_persist(pmemaddr, "Hello! PMEM", len);

/* delete the mappings */
pmem_unmap(pmemaddr, len);
```

Because directly accessing the storage media introduces new programming challenges, PMDK also offers applications with different interfaces, including libpmemobj for transactional object storage, librpmem for remote access to PMEM utilizing

RDMA-capable network interface cards (NICs), `libpmempool` for offline pool management, and so on.

4.4   Network Interconnects

During the past decade, the HPC field has witnessed a transition to commodity clusters connected with modern high-performance interconnects such as InfiniBand (32 Gb/s, 56 Gb/s, 100 Gb/s, or 200 Gb/s), Ethernet (10 Gb/s, 25 Gb/s, 40 Gb/s, 50 Gb/s, 100 Gb/s, or 200 Gb/s), iWARP (10 Gb/s, 25 Gb/s, 40 Gb/s, 50 Gb/s, or 100 Gb/s), RoCE (10 Gb/s, 25 Gb/s, 40 Gb/s, 50 Gb/s, 100 Gb/s, or 200 Gb/s), and Omni-Path (100 Gb/s). Figure 4.4 illustrates the various commodity high-performance interconnects and protocols that are being leveraged today. In addition, proprietary networks such as Cray Aries, Amazon EFA, and NVIDIA NVlink are being used on many HPC systems. Cray is working on their next-generation Slingshot architecture. We provide a short overview of these networking technologies and their features in this section.

4.4.1   Network Hardware Trends

The modern network interconnects are characterized by higher bandwidth, low latency, and low CPU utilization. On the commodity front, these technologies include InfiniBand, high-speed Ethernet, and Intel Omni-Path. On the proprietary front, these include Cray Aries, Cray Slingshot, and others.

High-speed Ethernet (1 Gb/s, 10 Gb/s, 25 Gb/s, 40 Gb/s, 50 Gb/s, 100 Gb/s): Traditionally, server connections in data centers used 1-Gb/s Ethernet (1-GigE) interconnect. With recent enhancements in CPU performance and I/O, the 1-GigE network has increasingly become the application and workload performance bottleneck. Recently, data center architectures are upgrading to a combination of 10-GigE connection and higher speed connections. Low latency is critical for many HPC applications including financial trading environments, and 10 GigE offers a compelling improvement in end-to-end latency. While 10 GigE is still paving its way into the data centers, the demand for higher bandwidth and faster data transfer rates has made way for 25 GigE, 40 GigE, 50 GigE, or 100 GigE. A few years back, 25 GigE was introduced to provide 2.5 times the performance of 10 GigE Ethernet, making it a cost-effective upgrade to the 10-GigE infrastructure. Many of these high-speed Ethernet infrastructures use hardware-accelerated versions of TCP/IP. These are called Transmission Control Protocol (TCP) offload engines (TOEs), which use the NIC hardware to offload the TCP protocol. The benefits of the TOEs are to maintain full socket streaming semantics and implement them efficiently in hardware to get reduced latency and higher bandwidth.

InfiniBand: InfiniBand (IBTA, 2021a) is an industry-standard switched fabric that is designed for interconnecting nodes in HPC clusters. Available as low-cost PCIe devices,

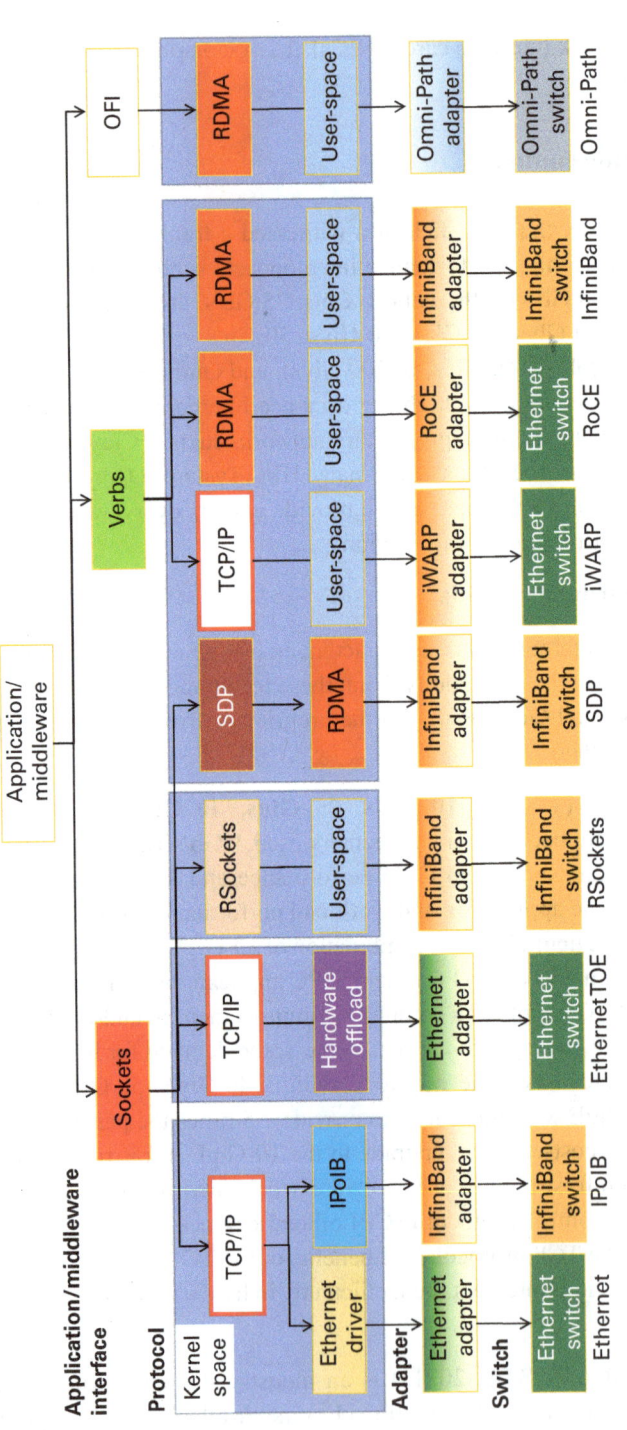

Figure 4.4
Overview of high-performance network interconnects and protocols. OFI (OpenFabrics Interfaces).

these interconnects provide not only high bandwidth (up to 200 Gb/s (Mellanox, 2018)) and low latencies ($< 1\mu$s), but they also help server scalability by using very little CPU intervention and reduced memory copies for network-intensive operations. In addition to this, they provide advanced features, such as RDMA, that enable the design of novel communication protocols and libraries. This feature allows the software to read memory contents of another process on a different node or server (remote process) without any software involvement at the remote side. It is being increasingly leveraged to design scalable communication engines for scientific HPC and data-intensive big data middleware. It has emerged as a popular interconnect for top supercomputing systems and data centers. The Top500 list released in November 2020 reveals that more than 31 percent of the overall Top500 systems (TOP500.org) use InfiniBand. Advanced designs with RDMA for big data middleware on InfiniBand are discussed in chapter 7.

Intel Omni-Path: Intel Omni-Path (Birrittella et al., 2015) is a part of Intel's Scalable System Framework that is designed to enable emerging HPC workloads to scale to tens of thousands of nodes. As the successor to Intel True Scale Fabric, this optimized HPC fabric is built on enhanced Internet protocol (IP) and Intel technology with multiple features: adaptive and dispersive routing, traffic flow optimization, packet integrity protection, and dynamic lane scaling to enable fault-tolerant application runtimes. During the last decade, Omni-Path was deployed on several top supercomputers. The Top500 list released in November 2020 reveals that 9.4 percent of the overall Top500 systems (TOP500.org) use Omni-Path. In 2019, Intel stopped its plan for the second-generation Omni-Path. However, in late 2020, a new Intel spin-out (Cornelis Networks (Cornelis, 2020)) was announced to continue with the design and development of the next-generation high-performance networking product as a successor to the Omni-Path product.

Aries: Cray Aries interconnects (Alverson, 2012), introduced as part of the Cray XC system, coupled with the Dragonfly network, were designed to provide cost-effective, scalable global bandwidth. These proprietary interconnects are being leveraged by several top supercomputing clusters, including Piz Daint (CSCS, 2021) at the Swiss National Supercomputing Centre and the National Energy Research Scientific Computing Center (NERSC) Cori (NERSC, 2021a) at the Berkeley lab.

Slingshot: Cray Slingshot is Ethernet-compatible, able to connect to third-party data storage, and can do it all so systems can run at exascale speeds. With sixty-four port switches capable of 25.6 Tb/s per switch, it is estimated to support up to a quarter-million endpoints in just three network hops, with only one optical cable. This is being designed to deploy the next-generation pre-exascale system NERSC-9 "Perlmutter" machine (NERSC, 2021b) and exascale machines such as the Argonne Aurora (Argonne National Lab, 2021), Oak Ridge National Laboratory Frontier (Oak Ridge National Laboratory (ORNL), 2021a), and Lawrence Livermore National Library El Capitan (Lawrence Livermore National Laboratory, 2020).

NVLink: NVLink (Foley and Danskin, 2017) is a high-speed, direct GPU-to-GPU interconnect. For multi-GPU configurations, it addresses interconnect issues by providing higher
bandwidth, more links, and improved scalability. A single NVIDIA Tesla V100 GPU can
support up to six NVLink connections for a total bandwidth of 300 GB/s, which allows a ten
times bandwidth gain over PCIe Gen-3. Servers such as the NVIDIA DGX-1 and DGX-2
are designed to leverage this technology for scaling distributed computing tasks.

SR-IOV: The single root I/O virtualization (SR-IOV) is a specification that allows a PCIe
device to appear as multiple separate physical PCIe devices. The SR-IOV specification is
an extension of the PCIe specification (PCI-SIG, 2021). It permits high-speed communication by enabling the network traffic to bypass the software switch layer of the Hyper-V
virtualization stack. It is being leveraged in HPC clouds such as Chameleon (Chameleon,
2021) and Microsoft Azure HPC instances (Microsoft, 2021a) to enable near-native network
performance in virtualized environments.

Amazon Elastic Fabric Adapter: Amazon EFA (Raja, 2019) is the next-generation network interface designed to provide enhanced networking specifically to Amazon's EC2
cloud instances, toward supporting network speeds of up to 100 Gb/s with a custom
interface in a virtualized environment.

4.4.2   InfiniBand Software Library and RDMA

InfiniBand (IBTA, 2021a) is an industry-standard switched fabric that is designed for interconnecting CNs and IONs in HPC clusters. We provide a short overview of its features in
this section.

Communication model and transports: Upper-level software uses an interface called
Verbs API to access the functionality provided by Host Channel Adapters (HCAs)
and other network equipment (such as switches), including connection establishment and
communication in InfiniBand. This is illustrated in figure 4.4.

InfiniBand follows a QP-based model. A process can queue up a set of instructions that
the hardware executes. This facility is referred to as a work queue (WQ). WQs are always
created in pairs, a QP, one for send operations and one for receive operations. In general,
the send WQ holds instructions that cause data to be transferred from one process memory
to another process's memory, and the receive WQ holds instructions about where to place
data that are received. The completion of work queue entries (WQEs) is reported through
CQs. Memory involved in communication through InfiniBand should be registered with the
InfiniBand network adapter. Registration is done using an InfiniBand Verbs call that pins the
corresponding pages in memory and returns local (lkey) and remote registration keys (rkey).

InfiniBand offers different transport modes: Reliable Connection (RC), Reliable Datagram, Unreliable Connection, and Unreliable Datagram (UD), Extended Reliable Connected, Dynamic Connected, and others. RC is a connection-oriented service and it requires
a distinct QP per communicating peer. It provides RDMA capability, atomic operations,

and reliable service. UD is connectionless and a mode of an unreliable transport. A single UD QP can communicate with any number of other UD QPs. However, UD does not offer RDMA, reliability, and message ordering. Moreover, messages larger than (MTU) size (8 KB in current Mellanox hardware) have to be segmented and sent in MTU-sized chunks.

Communication semantics and Verbs API example: InfiniBand supports two types of communication semantics in RC transport: channel and memory semantics.

In *channel semantics*, both the sender and receiver have to be involved to transfer data between them. The sender has to post a send work request entry (a WQE) that is matched with a receive work request posted by the receiver. The buffer and the lkey are specified with the request. The receive work request needs to be posted before the data transfer gets initiated at the sender. The receive buffer size should be equal to or greater than the send buffer size. This allows for zero-copy transfers but requires strict synchronization between the two processes. Higher-level libraries avoid this synchronization by preposting receive requests with staging buffers. The data are copied into the actual receive buffer when the target process posts the receive request. This allows the sender to proceed as soon as it is posted. There exists a trade-off between synchronization costs and additional copies. We illustrate the use of this two-sided SEND semantics with InfiniBand Verbs API with a simple example in listings 4.6 and 4.7.

As per the example in listings 4.6 and 4.7, the two peers register send/receive buffers (using the `ibv_reg _mr()` API) and exchange their QP information with each other using an out-of-band mechanism (e.g., TCP/IP). On receipt of a connection request, the target process, as shown in listing 4.7, preposts a receive request before exchanging its QP information with the host. On connecting, the host process, as shown in listing 4.6 posts a two-sided SEND message using `ibv_post_send()` and polls for the send completion using `ibv_poll_cq()`. On the target side, the peer polls on the CQ waiting for the message from its peer.

In memory semantics, RDMA operations are used instead of send/receive operations. These RDMA operations are one-sided and do not require software involvement at the target. The remote host does not have to issue a work request for the data transfer. The send work request includes the address and lkey of the source buffer and address and rkey of the target buffer. Both RDMA Write (write to the remote memory location) and RDMA Read (read from remote memory location) are supported in InfiniBand. We illustrate the use of RDMA with InfiniBand Verbs with a simple example in listings 4.8 and 4.9.

As per the above example, the two peers first register blocks of memory, exchange memory descriptors, and post read/write operations. Registration is accomplished with a call to `ibv_reg_mr()` that pins the block of memory in place and returns a `struct ibv_mr` containing a unique `uint32_t key`, thereby allowing remote access to the registered memory. This key, along with the block's address, is exchanged with peers through some out-of-band mechanism. The host then uses the remote key and address to prepare an RDMA write work request that it posts to `ibv_post_send()`.

Listing 4.6
InfiniBand Verbs send/receive example (host).

```
/* init IB device */
ibv_qp *qp;
ibv_cq *cq;
ibv_wc wc;
init_ib(qp, cq, ...);

char *buffer = malloc(1024);
struct ibv_mr *mr;
struct ibv_sge sge;
struct ibv_send_wr wr, *bad_wr;
uint32_t peer_key;
uint64_t peer_addr;
int flag;

mr = ibv_reg_mr(pd, buffer, SIZE,
    IBV_ACCESS_LOCAL_WRITE);

/* exchange qp with target peer */

strcpy(buffer, "Big Data!");
memset(&wr, 0, sizeof(wr));

sge.addr = (uint64_t)buffer;
sge.length = SIZE;
sge.lkey = mr->lkey;
wr.sg_list = &sge;
wr.num_sge = 1;
wr.opcode = IBV_WR_SEND;
wr.send_flags = IBV_SEND_SIGNALED;

/* post a send
 * poll for completion */
ibv_post_send(qp,&wr,&bad_wr);
while(!ibv_poll_cq(cq,1,&wc));

if (wc.status != 0)
    perror("send failure");

/* free and detach IB device */
cleanup(..);
```

Listing 4.7
InfiniBand Verbs send/receive example (target).

```
/* init IB device */
ibv_qp *qp;
ibv_cq *cq;
ibv_wc wc;
init_ib(qp, cq, ...);

int flag;
char *buffer = malloc(1024);
struct ibv_mr *mr;
uint32_t my_key;
uint64_t my_addr;
struct ibv_mr *mr;
struct ibv_sge sge;
struct ibv_send_wr wr, *bad_wr;

flag = IBV_ACCESS_LOCAL_WRITE;
mr = ibv_reg_mr(pd, buffer,
    SIZE, flag);

key = mr->rkey;
addr = (uint64_t)mr->addr;

/* exchange qp with host peer */

sge.addr = (uint64_t)buffer;
sge.length = SIZE;
sge.lkey = mr->lkey;

flag = IBV_SEND_SIGNALED;
wr.sg_list = &sge;
wr.num_sge = 1;
wr.send_flags = flag;

/* pre-post a receive request */
ibv_post_recv(qp,&wr,&bad_wr);
/* poll for completion */
while(!ibv_poll_cq(cq,1,&wc));

if (wc.status != 0)
    perror("receive failure");
```

Listing 4.8
InfiniBand Verbs RDMA Write example (host).

```c
char *buffer = malloc(1024);
struct ibv_mr *mr;
struct ibv_sge sge;
struct ibv_send_wr wr, *bad_wr;
uint32_t peer_key;
uint64_t peer_addr;

mr = ibv_reg_mr(pd, buffer, SIZE,
     IBV_ACCESS_LOCAL_WRITE);

/* get peer_key and peer_addr
   from target */

strcpy(buffer, "Big Data!");
memset(&wr, 0, sizeof(wr));

sge.addr = (uint64_t)buffer;
sge.length = SIZE;
sge.lkey = mr->lkey;
wr.sg_list = &sge;
wr.num_sge = 1;
wr.opcode = IBV_WR_RDMA_WRITE;
wr.wr.rdma.remote_addr = peer_addr;
wr.wr.rdma.rkey = peer_key;

/* post RDMA write */
ibv_post_send(qp, &wr, &bad_wr);
/* poll for completion */
while(!ibv_poll_cq(cq,1,&wc));

if (wc.status != 0)
  perror("write failure");
```

Listing 4.9
InfiniBand Verbs RDMA Write example (target).

```c
char *buffer = malloc(1024);
struct ibv_mr *mr;
uint32_t my_key;
uint64_t my_addr;
int flag;

flag = IBV_ACCESS_REMOTE_WRITE;
mr = ibv_reg_mr(pd, buffer,
     SIZE, flag);

key = mr->rkey;
addr = (uint64_t)mr->addr;

/* exchange qp, key and
   addr with peer 2 */

/* server bypass;
 * no polling or waits */

/*  ...  */
/*  ...  */
/*  ...  */
/*  ...  */
/*  ...  */

/* free to process
 * other operations */

/*  ...  */
/*  ...  */
/*  ...  */
/*  ...  */
```

The target peer is only involved in setting up the connection via QP and memory address exchanges, unlike the channel semantics. Thus, the host peer is said to perform a remote CPU bypassed or one-sided communication. Because the RDMA-capable NIC directly writes the data into the user's memory buffers and does not perform any explicit data copies to or from the communication buffers, this process is referred to as "zero-copy."

4.4.3 Network Protocols over RDMA-Capable Interconnects

InfiniBand is one the most widely used network interconnects in the Top500 supercomputers (TOP500.org). The majority of HPC applications and middleware (such as MVAPICH2 MPI library (Network Based Computing (NOWLAB), 2021b)) take advantage of InfiniBand either via native InfiniBand Verbs, RoCE, or iWARP interfaces. On the other hand, big data middleware (such as Hadoop and Spark) have been traditionally run over clusters with 10/40/100 GigE interconnects. While leveraging the RDMA capabilities of the underlying high-performance network interconnects is ideal, this may require potential redesign. Thus, to bridge the gap between HPC and big data, two alternative approaches to using native Verbs APIs have been introduced: (1) TCP/IP-compatible network protocols over InfiniBand (e.g., IP-over-InfiniBand) and (2) RDMA-capable communication over high-speed Ethernet networks (e.g., TCP/IP TOE). We elaborate on a few of these state-of-the-art network protocols in this section.

InfiniBand IP layer: InfiniBand also provides a driver for implementing the IP layer. This exposes the InfiniBand device as just another network interface available from the system with an IP address. Typically, these InfiniBand devices are presented as ib0, ib1, and so on. This interface is presented in figure 4.4 (second column from the left, named IPoIB). It does not provide operating system bypass. This layer is often called "IP-over-IB" or "IPoIB" in short. There are two modes available for IPoIB. One is the datagram mode, implemented over UD, and the other is the connected mode, implemented over RC. The connected mode offers better performance because it leverages reliability from the hardware.

InfiniBand Sockets Direct Protocol: Sockets Direct Protocol (SDP) (Goldenberg et al., 2005) is a byte-stream transport protocol that closely mimics the TCP socket's stream semantics. It is an industry-standard specification that utilizes advanced capabilities provided by the network stacks to achieve high performance without requiring modifications to existing sockets-based applications. This interface is presented in figure 4.4 (the fifth column from the left, named SDP). SDP is layered on top of the InfiniBand message-oriented transfer model. The mapping of the byte-stream protocol to the underlying message-oriented semantics was designed to transfer application data by one of two methods: through intermediate private buffers (using buffer copy) or directly between application buffers (zero-copy) using RDMA. Typically, there is a threshold of message size above which zero-copy is used, achieving an operating system–bypassed communication. SDP can also be used in the buffered mode.

RSockets: RSockets is a protocol over RDMA that supports a socket-level API for applications. This interface is presented in figure 4.4 (fourth column from the left, named RSockets). RSockets APIs are intended to match the behavior of corresponding socket calls. RSockets are contained as a wrapper library in the OpenFabrics Enterprise Distribution InfiniBand software stack, with near-optimal speed over InfiniBand networks. The wrapper library method can in principle be used on any system with dynamic linking. The performance improvements gained by scaling out to increase parallelism using RSockets over InfiniBand are significantly higher than what can be achieved using regular Ethernet networks.

Internet Wide Area RDMA Protocol: To improve bandwidth in data center environments that widely use 10/40 GigE, 10 GigE was standardized. To support increasing speeds of network hardware, it was not completely feasible to only offer a traditional sockets interface and there was a need to reduce memory copies while transferring messages. Toward that effort, the iWARP (RDMA Consortium, 2016) standard was introduced. iWARP is very similar to the Verbs layer used by InfiniBand, except iWARP requires a connection manager. This interface is presented in figure 4.4 (fourth column from the right, named iWARP). The OpenFabrics network stack (OpenFabrics Alliance, 2021) provides a unified interface for both iWARP and InfiniBand.

RDMA over Converged Ethernet: InfiniBand has started making inroads into the commercial domain with the recent convergence around RoCE (IBTA, 2021b). This interface is presented in figure 4.4 (third column from the right, named RoCE). There are two RoCE versions, RoCE v1 and RoCE v2. RoCE v1 is an Ethernet link layer protocol and hence allows communication between any two hosts in the same Ethernet broadcast domain. RoCE v2 is an Internet layer protocol, which means that RoCE v2 packets can be routed. In recent years, RoCE v2 is getting momentum due to its advanced features.

Example numbers of RDMA performance: As we just discussed, RDMA-capable networks are able to deliver very good performance to the end applications. Figures 4.5 and 4.6 show example numbers of RDMA performance over the years. These numbers are taken with the MVAPICH2 MPI library. Figures 4.5a and 4.5b show the one-way latency numbers of transferring small messages and large messages over different generations of InfiniBand and Omni-Path networks. As we can see, for transferring small messages, the RDMA-based MVAPICH2 communication library is able to achieve around 1 us of latency across two nodes. Figures 4.6a and 4.6b show the unidirectional and bidirectional bandwidth numbers of transferring various-size messages over different generations of InfiniBand and Omni-Path networks. As we can see, for transferring large messages, the RDMA-based MVAPICH2 communication library is able to achieve near-peak bandwidth.

4.4.4 RDMA over NVM

Both RDMA and NVMe/PMEM devices are designed to work over a PCIe bus. Therefore, NVM storage devices can be remotely accessed using any of the RDMA technologies, including InfiniBand, RoCE, and iWARP. RDMA over NVMe SSDs and PMEM devices are contrasted in figure 4.7.

NVM Express over Fabrics: NVMeoF defines a common architecture that supports a range of storage networking fabrics for NVMe block storage protocol over the storage or I/O nodes in the network fabric. This protocol can be used to transfer NVMe storage commands between the client nodes over InfiniBand or Ethernet networks using RDMA technology to the target nodes, as shown in figure 4.7a. It is designed to be a lightweight application that

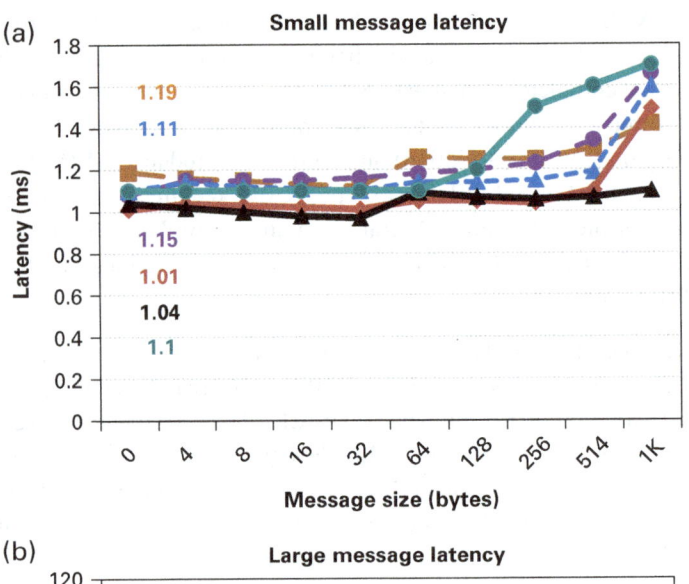

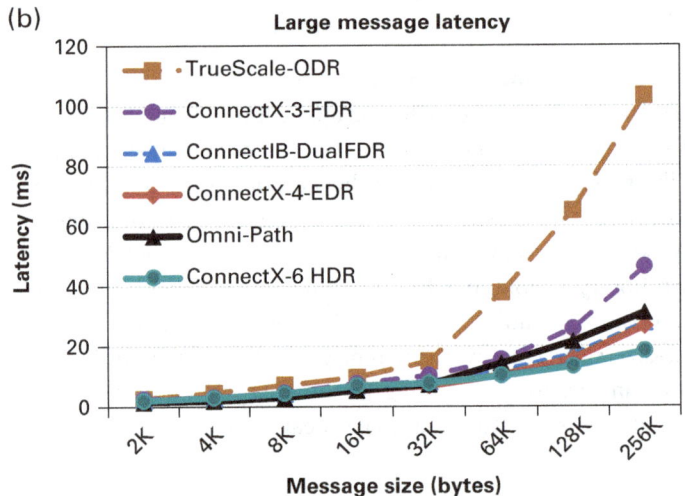

Figure 4.5
One-way latency: MPI over RDMA networks with MVAPICH2. (a) Small message latency. (b) Large message latency.

runs above the RDMA standard interface, like the Verbs API does. The NVMeoF protocol aims to standardize the wire protocol and drivers for efficient access over RDMA-capable networks with minimal processing required by the target node.

Persistent Memory over Fabrics: With Persistent Memory over Fabrics (PMoF), PMEM devices can be attached to the memory bus, as shown in figure 4.7b. With this architecture,

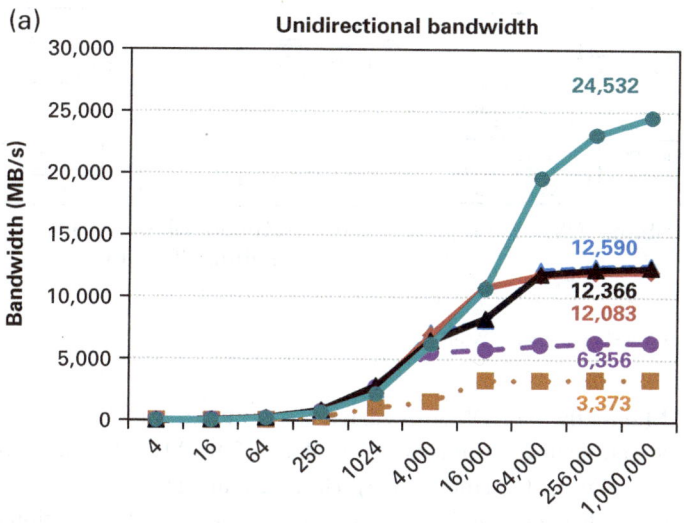

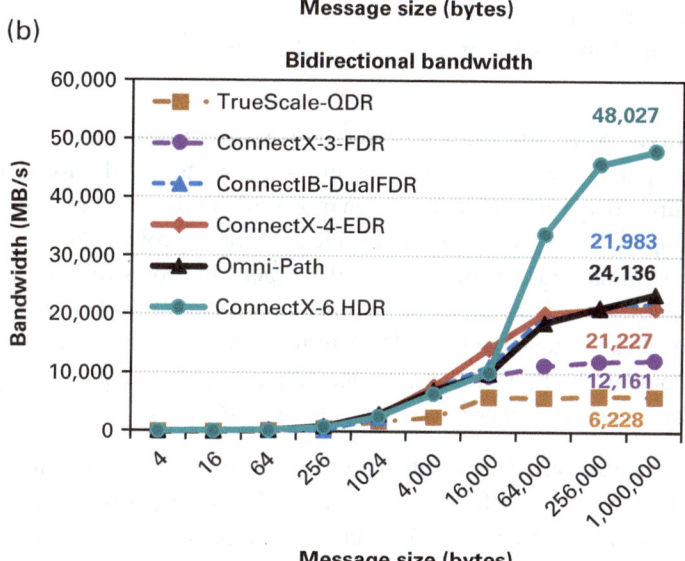

Figure 4.6
Bandwidth: MPI over RDMA networks with MVAPICH2. (a) Unidirectional bandwidth. (b) Bidirectional bandwidth.

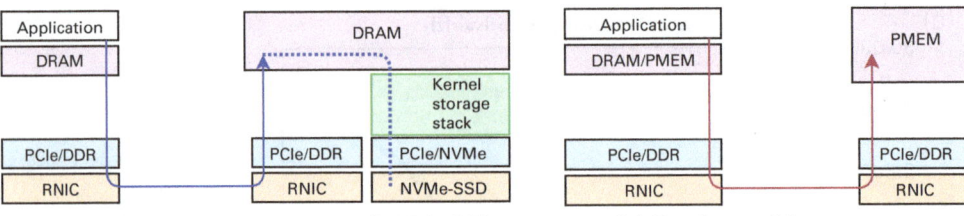

(a) NVM Express over Fabrics (NVMeOF)   (b) Persistent Memory over
                                                                    Fabrics (PMoF)

Figure 4.7
RDMA over NVM: contrasting NVMeoF and remote PMoF

we can access remote PMEM directly via RDMA C. Zhang et al. (2015). An RDMA Read from a target PMEM device can achieve performance similar to DRAM, with hardware optimizations such as Intel Data Direct I/O (Intel, 2012). However, an RDMA-based write into PMEM can only guarantee that the data have arrived at the remote side; additional steps are required to ensure durability.

There are currently two mechanisms being discussed in the community (see Douglas [2015] and Talpey [2015]):

Appliance-based durability: New hardware-assisted (HCA-assisted) commands, such as **RDMA Commit** (Talpey, 2015), with one-sided semantics similar to RDMA Read to support enhanced remote direct data persistence have been proposed (Talpey and Pinkerton, 2016). PCI extension to support RDMA Commit can enable the NIC to provide durability to the PMEM device, with minimal or no involvement of the target CPU cores for flushing the caches. Currently, RDMA Read is being studied to enable such a one-sided remote flush operation. RDMA Read can initiate the target HCA to force data to be flushed to the PMEM device via PCIe (Douglas, 2015). Currently, this technique, however, requires using nonallocating writes that bypass CPU caches for any local or remote writes.

General purpose server durability: A dedicated process on the target node can be explicitly notified by the host performing RDMA Write into PMEM to (1) flush cache-lines associated with the specified data (`clflush`) and invoke a persistent store barrier (`sfence`) and (2) send a message to notify the persist completion.

4.5 Summary

This chapter presents a landscape overview of state-of-the-art hardware capabilities that are being leveraged to design HPC clusters. These capabilities span the latest in computation, storage, and network technologies. To understand how these technologies make HPC possible when combined, consider the second-ranked Top500 (TOP500.org) system (based on

November 2020 ranking), Summit. The Summit (Oak Ridge National Laboratory, 2021b) cluster at the Oak Ridge National Lab consists of 4,356 nodes, each with two IBM Power9 CPUs, six NVIDIA V100 GPUs, and NVLink for high-speed CPU-CPU and CPU-GPU communications. It contains over half a terabyte of memory and a large DDN IME-based burst buffer for efficient I/O, connected via dual-rail Mellanox EDR (100 Gb/s) InfiniBand interconnect. It enables the performance of 148.6 petaFLOPS for scientific HPC application benchmarks.

While scientific computing workloads have demonstrated highly scalable application performance using the latest HPC technologies, Big data applications have yet to fully embrace these technologies. This opens up a lot of avenues for designing current and future big data middleware that explore efficient ways of converging big data and HPC technologies for high-performance data analytics. More discussions on accelerated designs for big data middleware with HPC technologies are presented in chapters 7–9.

5 Opportunities and Challenges in Accelerating Big Data Computing

Previous chapters provided an overview of parallel programming models and systems, parallel and distributed storage systems, and HPC architectures and trends. However, to design big data computing systems and applications with high-performance and scalability, there are many fundamental challenges as well as opportunities facing the big data community. This chapter presents a brief discussion on these challenges and opportunities in accelerating big data computing. In particular, we summarize six concrete challenges from the perspectives of computation, communication, memory/storage, codesign, workload characterization and benchmarking, and system deployment and management. Understanding these challenges will help the community to propose more advanced designs for next-generation high-performance big data computing systems and applications.

5.1 Overview

Data-intensive applications or big data applications have brought significant challenges with current-generation technologies, such as out-of-core data processing, significant I/O bottlenecks, bulk data transferring, and so on. In the last few years, accelerating big data processing and management middleware have become a hot research topic. One driving factor for this is the rapid power and capability growth in multi-/many-core architectures, emerging heterogeneous memory technologies (e.g., DRAM, NVM (Qureshi et al., 2009, Kültürsay et al., 2013), HBM, NVMe-SSD), and high-speed interconnects such as InfiniBand, Omni-Path, RoCE, and so on, as discussed in Chapter 4.

Advances in multi-/many-core processors and accelerators, networking, and heterogeneous memory/storage technologies are providing novel opportunities to design computation, communication, and I/O subsystems to significantly accelerate modern and next-generation big data frameworks. Based on the paradigm shifts seen in current big data processing frameworks, we envision that more and more high-performance big data computing systems, runtimes, and applications will be designed and developed in the future HPC and cloud computing infrastructures.

This chapter first introduces our envisioned architecture for next-generation HPC and cloud computing infrastructures. In the HPC community, such systems are typically called next-generation "high-end computing (HEC) systems." These advanced systems have been allowing scientists and engineers to tackle grand challenges in their respective domains and make significant contributions to their fields. Examples of such domains include astrophysics, earthquake analysis, weather prediction, nanoscience modeling, multiscale and multiphysics modeling, biological computations, computational fluid dynamics, and others. Besides the traditional scientific computing, the fields of big data analytics, machine learning, and deep learning are adding significant demands for HEC systems.

Even though big data applications on the HPC systems are very different in scale and data characteristics from big data applications in data centers or clouds, systems researchers and developers from both fields have common goals such as significantly improving the performance, scalability, resource utilization, manageability, and so on of big data applications. In this book, we discuss how to achieve these common goals for next-generation big data systems and applications on both HPC and cloud computing infrastructures, or the HEC system. Advanced designs and accelerations for big data systems and applications on clouds are discussed in chapter 11.

The HEC systems have been making steady progress during the last two decades. During the last decade, a lot of focus has been given toward designing and using exascale systems (The ASCAC Subcommittee on Exascale Computing, 2010). The community has been in a pre-exascale period since 2018. In June 2018, the Summit (Oak Ridge National Laboratory, 2018) system with a performance capability of 200 petaFLOPS was announced. In June 2020, the Fugaku (RIKEN Center for Computational Science (R-CCS), 2020) system was announced as the fastest supercomputer in the world. After the upgrade in November 2020, Fugaku's performance reached a new world record of 442 petaFLOPS. Multiple countries (the US, China, and Japan) and the European Union are or will be engaged in designing and developing exascale systems during the 2021–2023 period. In particular, three upcoming exascale systems from the US are being developed. The Aurora supercomputer (Argonne National Lab, 2021) is expected to be delivered to Argonne National Laboratory (ANL) by Intel and Cray (now Hewlett Packard Enterprise). It may be the first operational exaFLOPS supercomputer at ANL in 2022. The Frontier (Oak Ridge National Laboratory (ORNL), 2021a) supercomputer at Oak Ridge National Laboratory is anticipated to be operational in 2022 as well. Its performance would be greater than 1.5 exaFLOPS. Another supercomputer, El Capitan (Lawrence Livermore National Laboratory (LLNL), 2020), will be installed at the Lawrence Livermore National Laboratory and may become operational in early 2023. It is expected to have a performance of 2 exaFLOPS. As more HEC systems are deployed and used, there will be continuous demand to run the next-generation big data applications with finer granularity, finer timesteps, larger data sizes, and so on.

Based on the upcoming technological trends, we envision some of the next-generation HEC systems to have architectures as outlined in figure 5.1. The systems will have a large

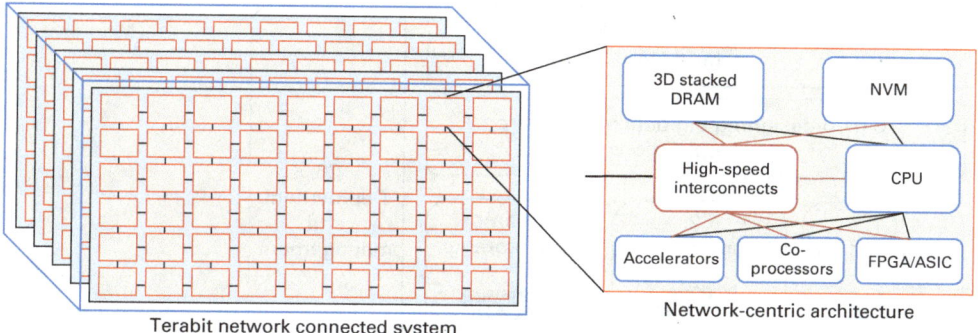

Figure 5.1
Envisioned architecture for next-generation HEC systems. Courtesy of Panda et al. (2018).

number of dense nodes interconnected with networks of terabits/second or higher (DOE Workshop Report, 2011). The dense nodes will be heterogeneous with a mix of CPUs, accelerators, coprocessors, and FPGAs/ASICs. The CPUs also could be a mix of strong and wimpy cores. The nodes will also have a large amount of memory with a combination of different technologies, such as NVRAM and 3D stacked memory. The components within a node or across nodes will be connected with extremely high-speed interconnects. This architecture will be able to provide a large amount of concurrent communication among different components (CPUs, accelerators, and memory) without contention. Because each node will be quite dense, the nodes will need to be connected to the overall terabits/second network with multiple adapters/ports. This will allow a good balance between internode and intranode transfers.

Efficiently running big data systems and applications on top of these envisioned HEC systems is a nontrivial task. As big data systems aim to process large volumes of data, it is critical to codesign the data placement, data movement, job scheduling, and other mechanisms in big data stacks to take advantage of the available NVRAMs, HBMs, SSDs, networks, GPUs, FPGAs, and multicore CPUs in an intelligent manner to reduce data access, management, and data processing time. The codesign approach requires a holistic understanding of the computation, communication, and I/O characteristics of big data frameworks and the interaction across different layers of applications, middleware, and hardware.

To understand the above-mentioned fundamental issues facing the big data community, this chapter tries to summarize the six concrete challenges (denoted C1–C6) in different layers as shown in figure 5.2.

C1: Computational challenges of achieving high-performance big data computing

C2: Communication or data movement challenges of achieving high-performance big data computing

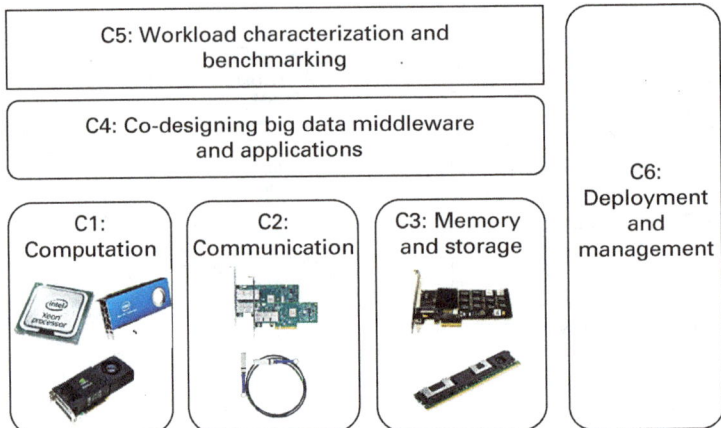

Figure 5.2
Challenges of achieving high-performance big data computing.

C3: Memory and storage management challenges of achieving high-performance big data computing

C4: Challenges of codesigning big data middleware and applications

C5: Challenges of big data workload characterization and benchmarking

C6: Deployment and management challenges of big data systems and applications

We will briefly discuss these challenges in detail in the following sections of this chapter.

5.2 C1: Computational Challenges

As discussed in chapter 4, modern computing platforms currently support approximately 32–128 processing cores per node. These platforms are rapidly changing and becoming heterogeneous with many-core processors and accelerators such as NVIDIA GPUs, AMD GPUs, FPGAs, ASICs, among others. With the emergence of multi-/many-core architectures, it is becoming critical to provide efficient onload and offload solutions for computation-intensive applications in big data domain. Such redesigns and optimizations can help big data middleware components to utilize multicore processors and accelerators for computation as well as to increase overlap capabilities with several other phases in the overall execution, thereby improving performance and scalability. However, to efficiently leverage these HPC resources, we need to solve many fundamental challenges for big data systems and applications.

Efficient NUMA-aware designs on CPUs: Nonuniform memory access (NUMA) is a common microarchitecture for modern CPUs. Under NUMA, cores in one CPU NUMA node can access their own local memory much faster than they can access remote memory (nonlocal) on another NUMA node. In this case, if data placement or access operations happen on remote NUMA memory, then the performance of big data applications and systems suffers significantly. Unfortunately, most of the modern big data middleware, such as Hadoop and Spark, is designed in an architecture-unaware manner. Thus, their computational engines cannot be fully NUMA-aware, and then there will be performance loss due to the remote memory access penalty of NUMA. There are many NUMA-aware studies in the literature (Ott, 2011, Lu, Shi, Shankar et al., 2017) that discuss how to optimize the performance of big data applications and systems on NUMA architectures. However, they are still not systematically exploited in modern big data systems because many technical challenges of NUMA have to be solved in different components and layers in these big data systems. More specifically, to achieve efficient NUMA-aware designs in big data systems and applications, we need to balance the workloads across cores on different NUMA nodes, achieve good locality for all the data accesses, avoid possible data movement and communication congestion overhead, and so on. All of these are fundamental challenges of designing efficient NUMA-aware schemes in big data systems and applications on modern and next-generation CPU architectures.

Efficient vectorization designs with SIMD on CPUs: On modern multicore CPU architectures (e.g., Intel Xeon processors) that support data parallelism via vectorization, there have been several studies directed toward leveraging CPU–single instruction, multiple data (SIMD) instructions for accelerating computation for data-intensive workloads. Using SIMD instructions to enable data parallelism has been studied for accelerating database operators such as scan, join, and aggregation (Polychroniou et al., 2015, Zhou and Ross, 2002), as well as Bloom filters (Polychroniou and Ross, 2014, Lu, Wan, et al., 2019). SIMD instructions have also been leveraged to enable data parallel key lookups over hash tables (Shankar et al., 2019b) for join operations (Ross, 2007, Behrens et al., 2018, Polychroniou et al., 2015). The modern Intel Skylake and Cascade Lake CPUs support SSE (128-bit), AVX2 (256-bit), and AVX-512 (512-bit) vector instructions. Specifically, with 512-bit extensions to the AVX SIMD instructions (AVX-512) for x86 ISA, we have the opportunity to operate on an entire cache-line in a single instruction. Thus, with this increasing vector-processing capability on modern hardware, we have the option to leverage different CPU-SIMD vector widths to enable different degrees of data parallelism for a given SIMD vectorization approach. These SIMD technology variants define important challenges: What kind of SIMD parallelism (hardware capability or CPU vector instructions) can we exploit to accelerate big data systems for a given application workload? Are there any design guidances or tools for SIMD to help researchers and developers to design their efficient vectorized computational engines in big data systems? Some preliminary SIMD benchmarking toolkits such as SimdHT-Bench (Shankar et al., 2019b) may be useful for answering these questions.

Accelerator-aware designs in big data stacks: Accelerators such as GPUs have been widely deployed on modern HPC and data center clusters. GPUs offer a magnitude higher peak performance than that of traditional CPUs. While GPUs have been mainly leveraged to accelerate computations involved in the scientific, machine, and deep learning applications, their use within big data stacks has become the focus of recent research. One such work (Al-Kiswany et al., 2013) focuses on offloading computationally intensive primitives based on hashing and introducing techniques to efficiently leverage the processing power of GPUs. Less computation-intensive middleware, such as KV stores, has also been exploring the use of GPU-aware designs. MemcachedGPU (Hetherington et al., 2015) proposed an accelerated KV store that can leverage the full system on contemporary hardware, by offloading User Datagram Protocol packet progressing and hash table access operations to the GPU. MegaKV (K. Zhang, Wang, et al., 2015) leverages the high memory bandwidth of modern GPUs and its data hiding capabilities to achieve higher server throughput. GPU-aware database and data warehousing systems (Govindaraju et al., 2004, Yuan, Yuan and Lee, Rubao and Zhang, Xiaodong, 2013) have also been proposed in many research studies. GPU-based designs in MapReduce-like systems (Yuan et al., 2016, Govindaraju et al., 2011) are also available in the literature. Typically, there are various ways to utilize GPUs or other accelerators in big data stacks. One approach is to use GPUs for accelerating big data workloads or libraries without changing their interfaces and programming models. With this approach, significant improvement needs to be made inside big data stacks to integrate GPUs seamlessly. The improvement cannot be fully compatible with their CPU-based counterparts. Another approach to leverage GPUs or other accelerators is to automatically generate CUDA code from the CPU code, which typically needs compiler-level support. In addition, some big data stacks choose to directly integrate some GPU-enabled application kernels or libraries, which typically are written with CUDA natively. All of these approaches need a lot of effort to redesign the computational engines, resource management, and task scheduling subsystems in big data stacks. Indeed these redesigns will create many technical challenges.

For other types of accelerators or processing units, such as FPGAs or ASICs, similar technical challenges exist, as we have discussed here. We will further discuss many advanced accelerations with multi-/many-core technologies for achieving high-performance big data computing in chapter 8.

5.3   C2: Communication and Data Movement Challenges

Modern and next-generation HEC systems will be equipped with high-speed interconnects. These advanced networking technologies will provide vast opportunities and critical challenges for designing efficient and scalable communication and data movement mechanisms.

Tamping heterogeneous networking technologies: Heterogeneous networking technologies can be used at different levels (intracore, intranode, intrarack, and interrack) based on their power-performance trade-off. For example, as discussed in chapter 4, the NVLink interconnect might be optimal at the intranode level to connect both CPUs and GPUs. Infini-Band or RoCE interconnect might be optimal at the internode and interrack levels. These advanced networking technologies will allow the establishment of high-dimensional topologies within intracore, intranode, intrarack, and interrack levels. These technologies will also allow having one-to-many, many-to-one, and many-to-many communication patterns in a flexible manner with excellent performance. Thus, the coexistence of these heterogeneous technologies on a given system will provide a lot of flexibility to the upper layers to design highly efficient communication protocols. Because next-generation high-speed networks will be operating at hundreds of gigabits or multi-terabit speeds, the communication mechanisms at different levels of the system need to be coupled together to provide the best performance for end-to-end communication across any two processes in the system. Otherwise, the benefits of these modern technologies will not be exploited properly. These are some of the challenges the designers of next-generation HEC systems need to consider and create innovative solutions for.

Efficient in-network communication and computation protocols: As high-performance networking technologies advance, more and more capabilities are being offloaded to the network, leading to an era of "in-network computing." Existing and emerging schemes such as CORE-Direct (Collectives Offload Resource Engine) (Mellanox, 2010), Erasure Coding Offloading (Shi and Lu, 2019, 2020), and Scalable Hierarchical Aggregation and Reduction Protocol (SHARP) (Graham et al., 2016) on modern high-speed networks are good examples of this trend. The CORE-Direct technology can offload the MPI collectives operations from the software library to the network, which can provide the capability for overlapping communications with computations (Mellanox, 2010). Similarly, the Erasure Coding Offloading capability on modern smart network interface cards (SmartNICs) can fully offload Erasure Coding computation tasks to the SmartNICs (Shi and Lu, 2019, 2020), which can minimize the CPU utilization and achieve better overlapping. The SHARP technology from Mellanox can further offload collective operations from the CPU to the switch network, which can improve the performance of MPI collective operations (Graham et al., 2016).

The trend of in-network computing is allowing different types of communication and computation operations such as point-to-point unicast, one-to-many multicast, many-to-one gather, all-to-all, encoding, and decoding operations to be implemented in the network efficiently. The above-mentioned techniques will become essential to achieve peak performance. Such capabilities and middleware capable of using these schemes will be essential to propel scientific and big data applications to post-exascale performance. As communication networks become large, the necessity for communication and synchronization protocols to be "network topology–aware" will become very critical to minimize network contention. In this context, networks at each level on next-generation HEC systems must provide topology information with low overhead, which can be used by algorithms and communication protocols from upper layers to provide network topology–aware communication and minimize network contention.

Integrated support with accelerators: As discussed in section 5.2, next-generation HEC systems will use accelerators and FPGA-/ASIC-based components in an extensive manner; therefore, the communication mechanisms and protocols also need to be redesigned to provide integrated support for these devices in a heterogeneous environment. Such protocols need to be designed to provide high-performance communication with high productivity. For example, data type support in communication libraries is a fundamental challenge. Communicating noncontiguous data is often simply implemented through packing the data into contiguous buffers on the sender side while the receiver unpacks the data into the user buffer in many big data processing libraries or frameworks. This approach takes away CPU cycles from the application and provides poor performance, especially for large messages, because it requires copying the data multiple times. To address the issue of unnecessary memory copies and lack of overlap opportunities in the data type processing, designs where the runtime offloads the data packing/unpacking to FPGAs and smart adapters are essential. Solutions such as BlueField (Mellanox, 2018) are excellent examples of such emerging trends. Furthermore, the advent of new fields such as deep learning has brought forward extreme and custom communication/computation requirements that runtimes must address. GPUs, FPGAs, and other ASICs, with the ability to express such complex communication and computation requirements with high performance, are seen as excellent fits for such requirements.

Fault-tolerance and QoS support: As HEC and post-exascale systems will be massive with millions of components, frequent failures will be common. Thus, networks at each level need to provide mechanisms for fault-tolerance so that the upper levels of the software can be designed to make the software infrastructure fault-tolerant. Emerging applications (such as deep learning and streaming) are interactive in nature. Furthermore, one needs to account for the interaction between multiple different applications that can potentially use the same interconnection network. For instance, if a checkpoint operation is in progress at the same time as a dense communication operation, they are very likely to interfere with each other, resulting in poor performance for both applications. In this context, providing quality of service (QoS) for different applications as well as for different communication operations in the same application will be important. It will also be critical for next-generation networking technologies to provide QoS mechanisms at the lowest level so that QoS-aware solutions for performance isolation across communication streams and jobs can be provided. Next-generation HEC systems will have a set of heterogeneous networks. Even though each network can provide good QoS support, achieving end-to-end QoS across heterogeneous networks will be challenging and will require innovative solutions.

Energy-aware communication schemes and protocols: High energy consumption will be a crucial challenge for next-generation HEC systems. In this context, the networking technologies must provide mechanisms for energy savings (such as an automatic reduction in speed in the absence of data transfer, variable speeds with different energy trade-offs). Furthermore, the higher-level communication and synchronization protocols used by programming models and their runtimes, which actually move the data around the HEC systems, must be designed with energy efficiency as one of the core objectives. This cannot be achieved through existing approaches that consider the runtime as a black box and use techniques such as CPU throttling and frequency scaling based on high-level energy and performance measurements. A holistic approach that encompasses a careful understanding of the power characteristics of the data movement phase, integrated design of the different communication protocols, and finally a codesign between the runtime and applications will be essential. Furthermore, it will also be essential for the different software components on the HEC environment (scheduler, application, and runtime) to have complementary and adaptive energy-conserving strategies to achieve maximum savings while minimizing the impact on performance.

Software-defined networking: During the past few years, the field of software-defined networking (SDN) is getting more prominence in the cloud computing area. As modern HEC systems move toward supporting interactive workloads with data analytics and visualization, the requirements for persistent connectivity (such as available in SDN with OpenFlow) are gaining more importance. The current-generation high-performance interconnects, such as InfiniBand, incorporate some SDN functionalities (in InfiniBand's case, through its OpenSM Subnet manager). However, per-flow resource management is not there yet. Researchers have been exploring how to incorporate such resource management for InfiniBand. During the coming decade, the SDN technology will mature and will be a critical desired feature for next-generation HEC systems. This will allow big data applications to take advantage of SDN features and functionalities. However, providing SDN functionalities across heterogeneous networks will be a challenge and will require innovative solutions.

Scalable communication protocols with heterogeneous memory and storage: As HEC systems will use not only DRAM but also other types of emerging memory and storage technologies, such as NVM and NVMe-SSD, the advanced features of NVM provide great opportunities to design novel high-performance communication and high-throughput I/O subsystems for big data applications. Technologies such as NVMeoF (NVMe Express, 2016) have enabled offloading of various computing tasks associated with I/O operations to the network. The birth of the NVMe standard has changed the storage landscape. The lower latency and high scalability offered by this standard provide unprecedented high performance. The emerging NVMe-oF standard allows low-latency access to remote flash devices using NVMe commands, offering the possibility to fundamentally redesign storage systems. The next-generation communication and I/O runtime need to be redesigned to take NVM and NVMe technologies into account. We envision that the NVM- and NVMe-aware communication and I/O protocols with RDMA can fundamentally change the landscape of the communication and I/O subsystem designs in next-generation HEC systems.

To address many of these communication and data movement challenges, we will further discuss different state-of-the-art designs for accelerating big data middleware and libraries over RDMA-capable networks for achieving high-performance big data Computing in chapter 7.

5.4   C3: Memory and Storage Management Challenges

Many recent designs for big data processing and management middleware to achieve high-performance have benefited from in-memory technologies (Li, Ghodsi, et al., 2014, Zaharia et al., 2012). However, for most of these designs, the data reside on DRAM and the I/O operations will read/write data from/to DRAM. While this design will improve the raw I/O performance of big data middleware, it also brings additional design challenges arising from the nonpersistence of data due to DRAM's volatile nature. To overcome this, big data middleware typically needs to "spill" data from DRAM into SSD or HDD, which causes extra I/O overhead. On the other hand, emerging NVM technology is being deployed on modern HPC clusters, and this technology will become more common on next-generation HEC systems. NVMs can offer byte-addressability with DRAM-like performance along with data persistence.

Similarly, NVMe-SSDs are being provided by most modern data centers. NVMe-SSDs offer low latency and highly parallel I/O processing along with good random read performance. Moreover, NVMe-oF (NVMe Express, 2016) provides the ability to access remote SSDs directly by using the NVMe standard. To highlight the potential impact of NVM technologies on designing next-generation big data analytics stacks, we can compare the performance of a random write benchmark on an Intel NVMe SSD using POSIX, NVMe, and NVMe-oF. We find that POSIX shows poor performance compared to NVMe and NVMf even though it uses the NVMe protocol underneath, highlighting the need for direct use of the NVMe standard. Unfortunately, current-generation big data stacks still heavily use POSIX-based I/O subsystems, which cause suboptimal performance and scalability. These performance trends and features of NVMe technologies provide great opportunities to design novel high-performance communication and high-throughput I/O subsystems for big data applications. Some of recent studies (Pelley et al., 2013, Huang et al., 2014, Arulraj et al., 2015, Apache Software Foundation, 2019) have shown that their initial designs with NVM and NVMe-SSD can benefit databases, KV stores, distributed file systems, and so on.

However, these recent studies still could not fully embrace all the features available on the emerging NVM architectures and propose highly optimized designs by combining NVM with high-speed networking and accelerator technologies. For instance, many of the existing designs such as (see discussions in Pelley et al. 2013; Huang et al. 2014; Arulraj et al. 2015), still focus on how to use NVM to achieve persistence for local data storing. These types of designs could not fully utilize all the NVM devices on the whole cluster to handle big data challenges. To use all the NVM devices efficiently across the data center, the current designs need to be able to achieve high performance and scalability by taking advantage of all the capabilities from the emerging NVM and NVMe technologies together with high-performance interconnects and accelerators.

All of the above-discussed trends raise two broad challenges: (1) What kind of architectural revolutions are going to happen for NVM-centric designs in big data analytics stacks during the coming years? (2) Can such revolutions benefit from scalable and resilient data processing schemes for designing next-generation NVM-aware big data processing? We envision that codesigning the emerging NVM and NVMe technologies together with the high-speed networking and accelerator capability delivered by next-generation HEC systems, the current-generation big data stacks no longer need to use the traditional POSIX-based I/O stack. Instead, the NVM- and NVMe-aware communication and I/O protocols can fundamentally change the landscape of the communication and I/O subsystem designs in the current-generation big data systems. Our vision is that NVM-aware designs with high-performance interconnect and accelerators can not only significantly improve the performance of these stacks, but they can also benefit the resiliency and scalability of these stacks.

To discuss more along with these research challenges, we will further present many advanced designs for accelerating big data systems and libraries with high-performance storage technologies for achieving high-performance big data computing in chapter 9.

5.5 C4: Challenges of Codesigning Big Data Systems and Applications

To exploit the high-performance capabilities facilitated by the advanced hardware components on next-generation HEC systems, it will be vital to codesign the upper-layer big data and AI applications and the corresponding systems with the underlying next-generation computation, communication, and I/O runtimes. The designs of the computation, communication, and I/O runtimes, as well as the co-designs of big data systems and applications, are not independent of each other. The research and development process for such codesigning on next-generation HEC systems will need to follow a spiral model. For instance, along with designing the advanced features of high-performance and energy-aware communication and synchronization protocols, heterogeneous memory, and storage support, it will be critical to understand and identify the performance governing patterns in the end applications, which will give more insights to the bottlenecks specific to these applications as they execute on next-generation HEC systems. For instance, tracing and profiling the data movement among different components (e.g., CPUs, accelerators, heterogeneous memory, and FPGAs/ASICs) will be used in designing the high-performance and energy-aware communication schemes. Furthermore, once the predictable characteristics are understood by the community, execution time estimation for different computation and memory access patterns will form the basis for designing the data placement techniques, as well as data management schemes, respectively.

As we can imagine, there are a lot of other types of open challenges that need to be carefully resolved when we codesign big data systems and applications with the components as discussed in the preceding sections. In the following chapters, especially chapters 7–10, we will present many different kinds of case studies to show how the community is able to propose many advanced designs and codesigns to handle these challenges.

5.6   C5: Challenges of Big Data Workload Characterization and Benchmarking

To efficiently codesign the big data systems and applications as discussed earlier, their innate performance characteristics need to be well understood. While there are third-party tools that help understand the execution behavior of some of the big data frameworks, not much work has been done in understanding their interactions with the system, applications, and underlying hardware architectures in a holistic manner. To address this lack of understanding, the community needs to propose new performance characterization frameworks or tools that will be capable of studying the interactions among the various components in figure 5.2 with a particular focus on their computation, communication, and I/O characteristics. The frameworks or tools need to be designed to capture vital metrics both at the system and software levels, such as interprocess communication patterns, block-level I/O request counters, physical memory usage, file system requests and caching behavior, network resources utilization, shared-memory usage, and so on.

In addition to statically characterizing the computation, communication, and I/O behavior of these applications for offline analysis, the frameworks or tools should also be capable of dynamically providing feedback to the big data processing and management systems and applications in the form of hints, which can help make informed optimization decisions at runtime. For instance, these hints can be used to dynamically tune the number of worker threads inside the big data systems or to dynamically decide on the thresholds to switch between the various network transport protocols based on the communication patterns of big data applications. The knowledge of workload characterization will guide and steer the appropriate designs and optimizations. Such workload characterization tools can be integrated into the spiral development model discussed in section 5.5. This will ensure that the interactions among various components of the system are taken into consideration and evaluated incrementally and that the benefits obtained are consistently increasing.

To help with performing meaningful big data workload characterization and performance evaluation tasks, the community needs more well-designed benchmarks and datasets. There exist many benchmarks that attempt to capture the behavior of various data-intensive computing environments. But most of these benchmarks are proposed to evaluate the system in an integrated manner, which means that even to run a simple benchmark, such as Hadoop TeraSort, we still need to deploy a heavy stack and make all the components run together to finish the benchmark. In this case, the performance numbers coming out from the benchmarks are affected by multiple components, which may prevent us from analyzing individual components. We believe that the big data community needs more microbenchmarks that can have very specific benchmarking goals, and these microbenchmarks should be able to run with individual components in an isolated fashion. Some preliminary microbenchmarks (Islam et al., 2012) have been proposed in the community to attempt to achieve this object. However, there are still many opportunities for more big data benchmarking research studies in the community.

To help understand what kind of benchmarks are available in the big data community, we will systematically summarize many different types of benchmarks in chapter 6.

5.7   C6: Deployment and Management Challenges

A typical big data processing and management framework has a pretty heavy stack that usually involves many single- or third-party components. These components may include application components, big data analytics frameworks, resource schedulers, distributed storage systems, and many other supporting libraries and runtimes. To deploy a fully functioning big data system, we need to solve many software dependency issues with a lot of engineering effort. There are different systems constraints for deploying and managing big data applications on HPC and cloud platforms. We need to deploy the big data systems and applications on these platforms and make their services become available as soon as possible. We also need to make sure the deployed systems and applications can run as efficiently as possible. In many cases, these tasks are not easy to accomplish because there could be many possible issues happening in different components or layers. To solve these challenges systematically, container-based or virtual machine–based application appliance techniques (i.e., encapsulate the software stack as the container or virtual machine images) seem promising. However, with such a virtualization layer, the application's runtime performance may get affected. How to deploy and manage big data systems and applications on modern and next-generation HEC platforms is still challenging. This needs a lot of new tools and designs to make big data systems and applications easy-to-deploy and easy-to-manage.

Regarding their performance, these big data systems and applications also need to automatically detect hosting environments and adjust their execution behaviors to deliver efficient performance on different platforms. Along with this research direction, in chapter 11, we will present many cloud-related studies for big data systems that are currently in the literature and community, especially focusing on discussing how to achieve high-performance designs with HPC cloud technologies.

5.8 Summary

In this chapter, we focus on discussing some concrete research challenges for achieving high-performance big data computing from the perspectives of computation, communication, memory/storage, codesign, workload characterization and benchmarking, and system deployment and management. These challenges will be very important for the community when designing or codesigning next-generation high-performance big-data computing runtimes, systems, and applications on top of next-generation HEC systems. In the following chapters, we will discuss more concrete design studies to show how the community has been trying to address some of these challenges from different angles.

6 Benchmarking Big Data Systems

Big data applications are reshaping both the business and scientific research domains. With the growth in available raw computational power, new and more advanced data processing systems are being developed every day. Therefore, studying and benchmarking the performance and capabilities of these frameworks are an integral part of developing new big data middleware. This chapter presents a survey on the different benchmarks being designed and developed to evaluate current and emerging big data middleware systems. The performance results of new designs using these benchmarks will be discussed in chapters 7–11.

6.1 Overview

As architecture, data processing frameworks, and application developers divert their attention toward innovative hardware/software codesign for big data and machine learning algorithms, there is a need for a systematic way to evaluate these evolving hardware and software systems. Benchmarking provides a means to these frameworks with complex, diverse, and dynamic workloads. These benchmarks must support scalability and portability studies, while they enable reproducibility and easy interpretation of performance data across the diversity of datasets and workloads.

Big data workloads span different application domains, including Internet services (e.g., search engines, social networks, e-commerce), AI, and medical sciences. As discussed in chapter 2, these workloads are managed and processed by different data middleware systems. Therefore, as shown in figure 6.1, big data benchmarks can be classified along the five types of data processing frameworks, as follows:

1. Offline data analytical benchmarks
2. Streaming benchmarks
3. Online data processing benchmarks
4. Graph benchmarks
5. Machine learning and deep learning benchmarks

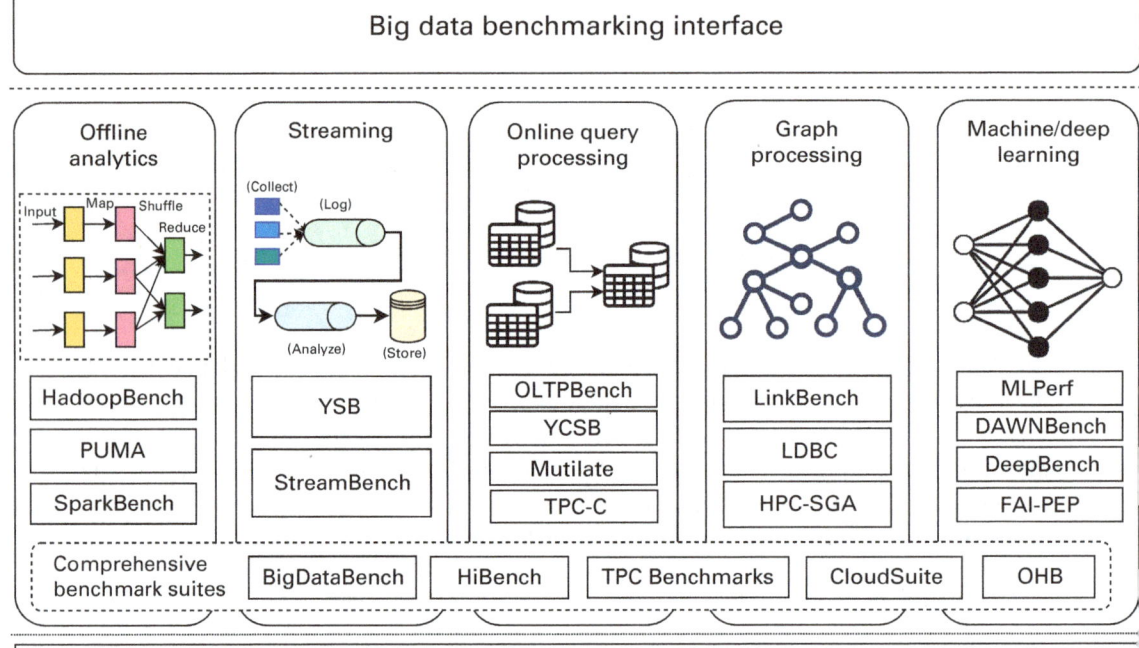

Figure 6.1
Big data benchmarks.

In this chapter, we will explore these different big data benchmarking domains and discuss popular benchmarks and their features along each of these directions and their features. We also explore comprehensive benchmark suites designed to benchmark big data ecosystems as a whole.

6.2 Offline Analytical Data Processing

Offline data analytics or batch processing involves iterative computations over large datasets. As discussed in chapter 2, these include frameworks such as Hadoop MapReduce and Spark. Some of the popular benchmarks for these data processing frameworks are discussed in the following sections.

6.2.1 Hadoop Benchmarks

Hadoop comes prepackaged with a set of microbenchmarks, HadoopBench, that can be used to test the fine-grained functionality of various components (Apache Software Foundation, 2021e). These include:

1. **TestDFSIO:** TestDFSIO is an HDFS benchmark and is available as part of the Hadoop distribution. It measures the I/O performance of HDFS. It is implemented as a MapReduce job in which each map task opens one file to perform a sequential write or read and measures the data I/O size and execution time of that task. A single reduce task is launched at the end of the map tasks that aggregates the performance results of all the map tasks.

2. **Sort:** The Sort benchmark uses the MapReduce framework to sort the input directory into the output directory. The inputs and outputs must be sequence files where the keys and values are stored. In this benchmark, the key and value lengths can be as small as 10 bytes to as large as thousands of bytes. The input of this benchmark can be generated by RandomWriter, which writes random-sized, KV pairs in HDFS. This benchmark is a very useful tool to measure the performance efficiency of MapReduce clusters with random KV sizes.

3. **TeraGen and TeraSort:** TeraGen is an I/O-intensive benchmark that generates the input data for TeraSort and stores it in the underlying file system. On the other hand, TeraSort is shuffle-intensive. It reads data from the file system, and then it performs a sort and writes the output back to the file system. These benchmarks have both Hadoop MapReduce and Spark implementations.

4. **WordCount and Grep:** The WordCount microbenchmark counts the appearance times of all words in the input dataset. This benchmark reads text files and counts the frequency of the words. Each mapper takes a line of the input file as input and breaks it into words. It then emits a KV pair of the word (in the form of (`word, 1`)) and each reducer sums the counts for each word and emits a single key/value with the word and sum. Similarly, the Grep microbenchmark extracts matching strings from input files and counts the appearance times of the strings.

6.2.2 PUMA Benchmarks

Purdue MapReduce (PUMA) benchmarks suite, designed by Ahmad et al. (2012), contains thirteen benchmarks developed by Purdue University. It represents a broad range of MapReduce applications exhibiting application characteristics with different computation and shuffle volumes. In other words, the PUMA benchmark suite includes a mix of computation-intensive, I/O-intensive, and shuffle-intensive benchmarks. It also contains three benchmarks from Hadoop distribution.

PUMA includes well-known search engine workloads, such as Grep and SequenceCount, that search for patterns in a large number of files and represent a class of programs that extract a small amount of interesting data from a large dataset. Similarly, the popular search engine indexing algorithm Ranked Inverted Index is used for full-text searches. The benchmark suite also contains the Adjacency List benchmark that generates adjacency and reverses adjacency lists of nodes of a graph for use by PageRank-like algorithms, whereas Self Join generates association among $k + 1$ fields given the set of k-field associations. MapReduce has also been used to parallelize existing data mining and machine learning algorithms (e.g., Apache Mahout machine learning library based on Hadoop MapReduce).

To evaluate such workloads, PUMA includes workloads such as the K-Means clustering algorithm used for knowledge discovery in data mining and adaptive machine learning algorithms that generate histograms for studying data trends (e.g., user ratings for a movies dataset).

6.2.3 SparkBench

SparkBench, by (M. Li et al., 2015), is a Spark-specific benchmark suite. It selectively embraces a set of representative applications to identify various performance bottlenecks and reveals the resource consumption behaviors across execution phases. It covers four critical usage patterns of Spark, including machine learning, graph processing, stream computations, and SQL query processing. It enables Spark users to study the performance, resource consumption, dataflow, and configuration parameter optimizations of different Spark application patterns. SparkBench can generate data according to many different configurable generators and store that data in different distributed or parallel storage systems such as HDFS, S3, and so on. Workloads are run as standalone Spark jobs that can be launched serially or in parallel.

6.2.4 AMPLab Big Data Benchmark

The University of California–Berkeley AMPLab introduced big data Benchmark (AMPLab, 2021) to compare and contrast big data systems that implement data-warehousing solutions on top of offline analytical frameworks such as Hadoop. These include analytic databases such as Impala (Apache Software Foundation, 2021i), SparkSQL (Apache Software Foundation, 2021s), and Apache Hive (Apache Software Foundation, 2021h). This benchmark measures the performance by employing relational queries, such as scans, aggregations, joins, and user-defined functions across different data sizes. For simplicity and reproducibility, it chooses to test platforms using a simple storage format, such as the compressed SequenceFile.

6.3 Streaming Data Processing

With the rise of web 2.0 and the Internet of things, it has become feasible to track information concerning fine-grained user activities and sensor data over time. Because these real-time application scenarios require that data be processed in a time-sensitive manner, users and developers have looked beyond the traditional batch-oriented approaches. This led to the emergence of distributed data processing systems, known as streaming frameworks, such as Apache Flink, Apache Storm, and others, which were discussed in chapter 2. Correspondingly, several benchmarks have been designed and released to qualitatively compare the different contenders. We discuss a few of these popular stream data processing benchmarks in the following sections.

6.3.1 Yahoo! Streaming Benchmark

Yahoo! Streaming Benchmark (YSB) (Chintapalli et al., 2016, 2021) is a stream processing benchmark that enables frameworks to evaluate a real-world use case. The benchmark is a simple advertisement application. It consists of several advertising campaigns and some advertisements for each campaign. The job of the benchmark is to read various JSON events from an event source such as Apache Kafka (Apache Software Foundation, 2021j), identify the relevant events, and store a windowed count of relevant events per campaign into a Redis server. These steps attempt to probe some common operations performed on streaming data streams. YSB currently supports Flink, Storm, and Spark streaming.

6.3.2 Stream Bench

With the increasing complexity of stream computing and the diversity in workloads leveraging these frameworks, benchmarking these systems presents newer and greater challenges. Toward providing standard criteria for benchmarking modern distributed streaming data frameworks, (R. Lu et al., 2014) present Stream Bench. Stream Bench proposes a message system functioning as a mediator between stream data generation and consumption. It also covers seven benchmarks intended to address typical streaming scenarios. In addition to evaluating performance with different data scales, it also takes fault-tolerance ability and durability into account.

6.4 Online Data Processing

Online processing is interactive; hence, it allows constant interaction between the user and the system. These systems include transactional SQL databases and high-performance semi-structured NoSQL database systems. Some popular benchmarks for online data processing applications running over these systems are presented in the following sections.

6.4.1 OLTPBench

OLTPBench, as presented by Difallah et al. (2013), is a "batteries included" test bed for benchmarking database management systems. It allows a user to run performance analysis experiments with tight control on the mixture, rate, and access distribution of transactions for its fifteen built-in workloads. The workloads include synthetic microbenchmarks, popular OLTP benchmarks, and real-world web applications. These include workloads such as AuctionMark, Wikipedia, TPC Benchmark C (TPC-C), Telecom Application Transaction Processing Benchmark, Stonebraker Electronic Airline Ticketing System, Twitter, Epinions.com, and many more.

OLTPBench is designed with a multithreaded load generator. The framework can produce variable rate, variable mixture load against any Java Database Connectivity–enabled relational database. It also provides data collection features, such as per-transaction-type latency and throughput logs.

6.4.2   Yahoo! Cloud Serving Benchmark

Cloud serving workloads are online data processing workloads that give higher priority to scalability, availability, and performance over the consistency of the data. Yahoo! Cloud Serving Benchmark (YCSB) (Cooper et al., 2010) is a typical benchmark for these cloud data processing applications. These services are often deployed using NoSQL-like distributed KV and cloud serving stores. It consists of an extensible workload generator that models a variety of data serving systems into five core workloads. The YCSB clients generate and load the data into the remote data storage layer. It generates an access pattern based on the chosen workload or input parameter selection and evaluates the underlying store to benchmark the performance and scalability of the cloud serving system. This comprehensive suite of standardized benchmarks can be used to evaluate the performance of different NoSQL databases, including, but not limited to, HBase, Cassandra, and MongoDB. These consist of six different batch-processing workloads for cloud systems, namely:

- Workload A: 50 percent read, 50 percent write
- Workload B: 95 percent read, 5 percent write
- Workload C: 100 percent read
- Workload D: New records are inserted and read
- Workload E: Short ranges of records are queried, instead of individual
- Workload F: Read a record, modify it, and write back the changes

6.4.3   Mutilate

Mutilate (Jacob, 2021) is a load generator designed for high request rates, good tail-latency measurements, and realistic request stream generation. It is used to benchmark distributed caching layers that are built based on KV stores like Memcached. Mutilate reports the latency (average, minimum, and various percentiles) for Memcached Get and Set commands, as well as for achieved queries per second (QPS) and network throughput. It enables mimicking and benchmarking real deployments of Memcached, which often handle the thousands of requests of and front-end clients simultaneously. The workload generator of Mutilate enables generating synthetic queries that are based on Facebook's real-world datasets and traces (Atikoglu et al., 2012).

6.5   Graph Data Processing

Graph data processing problems are representative of the fundamental core of traditional and emerging data-intensive applications, including complex network analysis, data mining, scientific computing, computational biology, as well as applications in security. Real-world networks such as the Internet, the worldwide web, social interactions, and transportation networks are analyzed by modeling them as graphs. To efficiently solve large-scale graph problems, it is necessary to design HPC systems and novel parallel algorithms, which were discussed in chapter 2. In this section, we review some well-known benchmarks for graph data processing and dedicated graph database systems.

6.5.1 LinkBench

Social network data can be modeled as a social graph, wherein the entities or nodes, such as people, posts, comments, and pages, are connected by links that model different relationships between the nodes. Different types of links can represent the friendship between two users, a user liking another object, ownership of a post, or any other relationship. The goal of LinkBench (Armstrong et al., 2013) is to emulate such a social graph database workload and provide a realistic benchmark for database performance on social workloads.

LinkBench is highly configurable and extensible and can generate a large synthetic social graph with key properties similar to the real graph. It contains traces from the production database of Facebook's social graph data, and the data generator produces synthetic data that has key properties of the workload to match production workloads. The benchmark covers standard insert, update, and deletion operations as well as key lookup, range, and count queries.

6.5.2 LDBC Graphalytics

The Linked Data Benchmark Council (LDBC) (LDBC, 2021) has developed two graph processing benchmarks: the Social Networks Benchmark (SNB) and the Semantic Publishing Benchmark (SPB). The SNB presents three different workloads: (1) an interactive workload that tests a system's throughput with relatively simple queries with concurrent updates, (2) business intelligence workload that consists of complex structured queries for analyzing online behavior of users for marketing purposes and touches more data as the database size grows, and (3) graph analytics workload that cannot typically be expressed using a query language like SPARQL. These offline graph analytics workloads cover algorithms such as PageRank, Clustering, and Breadth-First Search, which can produce large intermediate results. The SPB is a benchmark for testing the performance of Resource Description Framework (RDF) engines inspired by the media or publishing industry. The SPB performance is measured by producing a workload of create, read, update, delete operations that are executed simultaneously. The benchmark offers a data generator that uses real reference data to produce datasets of various sizes and tests the scalability aspect of RDF systems.

6.5.3 HPC Scalable Graph Analysis

HPC Scalable Graph Analysis Benchmark (HPC-SGA) (Bader et al., 2021) consists of four operations on a weighted directed graph with node degrees that follow a power-law distribution. This benchmark is based on the High Productivity Computing System Scalable

Synthetic Compact Applications graph analysis benchmark, which is characterized by integer operations, a large memory footprint, and irregular memory access patterns. This HPC-centric graph theory benchmark is representative of computational kernels in computational biology, complex network analysis, and so on, and is very unlike typical graph database workloads. The benchmark data generator produces nodes and edges with a positive weight value. The four types of operations supported include bulk insertions, retrieval of edges with the largest weight, extraction of a k-hop path from a given edge in an edge set, and extraction of a set of edges that identifies vertices of key importance along shortest paths of the graph.

6.6 Machine Learning and Deep Learning Workloads

Machine learning and deep learning have recently been gaining popularity. From the microarchitecture field to the high-level user applications, many research projects have been proposed in the literature to advance the knowledge of AI. Machine and deep learning benchmarking is thus a hot spot in the computing world today. There are several existing and emerging machine and deep learning benchmarks available to the community. We discuss a few popular and important benchmarks in this section.

6.6.1 MLPerf and MLCommons

MLPerf Inference Benchmark (Reddi et al., 2020) is a machine learning benchmark that measures how fast any system can perform machine learning inference. Each benchmark in MLPerf is defined by a model, a dataset, a quality target, and a latency constraint. They are executed via a load generator that issues queries to the machine learning model in one of several patterns, including offline, server, and stream modes. The MLPerf suite v0.5 consists of five benchmarks that span the English-German machine translations with the Workshop on Statistical Machine Translation (WMT) English-German dataset, two object detection benchmarks with the Microsoft Common Objects in Context (COCO) dataset (Lin et al., 2014), and two image classification benchmarks with the ImageNet dataset. MLPerf also proposes training benchmarks (Mattson et al., 2020) for image classification, object detection (lightweight), instance segmentation and object detection (heavyweight), translation (recurrent), translation (nonrecurrent), recommendation, and reinforcement learning. MLPerf is now part of the MLCommons Association (MLCommons, 2021).

6.6.2 MLBench

MLBench (MLBench-Team, 2021) is a framework for distributed machine learning. It is designed and implemented to enable improving transparency, reproducibility, and robustness, as well as to provide a means of fair performance comparison across reference implementations. MLBench has been helping with the adoption of distributed machine learning methods both in industry and in the academic community. MLBench is based on Kubernetes (Kubernetes Team, 2021) for easy distributed deployment on both public clouds and dedicated clusters. It supports several standard machine learning frameworks and algorithms, including decision trees, Support Vector Machines (SVM), logistic regression, decision forests, and so on.

6.6.3 Stanford DAWNBench

DAWNBench (Stanford DAWN Team, 2021) is an open-source benchmark and competition for end-to-end deep learning training and inference. This benchmark measures the end-to-end performance of training, such as time and cost, and inference, such as latency. DAWNBench gives researchers an objective judgment standard of different computation frameworks, hardware, hyperparameter settings, and optimization algorithms. DAWNBench can measure the end-to-end time and cost to train a deep learning model to a specific accuracy together with its inference time. Specifically, DAWNBench measures the inference time to 93 percent validation accuracy for different hardware, frameworks, and model architectures.

DAWNBench initially released benchmark specifications for image classification (ImageNet (Stanford Vision Lab, 2021) and CIFAR-10 (Canadian Institute for Advanced Research) (Krizhevsky, 2021)) and question answering (SQuAD (Rajpurkar et al., 2016)). The models in DAWNBench can be tested on both TensorFlow and PyTorch and on different types of CPU-GPU hardware platforms. The DAWNBench project also allows users to compete, by submitting their benchmarking results, toward promoting its development.

6.6.4 Facebook AI Performance Evaluation Platform

Facebook AI Performance Evaluation Platform (FAI-PEP) (Facebook AI Team, 2021) provides a means to compare the machine learning or deep learning inference performance metrics on a set of models over different back ends. This platform supports both mobile devices and server platforms, including CPU, GPU, digital signal processing, Android, iOS, and other Linux-based systems. Currently, FAI-PEP supports two platforms, namely, TensorFlow Lite and Caffe2. It includes sixteen popular machine learning models, including MobileNet (Howard et al., 2017), SqueezeNet (Iandola et al., 2016), ShuffleNet (X. Zhang et al., 2018), among others.

FAI-PEP provides a centralized model/benchmark specification, a benchmark driver for distributed execution, and a data consumption tool to compare the performance. The platform supports two modes of execution: standalone benchmark run and continuous benchmark run. The standalone benchmark run mode reports the results for one benchmark run. Continuous benchmark run mode repeatedly pulls the framework and runs the benchmarks. It can also accept any machine learning or deep learning models given by users and reports any metrics defined by users.

6.6.5 Deep500 and HPC AI500

Deep500 (Deep500-Team, 2021) is a customizable benchmarking infrastructure that enables a fair comparison of the plethora of deep learning frameworks, algorithms, libraries, and techniques. It supports distributed training on TensorFlow, Caffe, and PyTorch, and users can design different codes with its high-level APIs. Similarly, HPC AI500 (Jiang et al., 2019) is a benchmark suite for testing deep learning benchmarks on HPC systems. It selects several typical scientific fields to be the target scenes, for example, extreme weather analysis, high energy physics, and cosmology. Workloads in this benchmark are made up of state-of-the-art deep learning models and representative scientific datasets instead of standard deep learning models (e.g., *Visual Geometry Group* [VGG] and Long Short Term Memory [LSTM]) and datasets (e.g., Modified National Institute of Standards and Technology [MNIST] (LeCun et al., 2021) and ImageNet).

6.6.6 Baidu DeepBench

DeepBench (Baidu Research, 2021) from Baidu is an open-source benchmark covering both training and inference. DeepBench focuses on measuring the performance of basic operations in neural network libraries. It aims at determining the most suitable hardware for specific operations and communicating requirements to hardware manufacturers. The benchmark has been adapted to both server-side and edge-side platforms. For example, DeepBench can run on top of mainstream GPU devices as well as mobile devices, such as the iPhone. DeepBench focuses more on the basic operations of deep learning inference rather than complete inferences. The key aspect that it tries to determine is the hardware that can provide the best performance on the basic operations used in DNN.

DeepBench can measure operations and layers such as dense matrix multiplication, sparse matrix multiplication, convolution (Lawrence et al., 1997), LSTM (Hochreiter and Schmidhuber, 1997), and gated recurrent unit (Cho et al., 2014) layers. These operations and layers are widely used in applications such as DeepSpeech (DeepSpeech Team, 2021), language modeling (Bengio et al., 2003), machine translation, speaker identification (Reynolds and Rose, 1995), and others. Compared with the training benchmark, the inference benchmark has some distinct designs. It provides a batching scheduler to improve the performance issue led by individual requests and provides kernels with different precisions to adapt diversified terminal devices. Sparse kernels are employed for benchmarking the optimized neural network operations on mobile devices.

6.7 Comprehensive Benchmark Suites

Most of the state-of-the-art big data benchmarking efforts we have discussed target evaluating specific types of programming models or middleware system stacks. However, if we consider the broad use of big data systems, the benchmarks must include a diversity of data and workloads. Thus, the prerequisite for evaluating big data systems and architecture is a "comprehensive" benchmark suite that can span the different dimensions of application scenarios, algorithms, data types, data sources, software stacks, and application types. We discuss some popular big data benchmark suites in the following sections.

6.7.1 BigDataBench

Among the state-of-the-art comprehensive benchmark suites, BigDataBench and HiBench are two testing suites introduced from the systems perspective. BigDataBench (L. Wang et al., 2014) from the Institute of Computing Technology, Chinese Academy of Sciences is a comprehensive big data and AI benchmark suite. The core concept in BigDataBench is called "data motifs," which considers any big data and AI workload as a pipeline of one or more classes of computation units performed on different input datasets. The BigDataBench team has identified eight data motifs from a wide range of big data and AI workloads, including Matrix, Sampling, Logic, Transform, Set, Graph, Sort, and Statistic computation. These data motifs can reflect patterns from real workloads in terms of computation, memory access, communication, and I/O operations.

BigDataBench 5.0, the current version, provides thirteen representative real-world datasets and twenty-seven big data benchmarks. The benchmarks cover six workload types including online services, offline analytics, graph analytics, data warehouse, NoSQL, and streaming from three important application domains: internet services (including search engines, social networks, e-commerce), recognition sciences, and medical sciences. Specifically, for AI, BigDataBench can run on top of TensorFlow and Caffe, including model benchmarks such as convolutions, fully connected neural networks, and so on. The included data sources are text, graph, table, and image data. Using real datasets as the seed, the data generators in the Big Data Generator Suite generate synthetic data by scaling the seed data while keeping the data characteristics of raw data.

6.7.2 Intel HiBench

Intel HiBench (Intel, 2020a) is yet another comprehensive big data benchmark suite that helps evaluate different big data frameworks in terms of speed, throughput, and system resource utilization. This open-source benchmark suite consists of a total of nineteen workloads. They are divided into six categories: micro, machine learning, SQL, graph, web search, and streaming. These constitute a set of Hadoop and Spark workloads, including Sort, WordCount, TeraSort, Sleep, SQL, PageRank, Nutch indexing, Bayes, Kmeans, NWeight, and enhanced DFSIO, among others. It also contains several streaming workloads for Spark Streaming, Flink, Storm, and Gearpump.

6.7.3 TPC Benchmarks

TPC Benchmarks (Transaction Processing Performance Council) is the gold standard in database benchmarks. It is a trademark of the Transaction Processing Performance Council. It provides a suite of benchmarks spanning online transaction processing and decision support and has also been extended to big data processing systems such as

Hadoop and virtualized system environments. Some of the widely used TPC benchmarks include:

• TPC-C simulates a complete computing environment where a population of users executes transactions against a database. It is centered around the principal transactions of an order-entry environment and involves a mix of five concurrent transactions of different types and complexity either executed online or queued for deferred execution.

• TPC-E is an OLTP workload. It uses a database to model a brokerage firm with customers who generate transactions related to trades, account inquiries, and market research. It studies the scalability as transactions per second.

• TPC-DS is the industry-standard benchmark for measuring the performance of decision support solutions including, but not limited to, big data systems. It models the retail product supplier scenario with database schema and implementation rules designed to broadly represent modern decision support systems.

• TPC-H is a decision support benchmark. It consists of a suite of business-oriented ad hoc queries and concurrent data modifications. The queries and the data populating the database have been chosen to have broad industry-wide relevance while maintaining a sufficient degree of ease of implementation.

• TPCx-BB Express Benchmark BB (TPCx-BB) measures the performance of Hadoop-based big data systems. It measures the performance of both hardware and software components by executing frequently performed SQL analytical queries in the context of retailers with a physical and online store presence to benchmark Hadoop-based systems including MapReduce, Apache Hive, and Apache Spark MLlib.

6.7.4   CloudSuite

Since cloud computing has emerged as the dominant computing platform for providing scalable online services, it has been driven to enable operating on massive working sets with high degrees of parallelism and real-time constraints. To study cloud and data-centric computing, CloudSuite (CloudSuite Team, 2021, Palit et al., 2016) offers a benchmark suite based on real-world online services (e.g., web search, social networking, and business analytics). It consists of eight applications that have been selected based on their popularity in today's data centers. These include benchmarks that represent massive data manipulation with tight latency constraints such as in-memory data analytics using Apache Spark, a real-time video streaming benchmark, graph analytics, and web serving benchmarks that mimic multitier web server software stacks with a caching layer.

6.7.5   Ohio State HiBD Microbenchmark Suite

Ohio State High-Performance Big Data (HiBD) Benchmarks (OHB) (Network Based Computing Lab, 2021a) is a comprehensive benchmarking suite designed to evaluate the fine-grained performance of various data processing frameworks. Synthetic workloads are constructed and, together with their benchmarking application layers, are used to evaluate the standalone performance of various big data processing frameworks. The frameworks evaluated include Hadoop, Spark, Memcached, and others.

6.8   Summary

Benchmarking is defined as the quantitative foundation of any computer systems research. This chapter presents a birds-eye overview of state-of-the-art big data benchmarks that span different application domains and workloads. It also presents an overview of programming model-specific benchmarks and discusses state-of-the-art comprehensive benchmark suites used in practice. Many of these benchmarks are used in chapters 7–11 to demonstrate the strengths and weaknesses of new designs for accelerating big data processing.

7 Accelerations with RDMA

This chapter presents a brief overview of the challenges and choices involved in designing an RDMA-aware communication subsystem for big data middleware. This is followed by a comprehensive survey of various state-of-the-art RDMA-accelerated data processing and storage systems for big data proposed in the literature. Examples include accelerated stacks for Hadoop (including HDFS, MapReduce, and RPC), Spark, Memcached, Kafka, Mizan, Wukong, and others. An overview of the performance benefits achieved by these accelerated designs is presented.

7.1 Overview

Modern HPC systems and the associated middleware (such as MPIs and PFSs) have been exploiting the advances in networking technologies such as RDMA during the last decade. On the other hand, traditional big data processing and management middleware (e.g.. Apache Hadoop and Apache Spark) have relied on slower networking protocols such as TCP/IP (see chapter 4). However, with the convergence of big data and HPC, there has been a significant shift in how big data middleware is being designed for today's large-scale computing systems. Big data middleware is being optimized or redesigned to leverage RDMA-enabled networking for faster communication for both data processing tasks and I/O. Toward designing an RDMA-aware communication and I/O framework for data processing, there are two major challenges that need to be addressed:

1. How to integrate RDMA-based designs into the middleware?

2. Which RDMA operations to leverage to design the high-performance communication protocols involved?

Two distinct approaches have been explored toward integrating RDMA into big data middleware (Su et al., 2017). The first approach designs an independent RDMA-aware communication library that can replace the traditional TCP/IP-based communication substrate. This has been referred to as the "server-reply" approach, as it requires a server to respond or send the results to the clients. The second approach leverages the one-sided CPU-bypass communication semantics of RDMA to retrieve data directly from the server's memory. This is denoted as the "server-bypass" approach. Now, while this server-bypass approach ideally enables better performance and server scalability, it relies on middleware developers to design specific data structures and algorithms that make employing one-sided RDMA operations feasible. Therefore, several attempts have been made to leverage a hybrid server-reply/server-bypass mechanism to achieve a good trade-off involving end-to-end latency, server scalability, and CPU usage (Jose et al., 2011, Kalia et al., 2014, Su et al., 2017).

While modern RDMA hardware offers the potential for exceptional performance, the choice of RDMA operations employed and their usage can significantly affect the actual performance of the RDMA-accelerated middleware. The two-sided and one-sided RDMA operations (detailed in chapter 4) can be "inbound" or "outbound." It is vital to distinguish between inbound and outbound verbs as their performance characteristics are significantly different. One-sided RDMA Read/Write verbs and two-sided RDMA Sends are (1) outbound at the requesting process and (2) inbound at the responding process. Two-sided RDMA Recvs are always considered inbound. To elaborate more on the typical performance expectations, issuing a one-sided RDMA operation (i.e., outbound RDMA) has a much higher overhead than that of serving one (i.e., inbound RDMA). This is because the requesting side needs to maintain a certain context and involve both software as well as hardware to ensure the operation is sent and completed, while the responding side is purely handled at the NIC hardware. As evidenced by Kalia et al. (2016), the peak I/O (operation per second) of inbound RDMA is about five times higher than that of the outbound RDMA. thus, depending on the middleware's communication pattern, the choice of RDMA operations used to design the high-performance I/O protocol can be a game changer. This chapter explores RDMA-accelerated middleware along seven distinct directions:

1. Batch and streaming data processing systems that leverage RDMA to accelerate intermediate data shuffling stages.

2. Graph processing systems that employ RDMA-aware schemes for fast and distributed graph exploration.

3. RPC libraries that leverage RDMA for fast peer-to-peer communication substrates.

4. File systems that utilize RDMA for high-speed data movement for distributed fault-tolerance and remote data reads.

5. In-memory KV stores that leverage RDMA for high-performance client-to-server store and retrieve operations.

6. Database systems that employ RDMA for faster-distributed query processing.

7. Machine learning and deep learning frameworks that leverage RDMA for high-speed parameter convergence.

For each of these directions, the following sections discuss the architectures of various state-of-the-art high-performance designs proposed in the literature, including the choice of

RDMA-aware protocols (server-reply vs. server-bypass) and RDMA-accelerated primitives (inbound vs. outbound) employed for enabling high performance. We discuss RDMA-aware machine learning and deep learning frameworks in chapter 10.

7.2 Batch and Stream Processing Systems

Offline data analytical or batch computing frameworks involve an intermediate data shuffling phase, between the map and reduce phases. Similarly, message brokers, such as Kafka, used in the stream data processing pipeline act as a data intermediary to the streaming data computing system. Due to the communication-intensive nature of these intermediate data movement stages, these systems have demonstrated considerable benefits by exploiting RDMA over high-performance interconnects. This section discusses RDMA-accelerated designs available for popular batch and stream data processing frameworks.

7.2.1 Accelerating Hadoop MapReduce with RDMA

The default MapReduce framework employs bulk data transmission over the underlying interconnect and also a large number of disk operations during shuffle and merge phases that introduce a performance bottleneck on modern HPC clusters. Different enhancements have been proposed in the literature that address these issues and present RDMA-based shuffle inside MapReduce.

7.2.1.1 Hadoop-A

The two-phase MapReduce protocol in the Hadoop framework is made possible via a shuffle-and-merge intermediate stage (as discussed in chapter 2). Researchers, such as Wang et al. (2011) and Rahman et al. (2013), found that just replacing the network hardware with the latest interconnect technologies (e.g., InfiniBand and 10 GigE) and continuing to run Hadoop on TCP/IP (via HTTP (Apache Software Foundation, 2020c)) will not enable Hadoop to leverage the strengths of RDMA for the network-intensive shuffle phase. They found that the shuffle phase in turn can become a major bottleneck for Hadoop, especially when it is necessary to keep up with the advances of other processor technologies, storage, and interconnect devices that are being deployed to various HPC clusters and data centers. Wang et al. (2011) introduced a new protocol that directly builds the RDMA-based communication on top of the InfiniBand Verbs protocol and completely avoids the overhead of the Java Virtual Machine (JVM) stack for Hadoop data shuffling. They proposed two new components: the RDMA Server, which serves as a map output file supplier, and the RDMA Client, which requests the map outputs at the merge/reduce tasks. They also proposed new the RDMA Client sends a request along with the information of the available memory buffer, and the RDMA Server locates the data and writes it to the client buffer via a zero-copy RDMA Write operation. RDMA-aware protocols to support the TaskTrackers:

The merge phase at the reducers, which follows the intermediate shuffle phase in the default Hadoop MapReduce protocol, forms a "serialization barrier" that significantly delays the reduce operation and heavily relies on the local disk storage. Wang et al. (2011) proposed an alternative merge algorithm, referred to as "network-levitated merge," to merge data without repetition and disk access. These designs have demonstrated more than a 36 percent increase in Hadoop's data processing throughput while enabling significantly lower CPU utilization.

7.2.1.2 HOMR

While an RDMA-aware shuffle can enable better performance, several research works demonstrate that it cannot extract the full performance potential from modern HPC clusters without an efficient execution pipeline among different job phases in MapReduce. To address this, a high-performance design of MapReduce, referred to as Hybrid Overlapping in MapReduce (HOMR), has been proposed by Rahman et al. (2014). HOMR contrasts with the default MapReduce framework in the following ways:

• In the default MapReduce framework, shuffle uses HTTP for map output data transmission. In HOMR, this data shuffling phase is implemented over RDMA. RDMA Copiers send requests to the TaskTrackers for recently completed map output data. RDMA Responders in TaskTrackers receive those requests and respond to ReduceTasks. HOMR not only brings RDMA-based data transmission into the framework, but it also incorporates advanced features.

• In HOMR, the entire merge operation can take place in-memory, which reduces a significant number of disk operations in the ReduceTask. As in the RDMA-based design by Wang et al. (2011) data can be retrieved much faster. This creates the opportunity to transfer one map output file in multiple communication steps instead of one. By doing this, the merge phase can start as soon as some KV pairs from all map output files reach the reducer side. To complement this, HOMR implements an efficient caching technique for the intermediate data residing in map output files, while leveraging RDMA.

• Unlike initial RDMA-based designs proposed by Wang et al. (2011) and Rahman et al. (2013), HOMR employs bulk data transfer during the shuffle stage. Therefore, HOMR can start the reduce operation as soon as the first merge process generates a sorted output. Thus, it can overlap shuffle, merge, and reduce operations through efficient pipelining among these phases to enable Hadoop MapReduce applications to execute much faster than the default MapReduce framework on modern HPC clusters.

These advanced designs of HOMR are included in the RDMA–Hadoop 2.x software release, as a part of the HiBD project's software stack. The project is detailed in section 7.7.

7.2.2 RDMA-Accelerated Apache Spark

As mentioned in chapter 2, Apache Spark defines two types of dependencies between the distributed datasets (RDDs) that are specific to the map/reduce tasks: narrow dependencies and wide dependencies. When wide dependencies exist between RDDs, they will cause a global many-to-many data shuffle process in Spark, which is a communication-intensive process. The default Apache Spark design provides two approaches to perform data shuffle. The first is Java NIO–based data shuffle, and the second is based on the Netty (The Netty Project, 2021) communication substrate. In the latest Apache Spark (Zaharia et al., 2010), the Netty-based shuffle is the default approach. Both Java NIO and Netty still rely on a sockets-based send/receive two-sided communication model, which could not fully take advantage of the performance benefits provided by high-performance interconnects (Lu et al., 2014). This has become the major performance bottleneck for Apache Spark on HPC clusters.

To address this, a high-performance RDMA-based shuffle design has been proposed (Lu et al., 2014). However, without an efficient buffer and connection management and communication optimization in Spark, the RDMA-based shuffle design cannot extract the full performance potential from modern HPC clusters. The high-performance design of RDMA-based Apache Spark proposed by Lu et al. (2014) and Lu et al. (2016) involves the following advanced features.

SEDA-based data shuffle architecture: High throughput is one of the major goals in designing data shuffle systems. Staged Event-Driven Architecture (SEDA) (Welsh et al., 2001) is widely used for achieving high throughput. The basic principle of SEDA is to decompose a complex processing logic into a set of stages connected by queues. A dedicated thread pool remains in charge of processing events on the corresponding queue for each stage. By performing admission controls on these event queues, the whole system achieves high throughput through maximally overlapping different processing stages. Efforts on improving Hadoop components (e.g., HDFS, RPC) (Islam et al., 2014, Lu et al., 2013) have also shown that the SEDA approach is applicable for RDMA-based data processing systems. Using SEDA, the whole shuffle process can be divided into multiple stages and each stage can overlap with the others by using events and thread pools. In this manner, the design can achieve the default task-level parallelism in Apache Spark, as well as overlapping within block processing with the SEDA shuffle engine. In RDMA-based data shuffle engine, a pool of buffers is used for both sending and receiving data at any end point. The buffer pool is constructed from JVM off-heap buffers that are also mapped to the Java/Scala layer as shadow buffers.

Efficient RDMA connection management and sharing: As shown in RDMA-based designs for Hadoop (Islam et al., 2012, Rahman et al., 2013), the overhead of the RDMA connection establishment is a little higher than that of connection establishment with sockets. To alleviate this overhead, an advanced connection management scheme is needed to reduce the number of connections. Apache Spark uses a multithreading approach to support multitask execution in a single JVM, by default, which provides a good opportunity to share resources. This inspires an interesting idea to share the RDMA connection among different tasks that want to transfer data to the same destination. This connection sharing mechanism significantly reduces the number of total connections, which in turn enables reduced overall resource usage. Through nonblocking chunk-based data transfers, the designs by Lu et al. (2016) can efficiently work with the connection-sharing mechanism. This also facilitates a good trade-off between resource utilization and performance.

Nonblocking data transfer: With connection sharing, each connection may be used by multiple tasks (threads) to transfer data concurrently. In this case, packets over the same communication lane will go to different entities on both the server and the client sides. To achieve high throughput, a large data block can be divided into a sequence of chunks and be sent out in a nonblocking fashion. This can improve the network bandwidth utilization by offloading more packets to the NIC, as indicated by Lu et al. (2014).

7.2.3   RDMA-Accelerated Apache Kafka

Over the last decade, organizations have become heavily reliant on providing close to instantaneous insights to the end user, based on vast amounts of data collected from various sources in real time. To accomplish this task, a stream processing pipeline is constructed, which in its most basic form, consists of a Stream Processing Engine (SPE) and a Message Broker (MB). The SPE is responsible for performing actual computations on the data and providing insights from it. The MB, on the other hand, acts as an intermediate queue to which data are written by ephemeral sources and then fetched by the SPE to perform computations. Due to the inherent real-time nature of such a pipeline, low latency is a highly desirable feature for it. Thus, several existing research works in the community focus on improving the latency and throughput of the streaming pipeline. However, there is a dearth of studies optimizing the tail latencies of such pipelines. Moreover, the root cause of this high tail latency is still under investigation.

Javed et al. (2018) introduced Frieda, which presents a model-based approach to analyze in-depth the reasons behind high tail latency in streaming systems such as Apache Kafka. Having found the MB to be a major contributor of messages with high tail latencies in a streaming pipeline, an RDMA-enhanced high-performance MB, called Frieda, was designed and implemented with the higher goal of accelerating any arbitrary stream processing pipeline regardless of the SPE used. Experimental evaluations show a reduction of up to 98 percent in the 99.9th percentile latency for microbenchmarks and up to 31 percent for full-fledged stream processing pipeline constructed using YSB.

7.3   Graph Processing Systems

Large-scale graph exploration over a graph-based distributed in-memory store involves a large number of concurrent and latency-sensitive queries and can efficiently leverage RDMA to highly parallel and low-latency queries over big graph data. This section discusses two examples in literature that employ such high-performance designs for graph data processing over modern HPC interconnects.

7.3.1 Wukong

Many public knowledge bases are represented and stored as RDF graphs, where users can query the graph databases using SQL such as SPARQL. For instance, several large graph datasets, such as the Google Knowledge Graph (Singhal, 2012), PubChemRDF (NLM, 2020), and others, are represented as RDF format, which represents a dataset as a set of subject-predicate-object triples that form a directed and labeled graph. Early graph stores designed to cater to such RDF-based graph processing frameworks include RDF-3X (Neumann and Weikum, 2008), SW-Store (Abadi et al., 2009), and so on, as well as the more recent TriAD (Gurajada et al., 2014), Trinity.RDF (Semiodesk, 2021), and SHARD (Rohloff and Schantz, 2010). With massive queries over large and constantly growing RDF data, an RDF graph store must provide low latency and high throughput for concurrent query processing. Hence, designing a distributed graph-based RDF store that leverages RDMA-based graph exploration asserts great potential for high performance over the existing state-of-the-art RDF stores.

In this direction, Shi et al. (2016) present Wukong, a distributed in-memory RDF store that provides low-latency, concurrent queries over large RDF datasets. Unlike its predecessors, Wukong stores RDF triples as a native graph and enables fast and concurrent graph exploration queries. Its key techniques are based on leveraging one-sided RDMA for graph exploration in the following ways.

• **Graph model and storage:** To facilitate efficient use of RDMA, Wukong is built on an RDMA-friendly distributed hash table derived from DrTM-KV (Wei, Shi, et al., 2015). It simplifies the key/value store design adapted from DrTM-KV to remove unnecessary metadata for checking consistency and supporting transactions, but, Wukong employs DrTM-KV's nice features such as RDMA-friendly cluster hashing and location-based cache.

• **Full history pruning:** Wukong has the full history during graph exploration, which is forwarded by each step within and across machines (or CN) with RDMA Writes. With the full exploration history, the final join to filter out nonmatching results can be eliminated, which can drastically improve overall transaction processing performance.

• **Query distribution:** Wukong decomposes a query into a sequence of subqueries. It handles multiple independent subqueries simultaneously and adopts an RDMA communication-aware mechanism for each. Specifically, for small selective queries, it uses in-place execution that leverages RDMA Read to fetch necessary data so that there is no need to move intermediate data. Similarly, for large complex and nonselective queries, it uses one-sided RDMA Write to distribute the query processing to all related machines. It also provides a latency-centric work-stealing scheme to load-balance the subquery processing across its nodes.

Wukong significantly outperforms state-of-the-art systems and can process a mixture of small and large queries on a six-node InfiniBand cluster with Fourteen Data Rate

(FDR) 56 Gb/s interconnects and ten-core Intel Haswell CPUs at 269,000 queries per second.

7.3.2 Mizan-RMA

Mizan is an advanced version of Google's graph processing system Pregel (Malewicz et al., 2010) that utilizes online graph vertex migrations to dynamically optimize graph algorithm executions. The messaging module uses the MPI library to enable communication between workers. The Mizan graph processing framework uses the MPI two-sided programming model. On the other hand, MPI offers a remote memory access (RMA) model that separates synchronization from data movement. Because many modern high-performance interconnects offer RDMA features, the RMA programming model can be directly implemented over such network to deliver better computation/communication overlap than two-sided MPI Send/Recv operations do. One-sided communication has been seen as advantageous for applications with irregular communication patterns. Several studies have shown the benefits of the one-sided programming model for graph processing (e.g., Graph500 (2021) with MPI RMA (Li, Lu, et al., 2014)). Extending this concept, Li, Lu, et al. (2016) presents Mizan-RMA, which leverages MPI's one-sided programming model. Mizan-RMA leverages the CPU-bypass capabilities of RDMA by employing the MPI RMA routines in the RDMA-optimized MPI library MVAPICH2 (Network Based Computing (NOWLAB), 2021b) to achieve better computation/communication overlap and load-balancing for the workers for an overall performance gain of 2.8 times over Mizan.

7.4 RPC Libraries

By definition, an RPC is an interprocess communication that resembles a normal procedure call. While it resembles message passing between two processes, RPC is a client/server communication mechanism in which the "called procedure" takes the role of the server, and the caller takes the role of the client. Due to their communication-intensive nature, RPC libraries are being enhanced to leverage high-performance interconnect technologies such as RDMA.

7.4.1 Hadoop RPC with RDMA

Hadoop RPC is the basic communication mechanism in the Hadoop ecosystem. It is used with other Hadoop components such as MapReduce, HDFS, and HBase in real-world data centers, such as Facebook and Yahoo!. However, the current Hadoop RPC design is built on the Java sockets interface, which limits its potential performance. Lu et al. (2013) identify that the bottlenecks in the default Hadoop RPC design are not sensitive in socket-based communication over TCP/IP on the 1-GigE network because the data transfer on the network is the bottleneck.

They also put forth a redesigned data communication and the associated buffer management of Hadoop RPC when running on an RDMA-capable cluster equipped with InfiniBand, referred to as RPCoIB. RPCoIB utilizes native InfiniBand communication, JVM-bypassed buffer management, and message size locality to avoid buffer allocation and memory copy overheads in the data serialization and deserialization. It introduces RDMAInputStream and RDMAOutputStream classes and modifies the default Java-based IOStreams employed by the Hadoop RPC client/server to provide RDMA-based I/O streams. RPCoIB demonstrates a stand-alone performance gain of up to 50 percent in latencies and 82 percent in RPC throughput. When integrated with Apache Hadoop's HDFS and HBase components, RPCoIB demonstrates a 10 percent to 24 percent improvement in overall performance. RPCoIB is integrated as a part of the RDMA–Hadoop 2.x package, which is included as a part of the HiBD project (detailed in section 7.7).

7.4.2   RDMA-Based gRPC

gRPC (gRPC Authors, 2021) is the most fundamental communication mechanism in distributed machine learning frameworks, such as TensorFlow (Abadi, Agarwal, et al., 2016). It supports various channels to efficiently transfer tensors, such as gRPC over TCP/IP, gRPC InfiniBand Verbs, and gRPC + MPI. Similarly, an optimized gRPC library that replaces the inherent default TCP/IP-based communication channel with two-sided RDMA Send/Receive operations has also been proposed (CGCL-codes, 2017). Furthermore, Biswas et al. (2018) propose a unified approach to have a single gRPC runtime, referred to as AR-gRPC for TensorFlow, with adaptive and efficient RDMA protocols. It enables features such as hybrid communication protocols, message pipelining and coalescing, zero-copy transmission, among others, to make the runtime cater to different message sizes employed in distributed deep learning workloads.

7.4.3   RDMA-Based RFP

To solve the dilemma of whether server-reply or server-bypass can give the best performance (as discussed in chapter 1.1), Su et al. (2017) propose a new RDMA-based RPC paradigm called Remote Fetching Paradigm (RFP). In RFP, the server processes client requests sent from clients and maintains CPU usage that is similar to that of traditional RPC interfaces. RFP uses a hybrid mechanism that adaptively switches between repeated server-bypass (remote fetching) and server-reply to achieve higher performance based on the following two observations:

1. Because issuing a one-sided RDMA operation (i.e., outbound RDMA) has a much higher overhead than that of serving one (i.e., inbound RDMA), server-reply is suboptimal for scaling the server. Hence leveraging outbound RDMA at the client can be more beneficial.

2. Relying on only one-sided RDMA reads can lead to the need to coordinate multiple clients to avoid data access conflicts. By relying on the server to compute the results, the corresponding performance degradation can be avoided.

Su et al. (2017) demonstrate that, with these designs, RFP improves the throughput by 1.6–4 times under different workloads, compared to designs that rely solely on server-bypass or server-reply (e.g., Pilaf (Mitchell et al., 2013)).

7.5 Query Processing in Databases

Databases use RDMA as a fast data transfer mechanism from storage to database servers. This further extends transactions involving distributed join processing for faster convergence via fast data transfers between database servers. This section discusses some of the major related research works.

7.5.1 Accelerating Relational Databases with RDMA

Memory is a vital resource in relational databases or relational database management systems (RDBMSs). Insufficient memory scenarios in an RDBMS can force it to rely on slower storage devices such as HDDs or SSDs. This can not only significantly degrade workload performance, but it can also make suboptimal use of the high-speed network resources such as the RDMA-capable network adapters that are available in most data centers today. Specifically, symmetric multi-processing (SMP) RDBMSs, such as Amazon Relational Database Service, Microsoft Azure SQL Database, and so on, have memory demands that typically exceed the locally available memory. An RDBMS can potentially leverage RDMA to employ unused remote memory to significantly improve overall query performance.

Li, Das, et al. (2016) present an SMP RDBMS that uses available remote memory to accelerate memory-intensive workloads without a major rewrite. Their SMP RDBMS abstracts access to remote memory via RDMA and introduces corresponding lightweight file APIs. Li, Das, et al. (2016) propose several scenarios that can leverage remote memory abstractions:

• An SMP RDBMS relies on the procedure and buffer pool caches and generates a lot of temporary data. Instead of evicting entries when out-of-memory, the entries are cached in remote memory. The evicted entries can be accessed again, if necessary, rather than being fetched from SSDs/HDDs. Similarly, temporary data can use remote memory for extended capacities instead of slower-than-memory storage devices.

• Application-level semantic caches for RDBMS can create specialized redundant structures, keep them pinned in remote memory, and access them via RDMA when executing queries that benefit from them.

• Fast access to remote memory using RDMA can significantly reduce the impact of primary-to-secondary buffer pool swaps on application workloads.

Compared to using SSDs/HDDs when memory is insufficient, the RDMA-aware abstractions present improvements of three to ten times for TPC-H and two to a hundred times for TPC-DS workloads.

To achieve the performance gains on low-latency and high-throughput networks, such as InfiniBand and parallel transaction processing algorithms that are vital to designing high-performance database systems, several works have focused on exploiting RDMA primitives to speed up data movement across the machines. With the emergence of main-memory multicore join algorithms (see e.g., Kim et al. 2009; Albutiu et al. 2012), there has been increased scope for leveraging RDMA-aware data transfers for reducing the overall CPU usage and thus improve the overall performance.

Barthels et al. (2015) present a distributed radix join scheme demonstrating that the partitioning phase can be interleaved with the data redistribution over the network with RDMA. A relation is partitioned into either a local buffer, if the partition will be processed locally, or a designated RDMA buffer, if it has been assigned to a remote node. RDMA Write operations are executed at regular intervals to interleave the computation and communication. Similarly, Barthels et al. (2017) employ RDMA operations to implement radix hash join and sort-merge join algorithms via one-sided MPI RMA operations. Additionally, it also presents a study on the behavior of the enhanced distributed joins at a large scale and demonstrates a throughput of 48.7 billion input tuples per second over 4,096 cores.

7.5.2   HTM and RDMA

In-memory transaction processing is a crucial component for many online database systems used for web services and e-commerce. With the introduction of hardware transactional memory (HTM) via capabilities such as Intel's restricted transactional memory (Intel, 2021i), several researchers have investigated employing its atomicity, consistency, and isolation features for database transactions. Wei, Shi, et al. (2015) present DrTM, which is a fast in-memory transaction processing system that exploits advanced hardware features such as RDMA and HTM to improve overall performance by more than an order of magnitude compared to the state-of-the-art distributed software-centric transaction systems.

DrTM leverages the strong consistency between RDMA and HTM to ensure serializability among concurrent transactions across machines while enabling the offloading concurrency control within a local machine into HTM. It is built out of two independent components: transaction layer and memory store. The memory store builds an efficient hash table for DrTM by leveraging HTM and RDMA to simplify overall design and improve performance. Remote access is performed using one-sided RDMA operations for efficiency. In the transactional layer, the one-sided RDMA compare-and-swap (CAS) is used to lock remote records, which is working toward implementing the two-phase locking protocol. An RDMA Read fetches data into local memory and retries RDMA CAS if the lease is expired. Evaluations with DrTM using typical online transaction processing workloads such as TPC-C demonstrate about 17.9 times improvement over state-of-the-art systems such as Calvin (Thomson et al., 2012, Ren et al., 2014).

7.5.3 RDMA-Aware High-Speed Query Processing

Unlike the previous works that rely on enhancing specific components with RDMA-aware designs, Rödiger et al. (2015) present the design and implementation of a comprehensive distributed query engine based on RDMA that is capable of processing complex analytical workloads such as the TPC-H benchmark. With the state-of-the-art in-memory database system HyPer (Kemper and Neumann, 2011, Hyper, 2021) as the basis, it employs a hybrid approach, wherein NUMA-local processing is used for parallelism within the processing node. For distributed processing, it designs a new data redistribution scheme among servers that combines decoupled exchange operators and an RDMA-based communication multiplexer. This multiplexer is a dedicated network thread per server that transfers the data between local and remote exchange operators. It performs these data transfers by sending messages using a global round-robin schedule that employs two-sided RDMA Send/Receive operations (i.e., channel semantics).

By using RDMA to alleviate the overhead of TCP processing and the inflexibility of the classic exchange operator model, the low-latency network scheduling proposed by Rödiger et al. (2015), allows high single-server performance as compared to the default HyPer in-memory database running over a TCP/IP-based communication substrate, with up to six servers in the cluster.

7.6 In-Memory KV Stores

The performance of the in-memory KV stores continues to be of utmost importance as they go beyond the traditional object-caching workload. They have become a key infrastructure component for supporting distributed main-memory computation in data centers. The introduction of high-speed interconnects such as InfiniBand (IBTA, 2021a) has enabled a significant performance improvement for this high-performance middleware through the adaptation of RDMA-aware designs.

7.6.1 RDMA-Based Memcached

Memcached is a popular KV store designed to accelerate online data serving applications (see discussion in chapter 3). Several works (see, e.g., Atikoglu et al., 2012, Xu et al., 2014, etc.) have demonstrated that KV store workloads are read-dominated and latency-sensitive and, therefore, are ideal candidates to benefit from a low-latency communication substrate. Figure 7.1 depicts the RDMA-Memcached design proposed by Jose et al. (2011).

The RDMA-Memcached design can support all the existing Memcached applications without any changes. Shankar, Lu, Islam, et al. (2016) also proposed nonblocking APIs for accelerating offline data analytics toward facilitating applications beyond the traditional online database workloads to benefit from RDMA-optimized in-memory data stores. An RDMA-based communication substrate for the Memcached server and Libmemcached includes the following aspects.

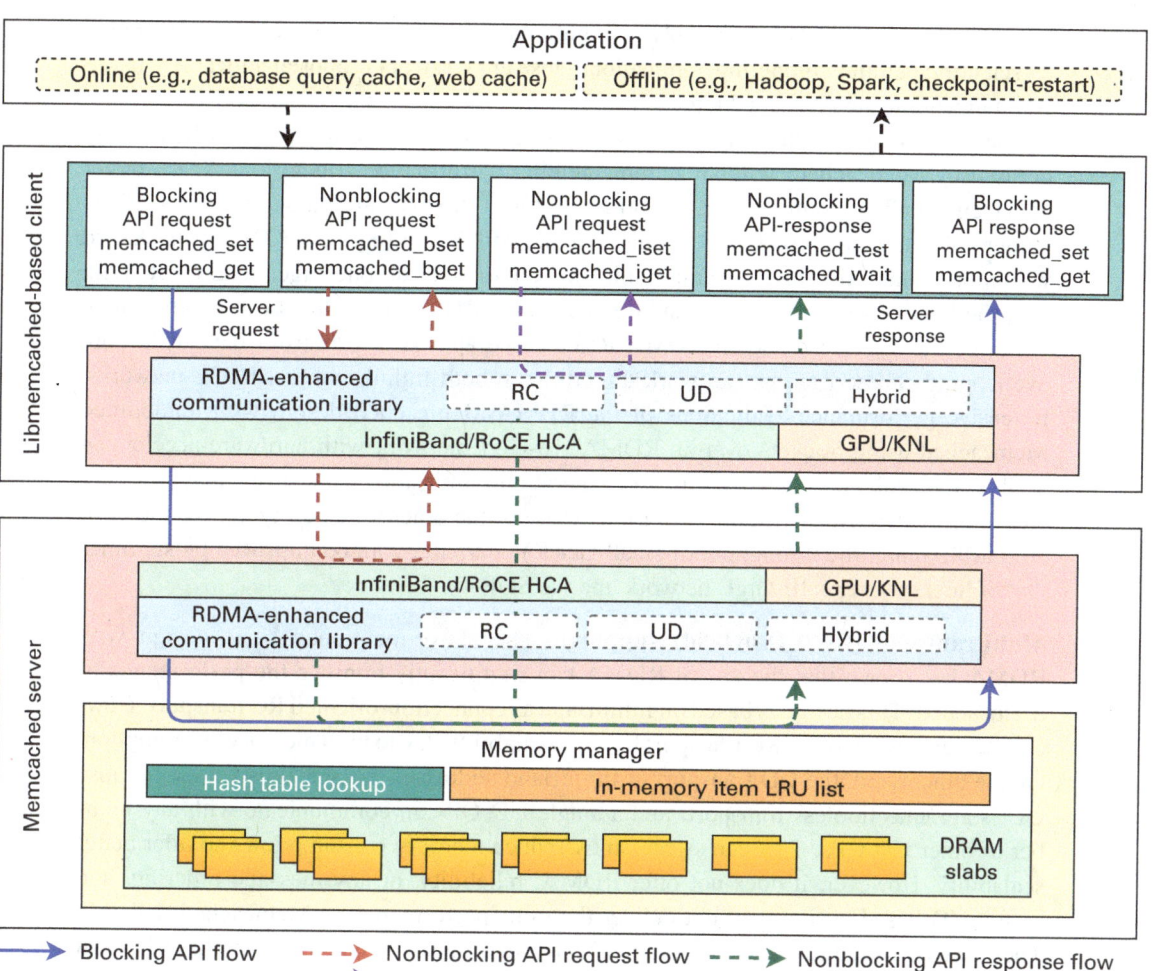

Figure 7.1
Design overview of RDMA-Memcached (Shankar, Lu, Islam, et al., 2016, Jose et al., 2011). KNL, kernel; LRU, least recently used.

RDMA-based communication substrate: The current-generation Memcached library is designed with the traditional Berkeley Sockets interface. While the Sockets interface provides a great degree of portability, the byte-stream model within the Sockets interface mismatches with Memcached's memory object model. Besides, Sockets-based implementations internally need to copy messages, resulting in further loss of performance. High-performance interconnects and their software APIs, such as OpenFabrics Verbs (linux rdma, 2021), provide RDMA capability with memory-based semantics that fits very well with the Memcached model. Inspired by the message passing library MVAPICH2 (Network

Based Computing (NOWLAB), 2021b) (highlighted in chapter 4.4.3), which is leveraged by scientific parallel computing applications, Jose et al. (2011) proposed an RDMA-based communication substrate for Memcached.

The communication substrate exposes easy-to-use Active Message–based APIs that can be used by Memcached, without reimplementing performance-critical logic (e.g., buffer management, flow control), and is optimized for both small and large messages. If the amount of data being transferred fits into one network buffer (e.g., ≤ 8 KB), it is packaged within one transaction and the communication engine will do one eager send for it. With this, the handshaking overheads that are needed for RDMA operations can be alleviated. For large messages, we use an RDMA Read–based approach to transfer data with more overlapping. These designs can work directly with both InfiniBand and RoCE networks. Extensive performance evaluations of the RDMA-Memcached design with unmodified Memcached using Sockets over an RDMA, 10-GigE network with hardware-accelerated TCP/IP and IPoIB protocols on different generation InfiniBand HCAs reveal that the latencies of RDMA-Memcached are better than those of the other schemes by up to a factor of 20. Furthermore, the throughput of small Get(K) operations can be improved by six times that of Sockets over a 10-GigE network and the SDP.

Multitransport (RC/UD/hybrid) support: The above-mentioned basic design with RDMA has shown that the use of RDMA can significantly improve the performance of Memcached. This design is based on InfiniBand as connection-oriented RC transport. However, the exclusive use of RC transport hinders scalability due to high memory consumption. On the other hand, the UD transport of InfiniBand addresses this scalability issue because UD is a connectionless transport, and a single UD QP can communicate with any number of other UD QPs. This can significantly reduce memory consumption and offer better scalability. However, it does not offer RDMA, reliability, or any message ordering, and a pure UD-based design may not deliver the same performance as Memcached with RC. These scenarios led to the introduction of a hybrid transport model (Jose et al., 2012) that leverages the best features of RC and UD to deliver both high performance and scalability for Memcached. The idea behind this is that the maximum number of RC connections can be limited to a specified threshold to achieve scalability.

7.6.2   One-Sided RDMA for Put(K, V) and Get(K)

Two well-known RDMA-aware KV store designs that exploit one-sided RDMA Reads or Writes include Pilaf and HERD.

Pilaf: As one of the earlier works proposed in this direction, Pilaf (Mitchell et al., 2013), a distributed in-memory KV store design, takes advantage of RDMA to achieve high performance with low CPU overhead. It processes write operations at the server (server-reply) and uses RDMA for read-only operations (server-bypass). The overall architecture involves the following RDMA-aware protocols for KV operations:

- **Get(K)** operations are executed independent of the server. The KV store clients perform RDMA Reads over multiple round trips to directly fetch data from the server's memory. To address the read/write races between the server and client, it introduces the notion of a "self-verifying data structure" by appending a checksum to every KV object that is cross-checked whenever a KV pair is read using RDMA.
- **Set(K, V)** protocol involves two-sided Send/Receive InfiniBand Verbs messaging to send the request to the server, which processes the local KV updates to its data structures.

Experiments by Mitchell et al. (2013) show that Pilaf can potentially achieve low latency and high throughput of around 1.3 million operations per second (90 percent gets) using a single CPU core compared with 55,000 for Memcached and 59,000 for Redis.

HERD: HERD, by Kalia et al. (2014), introduces a KV storage system that is designed to make the best use of an RDMA network. Unlike prior RDMA-based KV stores, HERD focuses its design on reducing network round trips while using RDMA primitives efficiently. Unlike Pilaf and RDMA-Memcached, which leverage the one-sided CPU-bypass allure of RDMA Reads, HERD attempts to substantially lower latency and saturates the throughput on modern RDMA hardware. The design employs a mix of RDMA Write and two-sided messaging verbs to enable using a single round trip for all Get(K) and Set(K, V) requests, as follows:

- For the request phase, the HERD KV store client writes the requests into the server's memory using unsignalled RDMA Writes. The server polls the memory instead of the network hardware and computes the corresponding responses.
- For the response phase, the server sends a reply over to the client via two-sided UD messages. Because KV store workloads deal with messages less than 4 KB, HERD can fit into a single datagram packet; thus remaining unaffected by the unordered message receipt with UD.

HERD demonstrates substantially lower latency guarantees with throughput saturation on modern commodity RDMA hardware. More precisely, by using a single round trip for all requests, this design supports up to twenty-six million KV operations per second with 5 us average latency on the InfiniBand FDR 56 Gb/s interconnects.

7.6.3 RDMA-Aware KV Stores for Analytics

In-memory data stores play a vital role in high-speed analytical systems. Among these systems, there are two prominent RDMA-aware KV stores presented in the literature.

FaRM: Fast Remote Memory (FaRM) (Dragojević et al., 2014) is a main-memory distributed computing platform that leverages RDMA to achieve an order-of-magnitude improvement in message rate and latency as compared to employing TCP/IP on the Ethernet network. It exposes the memory of all machines in the distributed cluster as a shared address space. With this, FaRM enables the server threads to use atomicity, consistency, isolation, and durability transactions with strict serializability to support collocation of objects and function shipping that enables the use of efficient single machine transactions. Transactions use optimistic concurrency control with an optimized two-phase commit protocol that takes advantage of RDMA as follows:

- It employs lock-free reads that are serializable with transactions and are performed using a single RDMA Read without involving the remote CPU. It relies on cache-coherent DMA to observe a consistent state of distributed data objects.

- It uses RDMA Writes to implement a fast message-passing primitive, coupled with a circular ring buffer and memory-based polling. This is in contrast to kernel-assisted communication as in TCP/IP.

FaRM achieves availability and durability using replicated logging to SSDs, but it can also be deployed as a cache. With this as the basis, FaRM demonstrates that its proposed techniques can achieve an order of magnitude performance improvement over TCP/IP-based implementation by building RDMA-aware KV and graph store systems.

HydraDB: HydraDB is a general-purpose in-memory KV middleware designed to comprehensively exploit the RDMA protocol to optimize various aspects of a general-purpose KV store, including latency-critical operations, read enhancement, and data replications for high-availability service, among others. Specifically, in terms of architecture, HydraDB utilizes RDMA as follows:

- A message passing interface driven by RDMA-Write, which is efficiently used for data replication, is capable of reducing the overhead incurred by offering high availability. HydraDB also introduces a sustained polling mechanism at both the clients and the servers to poll for requests and responses issued by RDMA.

- HydraDB exploits RDMA Read to allow clients to directly retrieve data from the server's memory in a single round trip. The first time a KV pair is accessed, the request is issued to the corresponding server using RDMA Write–based message passing, and these remote pointers are cached. For subsequently accessed KV pairs, an RDMA Read is issued on the cached pointer to retrieve data using a single one-sided operation.

In addition to RDMA, HydraDB employs a collection of state-of-the-art techniques, including continuous fault-tolerance, multicore-awareness, and so on, to deliver reliable high performance for cluster computing applications. HydraDB has been leveraged to accelerate cluster computing frameworks, including Hadoop, Spark, sensemaking analytics, and call record processing. Furthermore, Y. Wang et al. (2014) introduce several optimizations for efficient and reliable cache management, including lease-based KV management, which is integrated into HydraDB.

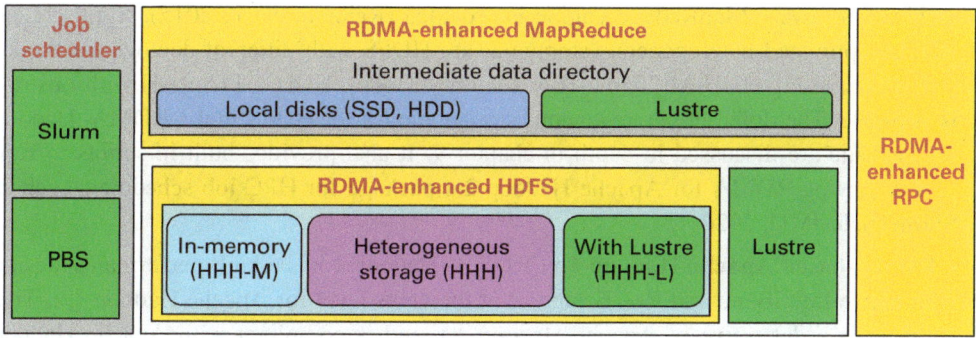

Figure 7.2
RDMA-based Hadoop architecture and its different modes. PBS, Portable Batch System.

7.7 HiBD Project

The HiBD project is carried out by the Network-Based Computing Laboratory of The Ohio State University. The main objective of the HiBD project is to design high-performance big data middleware that can leverage HPC technologies. It offers several types of HPC-optimized distributed data processing middleware. As of March 2022, the HiBD packages are being used by more than 340 organizations worldwide across thirty-eight countries to accelerate big data applications. More than forty-four thousand downloads have taken place from this project's site. The HiBD project currently contains the following types of accelerated big data middleware and the associated packages:

• **RDMA-based Apache Spark:** RDMA-Spark is a high-performance design with native InfiniBand and RoCE support at the verbs level for Apache Spark, based on the work of Lu et al. (2014) and Lu et al. (2016), which was presented in section 7.2.2. It supports multiple advanced features such as RDMA-based data shuffle, SEDA-based architecture, efficient connection management, nonblocking chunk-based data transferring, and JVM off-heap buffer management.

• **RDMA-based Apache Hadoop 2.x:** RDMA–Hadoop 2.x is a high-performance design with native InfiniBand and RoCE support at the verbs level for Apache Hadoop 2.x., based on the work by Rahman et al. (2013), Rahman et al. (2014), Rahman et al. (2015), which were presented in section 7.2.1. This software is a derivative of Apache Hadoop 2.x and is compliant with Apache Hadoop 2.x and Hortonworks Data Platform APIs and applications. Figure 7.2 presents a high-level architecture of RDMA for Apache Hadoop.

This package can be configured to run MapReduce jobs on top of HDFS and PFSs such as Lustre. More importantly, this package provides different modes designed to optimize performance for different kinds of applications and Hadoop environments. For instance, for an HPC environment with limited local storage and a Lustre file system, a Beowulf architecture–aware MapReduce engine is provided (Rahman et al., 2015). Similarly, an RDMA-optimized heterogeneous storage-aware HDFS with different deployment modes (HHH [Triple-H]. HHH-M [Triple-H In-Memory Mode]. HHH-L [Triple-H with Lustre]) is available. The different heterogeneous storage modes for HDFS in the RDMA–Hadoop 2.x package are discussed in-depth in chapter 9. It also provides built-in scripts to support deploying RDMA for Apache Hadoop 2.x package for HPC job schedulers such as Slurm (SchedMD, 2020a).

• **RDMA-based Apache Hadoop 1.x:** RDMA–Hadoop 1.x is a high-performance design with native InfiniBand and RoCE support at the verbs level for Apache Hadoop 1.x. This software is a derivative of Apache Hadoop 1.x and is compliant with Apache Hadoop 1.x APIs and applications. It supports multiple advanced features such as RDMA-based HDFS write, RDMA-based HDFS replication, parallel replication support in HDFS, RDMA-based shuffle in MapReduce, in-memory merge, advanced optimization in overlapping among different phases, RDMA-based Hadoop RPC, and others. Through all these advanced designs, the performance of Apache Hadoop 1.x is improved significantly compared to that of the basic Apache Hadoop version.

• **RDMA-based Memcached:** RDMA-Memcached is a high-performance design with native InfiniBand and RoCE support at the verbs level for Memcached and Libmemcached by Shankar, Lu, Islam, et al. (2016) and Jose et al. (2011), which was discussed in section 7.6.1. This software is a derivative of Memcached and Libmemcached and is compliant with Libmemcached APIs and applications. It supports multiple advanced features such as RDMA-based KV pair operations, SSD-assisted hybrid memory, and so on. Through RDMA-based design, the performance of KV pair operations gets significantly improved, while the SSD-assisted hybrid memory provides increased cache capacity.

• **Ohio State HiBD-Benchmarks:** Typically researchers use traditional benchmarks such as Sort, TeraSort, and so on, to evaluate new designs for the Hadoop ecosystem. However, all these benchmarks require the involvement of several components such as HDFS, RPC, MapReduce framework, and others, thus making it hard to isolate problems and evaluate the performance of individual components that are vital to the functioning of the entire system. To address this need for designing effective microbenchmarks and to evaluate the performance of big data middleware, the HiBD project proposed the OHB micro benchmarks to support stand-alone evaluations of HDFS (Islam, Sharmin, et al., 2012), MapReduce (Shankar et al., 2014), Hadoop RPC (Lu, Rahman, Islam, Panda,, 2014), Spark, and Memcached. This microbenchmark suite is highlighted in detail in chapter 7.5.

7.8   Case Studies and Performance Benefits

To further the argument that the communication stack of traditional big data frameworks need to be redesigned for modern high-performance hardware, this section presents a set of sample case studies from different RDMA-aware designs.

7.8.1   RDMA-Aware Data Analytical Frameworks

As discussed in section 7.2, RDMA can be leveraged to improve the performance of the data shuffling phase in map-reduce frameworks. In this section, we discuss the benefits of leveraging RDMA for Apache Spark and Apache Hadoop MapReduce 2.x.

7.8.1.1   RDMA for Spark

The plug-in-based approach with SEDA-/RDMA-based designs proposed by Lu et al. (2014) provides both high performance and high productivity. As shown in figure 7.3a, the RDMA-based design for Apache Spark (denoted RDMA-IB) improves the average job execution time of HiBench (Intel, 2020a) PageRank on sixty-four SDSC Comet (SDSC, 2021a) worker nodes by 40–60 percent compared to the execution time of IPoIB (56 Gb/s). These experiments were performed with full subscription utilization of cores so that on sixty-four worker nodes, jobs run with a total of 1,536 maps and 1,536 reduces. Spark is run in stand-alone mode. SSD is used for Spark local and work data. More detailed configurations and numbers can be found on the HiBD website (Network Based Computing Lab, 2022).

7.8.1.2   RDMA for Hadoop-2.x

The performance of Apache Hadoop 2.x is improved significantly and it becomes easy to use on HPC clusters as well. As shown in figure 7.3b, the RDMA-based design for Apache Hadoop 2.x (denoted RDMA-IB) improves the job execution time of Sort on sixteen SDSC Comet worker nodes by up to 48 percent compared to the execution time on IPoIB (56 Gb/s). These experiments were performed with a total of sixty-four maps and twenty-eight reduces. HDFS block size is kept to 256 MB. The NameNode ran in a different node of the Hadoop cluster and the benchmark is run in the NameNode. HHH data storage used 70 percent of the RAM disk.

7.8.2   RDMA-Aware Data Management Systems

As discussed in sections 7.5 and 7.6, RDMA can improve performance by significantly reducing the overhead of client-server communication and that of the communication among the nodes that make up the distributed server cluster. In this section, we present a case study for exploiting RDMA for in-memory data stores such as Memcached. Then, we discuss case studies that demonstrate the benefits of RDMA in database systems.

7.8.2.1   RDMA over Memcached

Leveraging a distributed KV store–based caching layer has proven to be invaluable for scalable data-intensive applications. To illustrate the performance benefits, we present two use cases that illustrate (1) online caching and (2) offline analytics.

Online caching: To illustrate the benefits of leveraging RDMA-Memcached to accelerate MySQL-like workloads, the hybrid microbenchmarks presented Shankar et al. (2015) is employed on an Intel Westmere cluster. Each node is equipped with 12 GB of RAM,

(a)

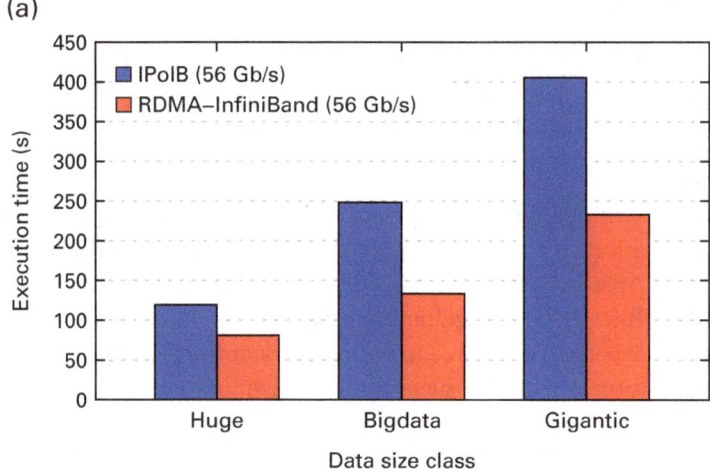

(b)

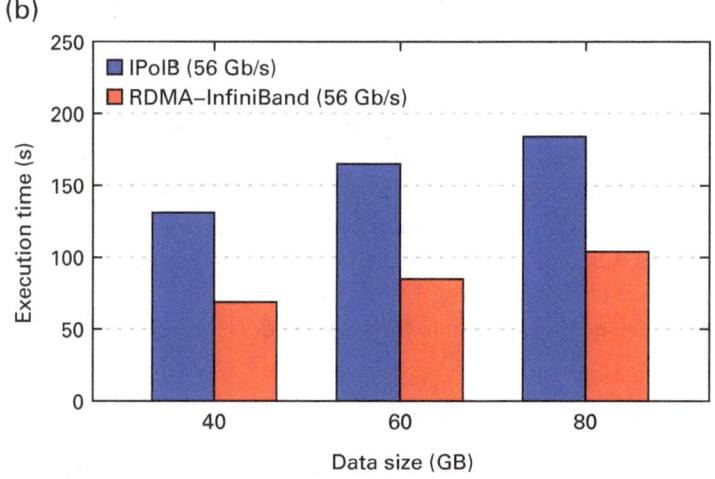

Figure 7.3
Performance improvement of RDMA-based designs for Apache Spark and Hadoop on SDSC Comet cluster. (a) PageRank with RDMA-Spark. (b) Sort with RDMA–Hadoop 2.x.

160 GB of HDD, and MT26428 InfiniBand QDR ConnectX interconnects (32 Gb/s data rate) with PCIe Gen-2 interface running Red Hat Enterprise Linux 6.1. A set of (eight to sixteen) CNs were used to emulate clients, running for both the BSD socket-based (Libmemcached v1.0.18) and RDMA-enhanced implementation (RDMA-Libmemcached 0.9.3). These experiments were run with sixty-four clients, one hundred percent reads, an aggregated workload size of 5 GB, a KV size of 32 KB, and an estimated average miss penalty of 95 ms.

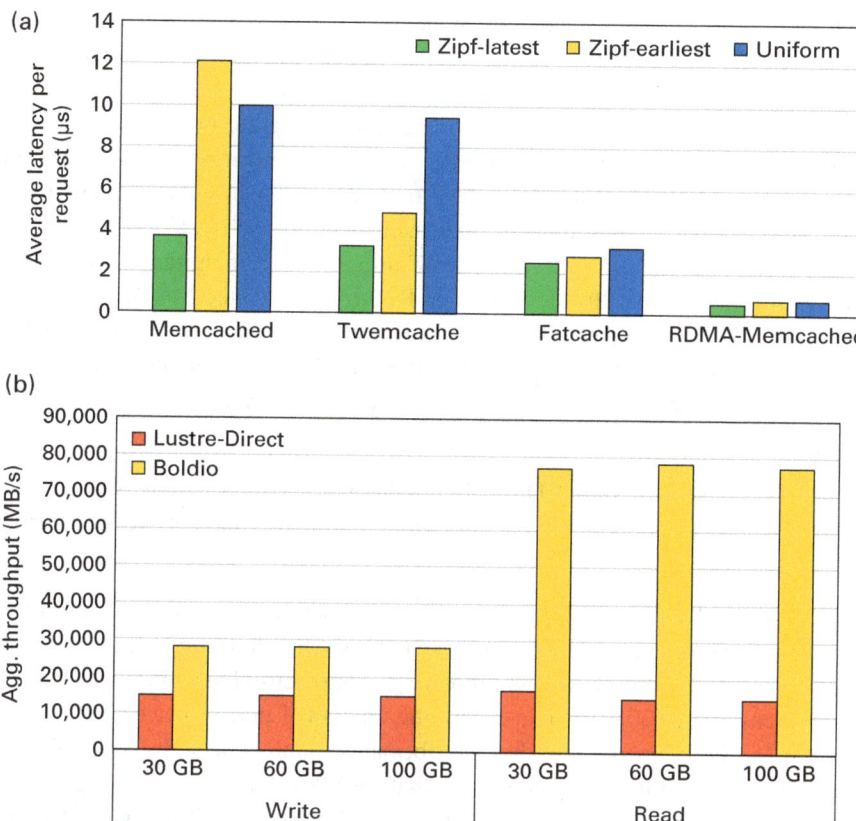

Figure 7.4
Performance benefits with RDMA-Memcached based workloads. (a) Memcached Set/Get over simulated MySQL. (b) Hadoop TestDFSIO throughput with Boldio.

We compared various Memcached implementations including default Memcached (Dormando, 2021), Twemcache (Twitter, 2019), Fatcache (Twitter, 2017), and RDMA-Memcached (Network Based Computing Lab, 2022). From figure 7.4a for the uniform access pattern, where each KV pair stored in the Memcached server cluster is equally likely to be requested, it can be observed that the SSD-assisted designs such as Fatcache and RDMA-Memcached give the best performance due to their high data retention. Specifically, RDMA-Memcached can improve performance by up to 94 percent over in-memory default Memcached designs (including Twemcache). The performance is studied with two skewed access patterns, namely, Zipf-Latest (latest is popular) and Zipf-Earliest (earliest is popular), which model KV pair popularity based on the most recently and least recently updated KV pairs, respectively. As compared to the other solutions in figure 7.4a, it can be observed that RDMA-Memcached can improve performance by about 68–93 percent.

Offline analytics: To further understand the benefits of employing RDMA to accelerate in-memory data stores, we illustrate how Memcached can be exploited to accelerate offline analytical applications like Hadoop's MapReduce-based batch processing workloads. A vital part of the Hadoop job involves retrieving data from HDFS for processing and writing results back on completion. This I/O phase is a potential performance bottleneck on HPC clusters as they involve accessing a single shared parallel file system, such as Lustre. Several recent studies have focused on extending the capabilities of the I/O subsystem for running Hadoop on HPC clusters with the help of in-memory data stores (Shankar, Lu, Islam, et al., 2016; Wang et al., 2014c). One such RDMA-accelerated Memcached-based burst-buffer system for Hadoop, referred to as Boldio (Shankar, Lu, and Panda, 2016), which can improve Write throughput by a factor of 3 and Read throughput by a factor of 7 over default Hadoop running directly over Lustre, is discussed in depth in chapter 9.5.2.

7.8.2.2   RDMA for databases

Modern database systems can exploit RDMA for low-latency and high-bandwidth data communication for client-to-server requests/responses and to exchange data between servers in the distributed database.

Server-to-server communication: Servers in a parallel database employ network-intensive protocols for various distributed operations such as data shuffling, replication, and distributed coordination of transaction execution (e.g., consensus algorithms).

For instance, Liu et al. (2019) validate how the analytical performance of a parallel database system can be improved over high-performance interconnects by leveraging an RDMA-based shuffle operator. They demonstrate that the TPC-H queries can be accelerated by up to two times by enhancing the shuffle operator to directly exploit two-sided RDMA Send/Receive operations over the UD transport service, as compared to using an external communication library such as RDMA-capable MPI, a sixteen-node Broadwell cluster with 100 Gb/s EDR InfiniBand interconnects. On the other hand, Zamanian et al. (2019) demonstrate that one-sided RDMA semantics are more suitable for ensuring high availability via replication. They propose an undo-logging-based replication protocol, known as Active Memory, that uses one-sided RDMA Writes, which outperforms replication protocols such as H-Store (Kallman et al., 2008) and Calvin (Thomson et al., 2012) that employ two-sided RDMA Send/Receive.

Similarly, Wang et al. (2017) demonstrate how RDMA can be used to accelerate communication among distributed servers for consensus algorithms that play a vital role in accelerating the performance of coordination managers such as ZooKeeper (Apache Software Foundation, 2021o) by presenting an RDMA-based Paxos protocol known as APUS (Cheng, 2019). Wang et al. (2017) show that the performance of a state machine replication-based system such as Calvin (Thomson et al., 2012) can be improved by up to 8.2 times by exploiting one-sided RDMA Writes as compared to ZooKeeper, implying an overhead of only 10.6 percent in Calvin's execution when replication is performed.

Client-to-server communication: Similar to the RDMA-based Memcached design discussed in section 7.8.2.1, Fent et al. (2020) proposed a high-performance communication layer, referred to as L5, that redesigns the database system's network stack to make optimal use of DMA features and intramachine communication techniques such as shared memory. They demonstrate that L5 can improve the performance for the YCSB-C workload by about 42 percent for RPC as compared to eRPC (Kalia et al., 2019), and that of the in-memory database Silo (Tu et al., 2013) by up to 1.8–15 times compared to that of MySQL (MySQL, 2020) and SQLite (Allen and Owens, 2010).

7.9 Summary

This chapter presented an overview of the RDMA-aware protocols for data movement and interprocess communication. A detailed outline of the different RDMA-aware designs that have been proposed for various high-performance middleware is discussed. Sample case studies are presented to demonstrate the performance benefits that can be harnessed with the new designs. Many of these designs are currently being used in production systems and provide a snapshot of how these designs can be accelerated using current and next-generation HPC technologies and systems.

8 Accelerations with Multicore/Accelerator Technologies

This chapter presents an overview of the common strategies used to leverage parallel and distributed computing systems. This is followed by a comprehensive survey of works in the literature that discuss accelerating different components of the big data ecosystem on current and emerging heterogeneous computing architectures. These recent technological advances include multicore CPUs with SIMD capabilities and coprocessing offload-based accelerators such as GPUs and FPGAs.

8.1 Introduction

Big data analytics has led to the evolution of a new type of workload, known as high-performance data analytics (HPDA), that is driving the research of big data middleware today. Conventionally, two primary approaches can be used by computational workloads to utilize parallel and distributed systems: (1) the "scale-out" approach, which adds multiple small nodes in distributed computing model, or (2) the "scale-up" approach, which focuses on enabling fewer but more powerful servers capable of running large computations. Traditional big data frameworks such as Hadoop focus on the scale-out approach by making use of low-cost and readily available commodity hardware. However, the evolution of modern hardware architecture has been continuously enabling data processing platforms to take advantage of the scale-up approach to create a hybrid scale-up/-out architecture that can meet today's real-time data processing needs.

The feasibility of employing the scale-up approach is made possible through the use of modern heterogeneous computing platforms composed of multicore CPUs with SIMD execution units, general-purpose computing devices such as GPUs (GPGPUs), and reconfigurable devices such as FPGAs (see detailed discussion in chapter 2). To efficiently handle the data computations involved in HPDA workloads on these heterogeneous computing architectures, many novel big data middleware designs are being continuously proposed to exploit the different hardware architectures, namely:

- Multicore CPU-optimized data processing

- GPU-aware big data middleware designs
- FPGA-/ASIC-accelerated big data algorithms and systems

Along each of the three directions, sections 8.2–8.4 discuss various works in the literature that address different components of the big data ecosystem, including online transaction/graph processing in databases, KV stores, and offline analytics (e.g., Hadoop MapReduce, Spark). Based on this, we present three simple use-cases for accelerating simple hash table operations in KV stores with emerging CPU, GPU, and FPGA architectures in section 8.5.

8.2 Multicore CPUs

Modern multicore processors exhibit multiple levels of parallelism through a wide range of architectural features that typically lead to the consideration of three levels of parallel programming: multithreading, distributed parallelism, and SIMD arithmetic support. These multicore processors follow a distributed-memory architecture, with each core being able to directly access its memory and shared main memory to form a NUMA architecture.

Thread parallelism enables creating and assigning threads to cores. This helps build cooperating threads for a single process communicating via shared memory and working together on a larger task. Multithreaded applications are an inherent part of most classical big data frameworks (Apache Software Foundation, 2016, Python, 2021). On the other hand, distributed parallelism and vectorization bring novel and newer features for the current big data applications to harness. Therefore, this section focuses on the NUMA and SIMD capabilities of multicore CPUs.

8.2.1 NUMA-Aware Designs

The NUMA-aware architecture is a hardware design that separates its cores into multiple clusters where each cluster has its local memory region and still allows cores from one machine to access the main memory in the system.

In a typical HPC cluster, a NUMA system classifies memory into NUMA nodes. All memory available in one node has the same access characteristics for a particular processor (i.e., one that is cache-coherent). Nodes have an affinity to processors and devices. These are the devices that can use memory on a NUMA node with the best performance because they are locally attached. Memory is *node-local* if it is allocated from the NUMA node that is best for the processor. As placement of data on the NUMA cores influences performance, several works have been dedicated to studying NUMA-aware algorithms and designs.

8.2.1.1 NUMA-aware DBMS

NUMA-aware database management system (DBMS) designs address two major dimensions, namely, data placement and scheduling tasks across sockets.

Research prototypes, such as HyPer (Leis et al., 2014), ATraPos (Porobic et al., 2014), and ERIS (Kissinger et al., 2014), explicitly control data placement to take advantage of the NUMA architecture. HyPer chunks all objects and distributes the chunks uniformly over the sockets. It leverages task stealing to schedule work local to the data chunks. ERIS employs range partitioning and assigns each partition to a worker thread core while dynamically partitioning to balance workload skewness. It uses one worker per core, which allows tasks to be optimized targeting a particular data partition. Along similar lines, ATraPos (Porobic et al., 2014) uses dynamic repartitioning for online transaction processing workloads, to avoid transactions crossing partitions, and to avoid interpartition synchronization.

While NUMA-aware data placement and scheduling have been basic approaches to accelerating DBMS on multicore machines, recent studies (Psaroudakis et al., 2015) have found that adopting these approaches without being aware of the workload access patterns and their impact on the caching behavior can hinder the performance by about 50–78 percent. To alleviate this impact on performance, task scheduling and data placement strategies that adapt to the workload have been envisioned (Psaroudakis et al., 2016, Kissinger et al., 2014).

8.2.1.2 NUMA-aware graph processing

While scaling graph data analytics on distributed clusters is desirable, enabling high performance per core or a single server machine has also led to several research works. Graph processing frameworks are typically comproed of two steps: construction and computation. The long-term graph data structures are allocated constructor threads on different NUMA nodes and the short-term data structures are allocated by the main thread. However, the computations are performed by different graph processing threads bound to such NUMA nodes at the computation stage. Recent studies (K. Zhang, Chen, and Chen, et al., 2015) have shown that such interleaved and centralized data allocation can significantly hamper data locality and parallelism. Additionally, sequential internode memory accesses (i.e., remote access across NUMA nodes/sockets) were found to have much higher bandwidth than both intranode and internode random ones do.

To overcome the overheads of NUMA data access on graph analytics, frameworks such as Polymer (Zhang et al., 2017) have been proposed. Such designs enable using graph-aware data allocation, memory layout, and access strategies that reduce internode NUMA memory accesses to the minimal or leveraged remote sequential NUMA access when necessary. This has proven to significantly boost performance on NUMA platforms and is comparable to state-of-the-art single-server graph processing systems, such as Galois (Lenharth and Pingali, 2015) and Ligra (Shun and Blelloch, 2013).

Similarly, cache optimized in-memory graph frameworks such as Cagra (Y. Zhang et al., 2017) have been proposed that eliminate random DRAM access with compressed sparse row segmentation and that uses a novel 1D segmentation and cache-aware merge schemes. On the other hand, from a programmer's perspective, domain-specific languages for graph processing, such as GraphIt (Y. Zhang et al., 2018), that can formulate NUMA-aware optimizations have also been designed.

8.2.2 SIMD Vectorization-Based Database Accelerations

Using SIMD instructions to enable data parallelism has been well-studied for accelerating big data computations. This has further advanced with the emergence of modern multicore CPU architectures that support data parallelism via vectorization support for up to 512 bits per vector register.

Using SIMD via vectorized CPU instructions (Intel, 2021c) to enable data parallelism has been studied for accelerating database operators such as scan, join, aggregation (Polychroniou et al., 2015, Zhou and Ross, 2002), and bloom filters (Polychroniou and Ross, 2014, J. Lu et al., 2019). SIMD instructions have also been leveraged to enable data parallel key lookups over hash tables for join operations (Ross, 2007, Behrens et al., 2018, Polychroniou et al., 2015, Gubner and Boncz, 2017). Similarly, network application scenarios, such as packet processing, that deal with batched hash table lookups (e.g., Cuckoo++ (Scouarnec, 2018), Intel DPDK library (Intel, 2021a)), also leverage high-performance CPU-optimized hash tables with SIMD-aware accelerations.

8.3   GPU Acceleration for Big Data Computing

GPU excels at performing simple repetitive operations on large amounts of data in many streams. In contrast to CPUs, offload-based accelerators such as CUDA-enabled GPGPUs (or just GPUs) are capable of offering much higher data accessing throughputs than CPUs because of the higher memory bandwidth and wider SIMD capabilities that enable fine-grained parallelism at the granularity of a GPU warp. For instance, an NVIDIA Volta GV100 GPU uses HBM (NVIDIA, 2017) with 900 GB/s of bandwidth, while the Intel Cascade Lake SP (Xeon Platinum 8280M) processor has an aggregated memory bandwidth of about 228.166 GB/s (Intel, 2019a).

The streaming semantics of GPUs can benefit different aspects of the database, including online data processing for SQL and NoSQL databases and offline data analytics. For online processing, there are two aspects of the GPUs that are widely leveraged: (1) leveraging the GPU's memory bandwidth for bulk data access operations (e.g., KV storage) and (2) accelerating computations in SQL and graph databases.

8.3.1   CUDA-Aware KV Stores

To take advantage of the massive amounts of available parallelism provided by the GPU, KV store requests are batched on the host, passed to the device on kernel launch, and processed in parallel on the device. One or more CUDA threads are dedicated to handling a single request in the Memcached CUDA kernel, with dedicated kernels per KV store operation (Set, Get, Delete).

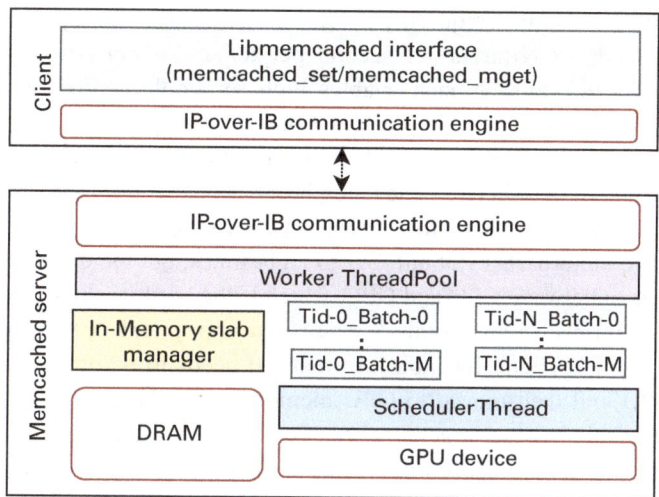

Figure 8.1
Architecture overview of GPU-aware hash table in Memcached.

Mega-KV (Zhang, Hu, et al., 2017, K. Zhang, Wang, et al., 2015) is one such GPU-aware Memcached design that maintains a hash table in GPU memory to achieve 1.4-2.8 times gain in throughput over highly optimized CPU-based KV store designs such as MICA (Lim et al., 2014). These GPU-centric designs point to two vital aspects.

The first aspect is the need for a GPU-friendly hash table layout and a set of corresponding access/update algorithms. GPU-optimized bucketed cuckoo hash table that stores 32-bit signatures to represent keys, and the 32-bit location information is one such example. With this layout, n threads in a warp (n = slots per hash table bucket) can be used to look up all n locations within a designated hash bucket in parallel. Based on this, tens of thousands of GPU threads launched can process batch sizes of up to a hundred thousand threads with a single CUDA kernel launch in a few hundred microseconds.

The second aspect is the end-to-end server workflow for processing request batches. The Memcached workflow is divided into preprocessing (batch incoming requests), postprocessing (responding to individual requests in a completed GPU batch), and GPU compute phases (offload and wait for GPU) that are pipelined to enable asynchronously processing multiple request batches. The overall design is illustrated in figure 8.1.

MemcachedGPU (Hetherington et al., 2015) design, on the other hand, presents an orthogonal approach to leveraging GPUs. Because the KV store operations are network-intensive, increasing server throughput can be enabled by offloading both network packet and application-specific data processing to the GPUs. MemcachedGPU employs the GNoM software framework to enable energy-efficient, latency bandwidth–optimized User Datagram Protocol network and Memcached Set/Get processing on GPUs; achieving 10GbE line-rate processing of about thirteen million requests per second with 17 percent improvement in cost-efficiency (thousands of requests per second per dollar) as compared to state-of-the-art CPU solutions. Such an approach enables high server throughput and energy-efficiency that matches optimized FPGA implementations.

Hetherington et al. (2012) present a CPU-GPU hybrid approach for porting Memcached to use the GPUs by offloading Get request batches and using host-visible memory to move data between the CPU and GPU for other KV store operations. This hybrid design outperforms their respective CPU counterparts by about four to eight times, but the design is affected adversely by moving data between CPU and GPU, which is the bottleneck.

Consequently, these GPU-centric designs require explicit data movement operations between the CPU and GPU. Each key (and location information) needs to be processed and buffered in batches at CPU and then moved to GPU memory: small batches of one thousand requests require moving 64 KB and larger batches of one million require approximately 8 MB. This needs to be performed every few microseconds to keep the pipeline busy. To alleviate this and enable a truly GPU-integrated design, the network and GPU I/O can be combined via RDMA over GPU memory (GPUDirect-RDMA (NVIDIA, 2021d)). A similar OpenSHMEM (Symmetric Hierarchical MEMory access) for GPU (NVSHMEM) implementation equipped with GDR for Mega-KV has been proposed by Chu et al. (2018).

8.3.2 GPU-Aware Online Data Processing

To facilitate database operations to benefit from the GPUs quickly, an early work (Bakkum and Skadron, 2010) implements a subset of the SQLite command processor directly on the GPU in a programmer-transparent manner. With a focus on offloading SELECT queries to a GPU, they demonstrated a thirty-five times improvement in the performance as compared to the SQLite on a single virtual machine. More recent works that leverage GPUs for online SQL operation processing include PG-Strom (HeteroDB, 2021) and BrytlytDB (Brytlyt, 2021). Both PG-Strom and BrytlytDB are GPU-accelerated databases based on PostgreSQL. They support the filtering, sorting, aggregating, grouping, joining tables, and other operations via GPU acceleration. BrytlytDB stores data in CPU memory and copies data back-and-forth from GPU memory, while supporting GPU-centric SQL joins with horizontal partitioning. On the other hand, PG-Strom enables the SSD-to-GPU Direct SQL mechanism that allows the GPU to direct access row data from the NVMe-SSD for SQL workloads.

Another example includes BlazingSQL (2021) that provides an extract-transform-load platform for massive datasets directly into GPU memory and the RAPIDS.ai (NVIDIA, 2021n) ecosystem. On the other hand, there are high-performance nonrelational GPU graph databases, such as the Blazegraph database (Blazegraph, 2021), that support RDF/SPARQL APIs for graphs with up to fifty billion edges on a single machine.

8.3.3 GPU Accelerations for Offline Analytics

One of the earliest works for employing GPUs in accelerating state-of-the-art big data systems is GPMR (Stuart and Owens, 2011), which is a stand-alone MapReduce library that leverages the power of GPU clusters for large-scale computing. HeteroSpark (P. Li et al., 2015) and Spark-GPU (Yuan et al., 2016) transform the general-purpose Apache Spark data processing system into a GPU-supported system, while ensuring minimal CPU-to-GPU data transfers and efficient GPU scheduling for suitable workloads. Continuing along these lines, GPU pioneer NVIDIA has released the XGBoost4J-Spark (Alarcon et al., 2019) package to enable GPU-accelerated end-to-end XGBoost pipelines on Apache Spark.

Another popular GPU-accelerated big data analytics platform is OmniSci (formerly MapD, or Massively Parallel Database). It is a framework that supports a SQL-based, relational, columnar database engine that can leverage the performance and parallelism of CPU/GPU systems to process and visualize large datasets without the need for CPU-centric optimizations such as indexing, preaggregation, and so on. Similarly, SQream DB (SQream, 2021) and Kinetica (2021) are distributed GPU-accelerated databases that utilize hybrid CPU/GPU machines to design a large enterprise-ready data warehouse for analytics.

Over the years, prominent GPU-aware generic data science frameworks have been developed, such as RAPIDS (NVIDIA, 2021n), that include a collection of libraries for executing end-to-end data science pipelines entirely on GPUs. In contrast to online and offline data processing, FASTDATA.io's PlasmaENGINE (FASTDATA.io, 2021) and IBM's Spark GPU (IBMSparkGPU, 2016) perform analysis on streaming data analysis or data transformation directly on the GPU.

8.4 FPGAs and ASICs

FPGAs (Xilinx, 2021) are semiconductor devices that are based around a matrix of configurable logic blocks connected via programmable interconnects. As discussed in chapter 4, FPGAs enable implementing fully pipelined dataflow architectures that enable us to fully exploit fine-grained task- and instruction-level parallelism to achieve high data processing rates. Along with low power consumption, they essentially eliminate shared-memory access overheads via locking mechanisms commonly seen in CPU-based designs. Similarly, ASIC chips are set to become the devices of choice for data processing applications in the near future. With this as motivation, there have been several research works dedicated to enabling big data frameworks to leverage FPGAs and ASICs in both HPC and data center environments. We discuss a few brief examples to give a taste of this growing technological direction.

8.4.1 FPGA-Aware KV Store and Databases

FPGA-aware designs for KV store designs address bottlenecks by rearchitecting the server while maintaining compliance with the standard Memcached-like semantics. The key bottlenecks we identified for CPU-based design are related to the network stack's processing overhead and its large latency, which is the result of the indirect access to the network via PCIe. The distances among the network interface, compute resources, and memory can be minimized by integrating both interfaces directly on the FPGA.

Blott et al. (2013) demonstrate that such a tight coupling enables achieving 10 Gb/s line-rate processing for all KV pair request sizes. Specifically, they show that FPGA-based implementation delivers a worst-case round trip time of 4.5 μs and achieves an increase of thirty-six times in requests per second per Watt over the best published x86 numbers with substantially low power usage. Lavasani et al. (2014) propose an FPGA-based accelerator with a CPU to handle complex Memcached applications by supporting more commands (or operations) through "inlining" critical operations and handling the rest at the CPU. An inline FPGA accelerator sits between the NIC and the CPU. The inline accelerator intercepts incoming packets going from the NIC to the CPU and can either process the packets completely without CPU involvement (i.e., fast path) or pass the packets through to the CPU without processing them (i.e., a complex slow path with CPU involvement). Based on the Memcached operation's latency needs, different types of user requests can be directed to different paths. This is especially suitable for cloud-based NoSQL work-loads that are read-heavy (i.e., dominated by read requests such as Get over update requests such as Set).

On the other hand, ready-to-use hardware such as Xilinx's FPGA-based KV store accel-erator, Alveo U200 (algo logic.com, 2020), implements an FPGA-based KV load/store processing for data center server CPUs integrated with a network accelerator stack and out-of-order memory controller.

8.4.2   Offline Data Analytics over FPGAs

Several research works have explored the feasibility of employing FPGA-based acceleration into Apache Spark. Ghasemi and Chow (2016) propose an FPGA-accelerated MapReduce implementation of the computation-intensive k-means clustering algorithm running over the Apache Spark framework in a distributed CPU-FPGA cluster. Similarly, Chen et al. (2016) presents an FPGA-as-a-Service framework that includes batch processing with Spark and presents a viable case study with a next-generation DNA sequencing application demonstrating gains of more than two times for both performance and energy efficiency. In essence, these works demonstrate how high performance can be achieved with the help of cost-effective and low-power programmable FPGA devices, despite employing hardware platforms that are tuned for performance.

Spark for unconventional cores (SparkCL (Segal and Margala, 2016)) is an extended version of Apache Spark built for use on heterogeneous clusters of conventional and unconventional computing cores such as FPGAs/GPUs/accelerated processing units. It en-hances Apache Spark APIs with accelerator-friendly OpenCL interfaces. To use SparkCL, one writes an algorithm in Java deriving from SparkCL-based classes and defines several accelerator support functions and the kernel that is meant to be accelerated. The Java to

OpenCL execution engine on each of the individual machines in the cluster is Aparapi-UCores (Segal et al., 2014). Any hardware that supports OpenCL-based accelerations can be supported, and the accelerator device selection can either be done automatically by the framework according to criteria such as speed or power efficiency.

8.4.3 Accelerating Deep Learning with FPGAs/ASICs

The recent growth of deep learning applications has improved the research and implementation of customized code and algorithms that can exploit programmable ASICs and FPGA devices. Specifically for convolutional neural networks, various FPGA-based accelerations (see, e.g., C. Zhang et al., 2015) have been developed to take advantage of the high performance and reconfigurability, while overcoming the under utilization of the logic resource, available memory bandwidth, and so on. A key advantage of the FPGA accelerator is that its performance is insensitive to data batch size, while the performance of GPU acceleration varies largely depending on the batch size of the data. For instance, Li et al. (2018) demonstrate that an optimized fully mapped FPGA accelerator architecture tailored for bitwise convolution and normalization, featuring massive spatial parallelism running over Virtex-7 FPGA, can outperform a Titan X GPU for processing online individual requests in small batch sizes by up to 8.3 times with a 75 times improvement in energy savings. Similarly, a Xilinx FPGA-based deep learning accelerator unit (Wang et al., 2016), exploiting three pipelined processing units to explore locality for deep learning applications, can achieve up to 36.1 times speedup compared to the Intel Core2 processors. Further details on deep learning over big data is discussed in depth in chapter 10.

8.5 Case Studies and Performance Benefits

To illustrate the benefits of the different HPC capabilities discussed, this section presents case studies for big data framework accelerations with GPUs, SIMD capabilities of CPUs, and FPGAs.

8.5.1 Hash Table Performance on CPU and GPU

This section presents two case studies for KV store accelerations with GPUs and CPU-SIMD.

To understand the potential of employing a GPU-aware hash table design into high-performance KV stores, it is crucial to study the stand-alone performance of insert and search operations on both CPU- and GPU-based hash tables. For the CPU platform, consider the chaining-based hash table design, used in popular in-memory KV stores such as Memcached. For the GPU platform, consider the GPU-optimized cuckoo hash table design from Mega-KV (described in section 8.3.1) (K. Zhang, Wang, et al., 2015). The hash table performance with a microbenchmark to simulate Multi-Set and Multi-Get operations in general-purpose KV stores is presented in figure 8.2. It demonstrates the hash table

(a)

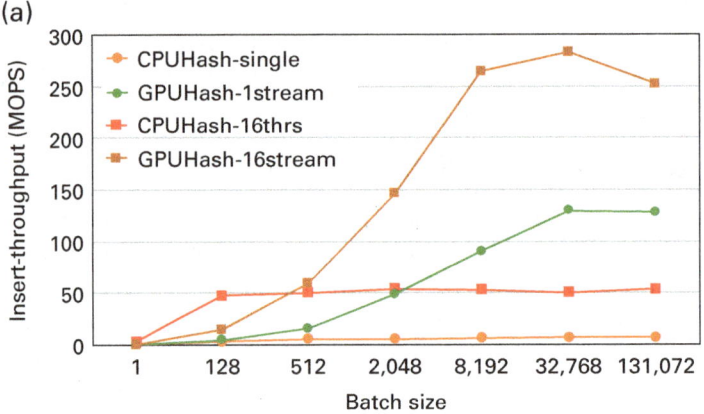

(b)

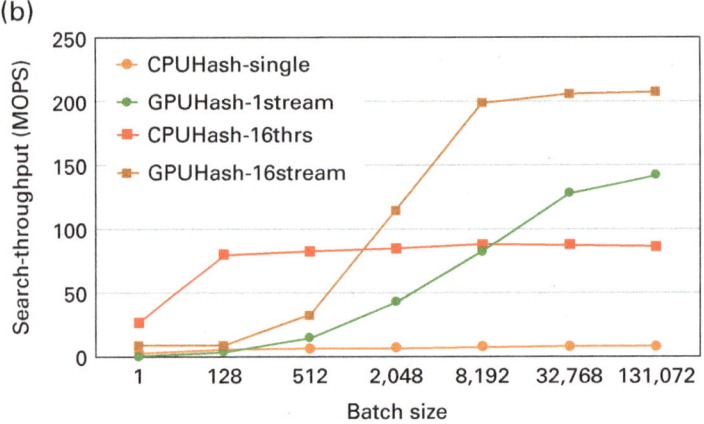

Figure 8.2
Stand-alone throughput with CPU and GPU-centric hash table (based on Mega-KV (K. Zhang, Wang, et al., 2015)). (a) Insert. (b) Search. MOPS, millions of operations per second; thrs, threads.

operation performance with varying batch sizes, wherein a batch size signifies the number of insert/search operations to perform per request. These experiments are performed on an HPC cluster equipped with nodes with Intel Broadwell (E5-2680-v4) dual fourteen-core processors, 128 GB of memory, a Mellanox InfiniBand, Extended Data Rate (EDR), host channel adapter (HCA), and a K80 NVIDIA GPU.

For the GPU-based hash table, compared with one CUDA stream, a 2.2–2.8 times performance improvement can be achieved with sixteen CUDA streams, by effectively overlapping data transfers with the CUDA kernel computation involved in inserting or searching KV pairs into slots in parallel in different hash table buckets. On the other hand, with the CPU-based hash table, employing sixteen concurrent threads for insert or search operations offers 8.2–9.8 times improvement in throughput, while the GPU-based hash table outperforms it by about 2.3 times for larger batch sizes. This confirms the observations presented in K. Zhang, Wang, et al. (2015) that indicate that the overall throughput of the KV store can be boosted by effectively utilizing the high memory bandwidth and latency-hiding capability inherent to offload-based accelerators, such as GPUs.

8.5.2   Hash Table Lookup Performance with CPU Vectorization

For this use-case, the focus is directed mainly toward read-dominated workloads. Hence, cuckoo hashing (Pagh and Rodler, 2004, Panigrahy, 2004) is used. It enables a KV pair to be located (i.e, a hash table lookup) using a constant number of memory accesses (unlike other collision resolution hash table schemes such as chaining, linear probing, and so on). The value stored corresponding to a key in the hash table is also referred to as a "payload." N-way cuckoo hashing is a well-known open-addressing–based hashing scheme that maintains N hash functions (h1, h2, . . . , hN), such that, a key can be found in exactly one of N locations (or hash buckets). It achieves this simplicity by shifting the complexity from lookup to insertion.

Considering the latest Intel CPUs that enable 512-bit vectors (AVX-512) (Doweck et al., 2017), for an N-way cuckoo hash table with 32-bit keys and payloads, one can lookup sixteen keys in parallel using one probe operation. Thus, one can lookup n keys in a maximum of $N \times n/16$ iterations, as compared to non-SIMD designs that may potentially perform $N \times n$ iterations in total. To evaluate this, the stand-alone performance of vertically vectorized three-way cuckoo hashing with AVX-512, that is, Cuckoo-Ver (AVX-512), is studied based on the algorithm presented in Polychroniou et al. (2015). Because the three-way cuckoo hash table enables a load factor close to 90 percent, it is contrasted with the state-of-the-art CPU-optimized non-SIMD MemC3 (Fan et al., 2013), the MemC3 (Scalar).

In this study, two key access patterns are investigated: a uniformly random pattern (Uniform) and a skewed access pattern (Skewed) based on Facebook's Memcached workload generator (Atikoglu et al., 2012). An input 32-bit column of one billion keys with about 90 percent selectivity (selectivity = percentage of the input that is likely to find a match in the hash table) is used, and the output is a 32-bit column with matching payloads, over a table with a load factor of 90 percent. This test is run on a shared hash table across all twenty-eight cores of a dual fourteen-core Intel Skylake node.

Figure 8.3 presents the hash table probing throughput for the two designs, for varying hash table sizes. From this figure, one can observe that for hash table sizes that can fit into the L2 cache (512 KB), Cuckoo-Ver outperforms MemC3 by about 2.7–6.6 times. When hash table size exceeds the L2 cache, one can observe that, while Cuckoo-Ver performance drops by 3–5 times, it still maintains a gain of 1.63–2.6 times over the MemC3 hash table. From the above-mentioned experiments, one can observe that, while the performance of the vertical-vectorized SIMD-aware hash table is bound by memory, it can maintain a consistent improvement over cache-optimized non-SIMD hash table designs.

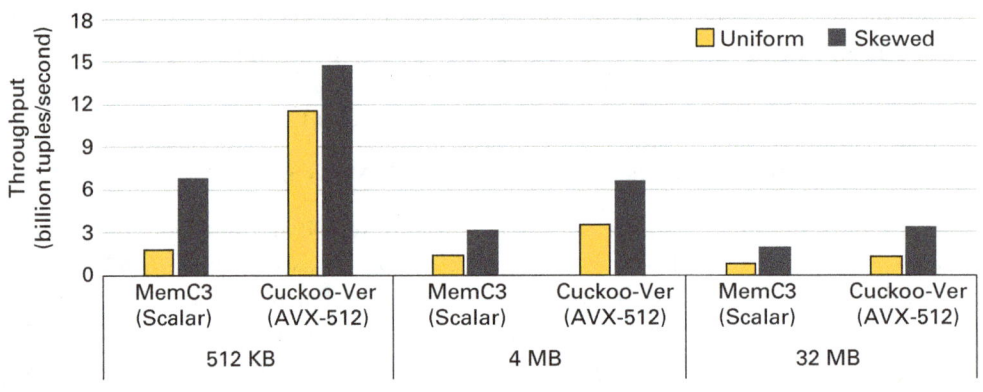

Hash table size vs. cuckoo hash table designs

Figure 8.3
Stand-alone hash table probing performance on the twenty-eight–core Intel Skylake CPU, over a three-way cuckoo hash table versus non-SIMD CPU-optimized MemC3 hash table with 32-bit key/payload (Shankar et al., 2019a).

8.5.3   Accelerating Spark with GPUs

As discussed in section 8.3, the increasing number of CPU-GPU systems have led several researchers to study how best to leverage the GPU's massive compute capabilities in existing offline analytical frameworks. Spark-GPU (Yuan et al., 2016) is one such system that attempts to address differences between Spark's Java-based network-centric execution model and GPU's streaming architecture. It introduces the notion of GPU-RDDs that mimic Spark's RDD notion for the application developer while enabling it to reside and be processed in the GPU's memory. With GPU-optimized task management and data transfers, Spark-GPU supports scan, broadcast join, hash join, aggregation, and sort operators. It demonstrates significant improvements specifically for computation-intensive data mining tasks such K-Means and logistic regression by up to 16.13 times. On the other hand, it shows a comparably less, but considerable, improvement of about 4.8 times for data warehousing workloads such as TPC-H benchmark (TPC, 2021).

8.5.4   Processing Real-Time Data on GPUs

Several works, such as cuStreamz (NVIDIA, 2020) and Flink with GPU support (Apache, 2020), envision leveraging GPU for streaming data, enabling them to scale cost-effectively. In particular, cuStreamz is a streaming library that accelerates the Python streaming library Streamz (Streamz, 2021) by leveraging RAPIDS cuDF (RAPIDS, 2021) to perform fast computations on streaming data on GPUs. It reads data directly from Kafka into GPU memory to enable processing high-velocity data efficiently. Experiments with cuStreamz (Dandu et al., 2020) on AWS EC2 with g4dn.xlarge (GPU) instances demonstrate that the total cost of ownership is improved by about 1.2 times for stateless streaming with simple aggregations and by up to 4.5 times for complex aggregations. It also improves stateful use-cases with short time windows by about three times.

8.5.5 GPUs and FPGAs for Graph Analytics

Recent advances in the computational capabilities of GPUs make them attractive for running graph applications. However, graph workloads typically require out-of-core computing (data that are too large to fit into DRAM) (Zhu et al., 2015). It, therefore, requires the support of a secondary storage extension such as NVM. One such recent work that focuses on leveraging an array of SSDs with a GPU on a single node to process large graphs is Garaph (Ma et al., 2017). It employs all CPU and GPU cores using techniques such as vertex replication for maximizing GPU utilization and edge-based partition for balancing the CPU workloads. Garaph achieves a 2.5–5.36 times improvement over other GPU graph frameworks such as CuSha (Khorasani et al., 2014) and CPU shared memory systems such as Ligra (Shun and Blelloch, 2013) for single-source shortest path and connected components graph algorithms over the Twitter 2010 dataset (Kwak et al., 2010) on a node equipped with Haswell CPUs and an NVIDIA Pascal GPU.

On the other hand, many works have been exploring leveraging FPGAs for graph workloads (Besta et al., 2019). ThunderGP (Chen et al., 2021) is a recent HLS-based graph processing framework for FPGAs that is developed on the gather-apply-scatter model to build highly parallelized and memory-efficient templates. ThunderGP can achieve up to 6,400 million traversed edges per second on Xilinx multiple super logic region–based FPGAs and about 2.9 times improvement of state-of-the-art FPGA graph frameworks such as HitGraph (Zhou et al., 2019).

8.6 Summary

This chapter presents an overview of the different hardware-centric optimizations introduced to accelerate computations for various high-performance big data middleware in use today. It discusses designing big data processing systems, such as databases, offline analytical engines, and KV stores, that leverage multicore CPUs via NUMA and SIMD, GPGPUs, FPGAs, and customizable ASICs. As technology improves, more acceleration-based solutions that can leverage the scale-up capabilities of modern, heterogeneous computation hardware will be feasible.

9 Accelerations with High-Performance Storage Technologies

This chapter presents a brief overview of the different architectural and design choices involved in building a file system or a data store using high-performance storage technologies for big data middleware. This is backed by an extensive survey of various state-of-the-art NVMe/NVMEoF and NVRAM-accelerated single-node and distributed storage systems and runtimes proposed in the literature.

9.1 Overview

Most traditional HPC systems are influenced by the Beowulf cluster architecture (Sterling et al., 2003b), where the computation and storage nodes constitute separate layers that are connected via high-speed interconnects. As discussed in chapter 4, the architecture of these clusters opens up the possibility of keeping the CNs lean, with a lightweight operating system and limited storage capacity. The subcluster of dedicated I/O nodes with enhanced PFSs, such as Lustre, provides access to large datasets. While these HPC clusters provide the computing power necessary for scaling big data applications, the limited local storage poses a significant performance overhead, as it necessitates moving the data to and from the CNs. To overcome this challenge, modern HPC architectures are being equipped with heterogeneous storage devices such as DRAM, NVRAM (i.e., byte-addressable NVM), SATA-/PCIe-/NVMe-SSD, and HDD. It is, therefore, becoming critical for big data middleware to design storage solutions that can take advantage of the heterogeneous storage options in the most optimized manner for enabling HPDA applications.

Designing new file systems and storage runtimes for PMEM (e.g., NVRAM) and NVMe-SSDs has been an active topic of research for the past decade. Several storage systems have been proposed for different storage technologies and purposes. In general, these systems can be broadly classified based on the kind of technology they use (NVRAM, SSD, or hybrid/tiered), their scalability (single-node or distributed), and the interface provided (KV or file system). Thus, to perform big data analytics at scale, it is necessary to answer the following questions.

What kind of storage devices can we leverage to meet the fast I/O needs of HPDA applications? The majority of the big data storage systems today are looking to leverage ongoing advancements in memory and storage technologies to provide distributed in-memory computing performance with the cost and durability of disk storage. These advancements include flash-based NVMe-SSDs and emerging PMEM technologies that are largely discussed in chapter 3.

How do we exploit the different storage devices available in the best possible manner? Typically, two architectural approaches have been adopted to design large-scale storage solutions depending on the application's data access pattern, namely, scale-up and scale-out. In the scale-out approach, multiple new hardware types (e.g., small servers) can be added and configured as the need in a distributed computing model. The scale-up system presents a vertical scalability approach that increases the capacity of a single system to define a more powerful server.

In this chapter, we explore high-performance storage middleware that leverages current and emerging high-speed I/O devices, along with the following three directions:

1. File systems and data stores that leverage emerging NVMe and PMEM (NVRAM) technologies.

2. Hierarchical file systems that exploit the growing local storage in conjunction with large PFSs for HPDA.

3. Burst buffer systems that aim to provide high-performance while maintaining the basic semantics of the traditional Beowulf architecture.

9.2   Exploring NVM-Centric Designs

NVMe and NVMEoF on RDMA-capable networks are a novel network storage technologies that are widely being explored for processing big data in data center environments. Built from the ground up, this technology is set to dramatically improve the performance of existing storage network applications and accelerate new and future computer technologies such as scale-out software-defined storage and storage disaggregation.

In this section, we discuss the different research prototypes and out-of-the box commercial solutions that have been changing the storage landscape, by exploiting NVMe/NVMEoF and PMEM for HPC and data center storage systems.

9.2.1   NVMe-Centric Storage Middleware

As discussed in chapter 3, the NVMe specification (NVMe, NVM Express, 2021) is a recent innovation that has significantly impacted research in storage systems. NVMe-based SSDs have been emerging as the latest storage technology bridging the dreaded performance gap between hard disks and memory. These new devices are built for extremely low latency and achieving high degrees of parallel I/O. Several user-space I/O engines have been proposed recently to allow programmers to directly access devices from user-space. These include Intel SPDK (Intel, 2021j), NVMeDirect (Kim et al., 2016), and POSIX asynchronous I/O (Linux, 2021a).

On the other hand, NVMEoF defines a common architecture that supports a range of storage networking fabrics for NVMe block storage protocol over a storage networking fabric. Two types of fabric transports for NVMe are currently under development: NVMEoF using RDMA and NVMEoF using Fibre Channel.

NVMe technologies can be used to boost the performance of storage clusters via scale-up NVMe-SSD arrays. The emerging high-capacity, ultra-performance NVMe-SSDs can be leveraged by almost every component of the front-end and back-end tiers of a multitier data center architecture. We describe some concrete examples for NVMe for each component in this section.

9.2.1.1   NVMe-SSDs for components of the front-end tier (online data analytics)

Memcached distributed caching layer in the multitier data center architectures can leverage NVMe-SSDs to boost their server capabilities while trying to achieve near in-memory speeds, toward enabling both high-speed I/O and end-to-end performance improvements for online data processing workloads. Examples include NVM-assisted designs for Memcached such as Fatcache (Twitter, 2017), MyRocks over NVM from Facebook by (Eisenman et al., 2018), and so on.

For other components of the front-end tier, some prominent examples of how NVMe-SSDs can provide performance enhancements are as follows.

In-memory databases: In-memory databases such as Redis (Redis Labs, 2021a) leverage DRAM capabilities to enable optimal performance, but they are limited by DRAM capacities. NVMe devices can enable in-memory storage systems with weak consistency needs to extend their data availability and reliability features at low costs, for example, while maintaining critical data in DRAM.

Such a Redis on Flash design for x86 servers is a prominent example of NVMe for memory-centric computing for online big data analytics. Redis on Flash allows Redis to run optimally with flash memory (e.g., Intel NVMe Flash, Samsung NVMe-Flash) used as a RAM extender. It enables memory-like performance at one-tenth of the DRAM cost (e.g, three million database operations per second at under 1 ms of latency, while generating over 1 GB NVMe throughput, on a single server with Redis on Flash and Intel NVMe-based SSDs).

Durable NoSQL databases: Cassandra is an open-source, NoSQL data store that has been primarily designed and optimized for fast writes and to scale linearly with an increase in cluster size. Cassandra writes to disk (SATA-/NVMe-SSD) when its in-memory data tables are full, and reads missing in the memtable cache are directed to NVMe. With this architecture, as Xu et al. (2015) demonstrates, a single NVMe-SSD can perform on par with four SATA-SSDs for YCSB (Cooper et al., 2010). For network-attached storage systems prevalent in data centers, Balakrishna (2016) presents a use-case of NVMEoF for Cassandra to demonstrate that efficient NVMEoF schemes can be designed to have the same benefits as direct-attached local NVMe-SSDs. Such designs can also be extended to any NoSQL database (e.g, for RocksDB (Guz et al., 2017)).

SQL/OLTP platforms: High-performance database platforms manage high-capacity, high-bandwidth transaction-based applications for OLTP workloads (e.g., session management, real-time data analysis, e-commerce). These systems need to enable fast access to mission-critical data, enabling transaction processing with ultra-low and consistent latency, where access delays can be extremely costly. Using NVMe-SSDs enables fast transaction processing and fast, consistent response times. From the TPC-C benchmark study by Samsung (Lagrange, Veronica and Choi, Changho and Balakrishnan, Vijay, 2016), a comparison of tpcc-mysql on Serial Attached SCSI (SAS)-HDD, SATA-SSD, SAS-SSD, and NVMe reveals that NVMe-SSDs can enable 180 times the benefits of traditional HDDs and 2.4 times improvement over SATA-SSDs.

Streaming applications: For streaming applications, pipelined frameworks that consist of a message broker (e.g., Kafka (Apache Software Foundation, 2021j)) coupled with a search storage engine (e.g., Elasticsearch (Elasticsearch B.V., 2021)) are employed to run Exact Search, Range Query, Term Match, and other operations. While the workflow utilizes the compute, memory, and networking to its capacity, efficiently using fast NVMe and NVM-EoF drives for I/O can enable further optimizing different service-level agreements (SLAs) targeted for high-throughput or low-latency scenarios for real-time decision making.

9.2.1.2   NVMe-SSDs for components of the back-end tier (offline big data analytics)

The back-end tier is dominated by large-scale I/O-intensive workloads over distributed processing systems, such as Apache Spark, and machine and deep learning frameworks, such as Tensorflow, and so on, over distributed storage systems. These middlewares can be accelerated with NVMe-SSDs. Some examples are included here.

Intermediate data shuffling in Hadoop/Spark: Hadoop MapReduce and Spark applications involve a network-intensive data shuffling phase that follows the map phase. Once map tasks finish processing their designated data partition, the intermediate data are written to disk before shuffling them over the network for the reduce phase. Because the performance of this intermediate data shuffling phase involves I/O, employing NVMe-SSDs with NVM-EoF to shuffle data over the network can reduce the overall Spark job time.

Distributed storage systems: Heterogeneity-aware HDFS designs can also benefit from NVMe-SSDs to enable hybrid storage placement strategies that focus on NVM-based strategies (Islam et al., 2016b, Li, Ghodsi, et al., 2014). Disaggregated storage such as IBM Crail (Stuedi et al., 2017) is also heavily invested in designing NVMe-oF based solutions toward a scalable storage layer that can support all big data analytics. Similarly, scale-out object-based storage systems, such as Ceph, can be potentially codesigned with NVMEoF to enable performance with NVMEoF schemes to access remote NVMe to perform on par with local NVMe accesses (Tang, Haodong and Zhang, Jian and Zhang, Fred, 2018).

Accelerating machine and deep learning: Any data center infrastructure that supports a machine or deep learning workflow needs to be able to handle a large number of files and must have a large amount of data storage with high-throughput access to all the data. Legacy file systems cannot supply high throughput and high file I/O operations per second with HDDs and are not suitable for the low-latency, small-file I/O, and metadata-heavy workloads that are common in AI and analytics. To overcome this I/O starvation, massively parallel shared file systems with NVMe-SSD and NVMEoF optimized schemes are being built (e.g., Matrix PFS (Murphy, Barbara, 2018)).

9.2.1.3   NVMe-based file systems

To get the best performance from the NVMe devices, it is vital to design either a stand-alone single-node or distributed file systems that can exploit the throughput capabilities of the device. Along this direction, several file system designs have been proposed in the literature.

ReconFS by Y. Lu et al. (2014) is a kernel file system designed for flash-based SSDs. It reduces metadata overhead by decoupling the volatile and persistent directory tree maintenance. DevFS by Kannan et al. (2018) is a direct access file system for NVMe-SSDs embedded completely inside the storage device. It utilizes a reverse caching mechanism to reduce host memory overhead. F2FS is a kernel file system designed by Lee et al. (2015) with the characteristics of flash storage in mind. It uses append-only logging and a flash-friendly on-disk layout to improve performance. Moneta (Caulfield et al., 2012) uses a novel architecture that allows it to avoid entering the kernel and making permission checks. This allows Moneta to overcome the two major sources of overhead in flash-based file systems. These file systems are all designed by considering the performance characteristics of NAND flash memory. However, as presented in chapter 4, with newer SSDs built using 3D XPoint (3DXP) memory, the performance characteristics are significantly different. For example, 3DXP memory does not suffer from the write amplification effect and also has symmetric read-write performance (for both sequential and random I/O). Therefore, many design decisions taken by these systems, such as the use of log-structured or append-only designs, are not optimal for this new class of SSDs.

Reflex by Klimovic et al. (2017), Ceph, 2021, and PolarFS by Cao et al. (2018) are three storage runtimes, specifically designed for NVMe-SSDs. Reflex enables fast remote access to SSDs using DPDK, which bypasses the kernel, avoiding expensive data copies and context switches for high performance. PolarFS and Ceph on the other hand, utilize the user-space Intel SPDK runtime for direct device access and RDMA for low-latency remote I/O.

9.2.2 PMEM-Based Data Storage

A number of prior works have developed new data stores and file systems for nonvolatile or heterogeneous memory systems.

9.2.2.1 KV stores

RocksDB and PebblesDB (Raju et al., 2017) are two log-structured merge (LSM-tree–based) KV stores. PebblesDB uses a fragmented LSM tree that was specifically designed to reduce write amplification in flash storage. WiscKey by L. Lu et al. (2017) is based on the same principles as RocksDB but separates the storage of keys and values for better performance on flash storage. Aerospike (Aerospike, Inc., 2021) and Scylla (Scylla, 2021) are two NoSQL stores providing KV semantics. Their designs have been specifically optimized for NVMe-SSDs. uDepot by Kourtis et al. (2019) is another such system that improves performance by keeping the index in DRAM and using logging for resilience. KVell by Lepers et al. (2019) achieves maximum bandwidth for NVMe-SSDs by avoiding synchronization and avoiding sequential access. In contrast to these systems, several others focus primarily on NVRAM-based KV stores. For instance, Bullet by Huang et al. (2018) is a persistent KV store on NVRAM that separates volatile and persistent domains and proposes techniques to keep the two consistent. The volatile front end in Bullet only serves as a cache with cache misses being served from the persistent back end. MyNVM by Eisenman et al. (2018) is another NVRAM-based KV store that aims to improve the memory footprint of RocksDB by using NVRAM as a second-level block cache. Pmemkv (2018) is a collection of NVRAM- and DRAM-based KV stores that utilize PMDK (Intel, 2019e) for guaranteeing consistency and persistence. HiKV by Xia et al. (2017) is a hybrid memory solution that maintains two indices (hash index in NVRAM and B+ tree in DRAM) to improve performance. Faster is a concurrent solution, which uses a resizable in-memory hash table and an append-only log to store data, designed by Chandramouli et al. (2018).

9.2.2.2 Single-node NVRAM file systems

NOVA by Xu and Swanson (2016) is an NVM file system that provides strong consistency and durability while maintaining performance. It does so through the use of novel techniques, such as per-inode logging and in-memory indexes. BPFS by Condit et al. (2009) is another storage class memory file system that uses shadow paging. It uses short-circuit shadow paging and fine-grained copy-on-write to improve performance. SCMFS presented by Wu and Reddy (2011) is another such system that improves performance by leveraging the virtual address space for continuous file addressing. The PMFS NVRAM file system, introduced by Dulloor et al. (2014), provides metadata atomicity through journaling. XFS DAX and fourth extended file system (ext4) DAX are extensions of the XFS and ext4 file systems that provide optimized support for NVRAM by bypassing the page cache. EvFS by Yoshimura et al. (2019) is a user-level, event-driven file system for NVM. It enables asynchronous processing through the use of a page cache and direct I/O. Similarly, HiNFS by Y. Chen et al. (2018) solves the tail latency problem of NVRAM by using a cache-line–sized page cache. Other recent works on PMEM file system designs include Aerie by Volos et al. (2014) and SplitFS by Kadekodi et al. (2019). These NVRAM-only file systems can deliver desirable performance. However, given that NVRAM is expected to have a lower capacity and higher cost than flash, these systems are unable to satisfy all use cases.

9.2.2.3   Distributed NVRAM file systems

NVFS (Islam et al., 2016b) is an optimized version of HDFS that leverages NVRAM and RDMA to design a low-latency runtime. Orion is a distributed storage system also designed with NVRAM and RDMA by C. Yang et al. (2019). Through a clean slate design and by utilizing the byte addressability of NVRAM, Orion delivers fast metadata and data access. Octopus by Y. Lu et al. (2017) is another such storage runtime that improves performance through techniques such as client-based replication and self-identified RPC. CDBB, introduced by Fan et al. (2018), is a collaborative distributed burst buffer that utilizes node-local NVRAM to absorb checkpoint I/O in HPC clusters. These solutions represent the current state-of-the-art distributed PMEM file systems. In general, though, we find that there is a dearth of such solutions. This is likely because NVRAM technology is new and expensive, and its integration with networking standards such as RDMA is still evolving and not yet complete.

9.2.3   Distributed Storage with NVMe/NVMEoF for Next-Generation Data Centers

For edge-computing and rack-scale architectures commonly employed to scale data center storage, NVMe technologies can be used to boost I/O performance predominantly along two important directions scale-out storage and scale-up storage.

Scale-out storage: NVMEoF changes the game for intra-SN communications. Instead of an IP, it uses a storage protocol designed specifically for memory transfers (i.e., RDMA). This will mean that high-speed storage devices can essentially sit anywhere on the network and be treated as one large pool of high-performance storage without compromises or bottlenecks. Many vendors such as Xilinx, Mellanox, and others, are enabling NVMEoF scale-out solutions over distributed NVMe-SSDs or SSD arrays.

Scale-up storage: This refers to scaling storage in conventional arrays that were based on SATA or SAS-based all-flash arrays that were accessed over a network, via a single controller head. This architecture has been leveraged in direct-attached storage to SAN storage for decades. This provides a performance boost because the media (NVMe-SSDs) are faster. By leveraging the NVMEoF path, the network I/O bottleneck in this traditional share storage architecture can be alleviated. Some solutions along this direction are Axellio (X IO, 2021), E8-D24 Rack Scale Flash (E8 Storage, 2021), and so on.

With these two approaches, the emerging high-capacity and low-latency NVMe-SSDs are steadily enabling data center and even high-performance computing systems to serve data for performing valuable big data stacks much further.

9.2.4   Application-Specific NVM Devices

There have been three recent efforts to design SSDs that provide interfaces that better match with application semantics and requirements.

Open-Channel SSD: Open-Channel SSD (OCSSD) (LightNVM, 2018) is a nonstandard SSD that breaks away from traditional NVMe protocol. OCSSD does not implement a flash translation layer, but rather it exposes the internal physical storage to the operating system or user-space runtime to directly program with. The intuition behind this approach is to expose the internal parallelism and to enable low and predictable I/O latency. The SPDK runtime has also added support for OCSSD. Furthermore, several KV stores (e.g., NVMKV by Marmol et al. (2014) and LOCS by P. Wang et al. (2014)) and caches (e.g., FlashTier by Saxena et al. (2012) and DIDACache by Shen et al. (2018)) have been designed to leverage the features of OCSSDs and provide high throughput. Unfortunately, there is not yet a commercially available OCSSD that can be widely deployed. This remains an active research area.

KV-SSD: KV-SSD (Kang et al., 2019) is another nonstandard SSD, introduced by Samsung, that provides a KV interface instead of a block-based interface. The benefits of a KV storage interface for designing KV stores are several. First, there is no need for maintaining metadata and worrying about its consistency because that task can now be offloaded to the device. Second, no additional information is required to be stored for each KV pair. This results in an ideal read and write amplification factor of 1. Finally, the host memory consumption now is independent of the number of keys in the system, significantly lowering the memory footprint. Crail-KV by Bisson et al. (2018) is an active distributed KV system, which is an extension of the Crail file system designed using a KV-SSD. The KV interface exposed by Crail clients can be preserved down to the SSD, which significantly simplifies metadata management and improves latency.

Multistreamed SSD: Multistreamed SSD, introduced by Kang et al. (2014) and manufactured by Samsung, is a new type of NVMe-SSD that provides the ability to write to flash pages using streams. Each stream writes to a separate pool of hardware pages. By utilizing this feature, a single request can be assigned a dedicated stream, ensuring that related data are resident in physically contiguous locations. Consequently, the SSD will never have only a fraction of invalid pages within an erase block, rather only an all or nothing effect, as discussed in detail by Rho et al. (2018). This effect significantly reduces write amplification and garbage collection overhead. This feature can be incredibly useful for designing file systems. Each file system thread can be mapped to a separate stream, which will improve the locality of data accesses and performance as a consequence.

9.3 Hybrid and Hierarchical Storage Middleware

The high-performance of PMEM and NVMe-SSDs introduces a significantly high cost, in terms of the average price per byte, compared with the cost of traditional SSDs and HDDs. As a consequence, the available NVM devices come with limited capacities. To overcome this, hierarchical or tiered architectures have been widely explored for designing large-scale storage systems. The storage systems manage a hierarchy of heterogeneous storage devices and place data in the storage layer that best matches the performance requirement and the future access pattern of the application.

9.3.1 Hybrid/Tiered File Systems

HMVFS (a hybrid memory versioning file system), by Zheng et al. (2016), is a tiered file system that exploits the byte addressability of NVRAM for performance and the large capacity of block-based storage. Ziggurat, by Zheng et al. (2019), is a storage system that combines NVRAM and slow disks to create a solution that has NVRAM-like latency with disk-like capacity. Strata is a cross-media file system that uses a log-structured design and a user-space file cache for improved performance Kwon et al., 2017. NV-Booster, by Q. Wei et al. (2015), is a PMEM file system that stores metadata in NVRAM, but data on disks. Their results show that NV-Booster can be used to accelerate the performance of Ceph by as much as ten times. TridentFS, by Huang and Chang (2016), is another hybrid file system that improves performance by keeping hot data in NVRAM and evicting cold data parallelly to flash and disk storage. IBM Crail (Stuedi et al., 2017) and Pocket, by Klimovic et al. (2018), are two-tiered distributed storage systems that are designed for data center and cloud environments. They support DRAM, flash, and disk storage media. While Crail leaves the choice of which data tier to use to the application, Pocket leverages application-provided hints and service agreements to select the appropriate tier to use. Along similar lines, ONFS, by Liu et al. (2017), is a tiered HPC-oriented system that offers high bandwidth, low latency, and high capacity. Initial results show up to 6.5 times higher application I/O bandwidth as compared with that of Lustre. In general, hybrid systems offer a better performance-to-cost ratio than other types of systems do. With the storage hierarchy getting deeper each year, the importance of these systems is increasing too.

9.3.2 Accelerating I/O with Heterogeneous Storage for HPC

On modern HPC clusters, each node may be equipped with different types of storage devices such as RAM disk, SSD, and HDD. Specifically, for big data workloads running in the Hadoop environment, a high-performance HDFS design should be able to fully take advantage of these available storage resources in the most efficient manner. A prominent example of a heterogeneity-aware storage system for big data is Triple-H (HHH), presented by Islam et al. (2015), that proposes high-performance optimizations to default HDFS. While the core idea in HHH is to exploit RDMA over InfiniBand interconnects to reduce the total time spent in I/O, it also presents key designs that enable it to leverage different tiers of storage efficiently.

9.3.2.1 RDMA-based data transfers

As discussed in chapter 2, HDFS involves a communication-intensive framework because of its distributed nature. For instance, network communication is involved in replicating HDFS data blocks for fault-tolerance. All existing communication protocols of HDFS are layered on top of sockets over TCP/IP. Due to the byte stream communication nature of TCP/IP, multiple data copies are required, which results in poor performance in terms of both latency and throughput. Consequently, even though the underlying system is equipped with high-performance interconnects such as InfiniBand, HDFS cannot fully utilize the hardware capability and obtain peak performance. Therefore, the high performance HDFS design, presented by Islam et al. (2012, 2013, 2014) incorporates various RDMA-aware designs detailed in this section.

RDMA-based communication: The high-performance HDFS design (Islam et al., 2012) uses RDMA for HDFS write and replication via the Java Native Interface (JNI) mechanism, whereas all other HDFS operations go over Java Socket. The JNI layer bridges Java-based HDFS with a communication library written in native C for RDMA operations. This design supports dynamic connection creation along with connection sharing for ensuring low over-head in connection management. In this design, the existing HDFS architecture is kept intact.

Parallel replication: By default, HDFS supports pipelined replication to maximize throughput. HDFS was designed to run on commodity hardware over slow networks. In the presence of high-performance interconnects, the network bandwidth is no longer a limita-tion. Therefore, the high-performance HDFS design incorporates parallel replication (Islam et al., 2013) that can send all three replicas in parallel from the client to the DataNodes. Sim-ilar RDMA-based replication approaches are also being leveraged to design replicated state machines to offer consistent services on such unreliable systems (Poke and Hoefler, 2015) and fault-tolerance in in-memory systems (Taleb et al., 2018).

SEDA-based overlapping in different stages of HDFS write: HDFS write operation can be divided into four stages: read, packet processing, replication, and (4) I/O. After data are received via RDMA, the data are first read into a Java I/O stream. The received packet is then replicated after some processing operations. The data packet is also written to the disk file. In the default architecture of HDFS, all these stages are handled sequentially by a single thread per block. In contrast, SOR-HDFS (Islam et al., 2014) incorporates a SEDA–based approach (Welsh et al., 2001) to maximize overlapping among different stages of HDFS write operation at the task/packet level and the block level, by assigning different parallel thread pools to the various stages.

9.3.2.2   Exploiting heterogeneous storage

Each HDFS DataNode in HHH runs a RAM disk and SSD-based buffer-cache on top of the disk-based storage system. It also uses the PFS (i.e., Lustre) installation in HPC clusters for storing data from Hadoop applications. It introduces several features to optimally exploit the available storage layers, including the following.

Buffering and caching through hybrid buffer-cache: In the HHH design, the data written to (or read from) HDFS are buffered (cached) in RAM disk, which is the primary buffer-cache in HHH. The size of the buffer-cache is enlarged by using SSD. HHH hides the cost of disk access during HDFS reads/writes by placing the data in this buffer-cache.

Storage volume selection and data placement: HHH calculates the available space in the storage layer based on the placement policy selection. If there is sufficient space available for the file, it is stored in the requested medium. Otherwise, the weight of the replica is adjusted and placed in the next level of the storage hierarchy. However, for each replica, it remembers the weight assigned by the Placement Policy Selector so that whenever there is space available in the requested level, the file can be moved there.

Fault-tolerance of RAM disk data through SSD-based staging: Data stored in RAM disk cannot sustain power/node failures. Therefore, all data placed in the RAM disk is asynchronously flushed to an SSD-based staging layer.

Data movement through eviction/promotion manager: The accelerated HDFS design integrates a dedicated module that is responsible for evicting cold data and making space for hot data in the buffer-cache. It also promotes the hot data from SSD/HDD/Lustre to RAM disk/SSD/HDD. Because the DFSClient reads each file sequentially, blocks of the same file are moved during the same window of data movement.

Efficient data placement policies: Data placement policies play a very important role in achieving high performance for HDFS. HHH has three types of placement policies:

1. **Greedy placement:** In this policy, the incoming data are greedily placed in the high-performance storage layer by the DataNodes. So, as long as there is space available in the RAM disk–based primary buffer-cache, the DataNode writes all the data there. Then the DataNode switches to writing data to the next level of storage based on SSD and so on. HHH follows a hybrid replication scheme through which the first replica of a file is placed in the RAM disk–based buffer-cache in a greedy manner while the other two are stored in the SSD-based secondary buffer-cache. Alternatively, two replicas can be placed to RAM disk and the other one in SSD.

2. **Load-balanced placement:** This policy spreads the amount of I/O across multiple storage devices in a balanced manner. The I/O load is balanced between RAM disk and SSD. Both greedy placement and load-balanced placement are performance-sensitive policies.

3. **Storage-sensitive placement:** With this policy, one replica is stored in local storage (RAM disk or SSD) and one copy is written to Lustre through the PFS. This policy reduces local storage requirements.

9.3.2.3   Heterogeneity-aware HDFS in HiBD software stack

As described in chapter 7, the RDMA-based Apache Hadoop 2.x package of the HiBD software stack incorporates an RDMA-optimized heterogeneous storage-aware HDFS with different deployment modes. This enhanced HDFS design is based on the HHH design we have discussed here. For heterogeneity-aware data storage within HDFS, four different modes are included:

1. **HHH:** Heterogeneous storage devices with hybrid replication schemes are supported in this mode of operation to have better fault-tolerance as well as performance. This mode is enabled by default in the package.

2. **HHH-M:** A high-performance in-memory-based setup has been introduced in this package that can be utilized to perform all I/O operations in-memory and obtain as much performance benefit as possible.

3. **HHH-L:** With PFSs integrated, the HHH-L mode can take advantage of the Lustre available in the cluster.

4. **HHH-L-BB:** This mode deploys a Memcached-based burst buffer system to reduce the bandwidth bottleneck of shared file system access. The burst buffer design is hosted by Memcached servers, each of which has a local SSD.

Along similar lines, for the Hadoop ecosystem, a novel byte-addressable NVM-assisted shuffle approach for the MapReduce framework has also been proposed (Islam et al., 2013).

9.4   Burst Buffer Systems

A burst buffer system is a layer of high-performance storage, such as NVRAMs or SSDs, that sits between the CNs and the PFS as a staging layer to absorb application I/O requests and alleviates the load on the underlying shared storage system. Several storage system vendors, such as DDN, Cray and so on, have followed a hardware approach to design integrated burst buffer systems (e.g., DDN Infinite Memory Engine (IME) (DDN, 2021), and Cray DataWarp (Cray, 2021)), to handle both checkpoint and recovery in addition to boosting application performance and efficiency. Traditionally, KV stores were introduced to accelerate online data processing workloads by reducing the load on the underlying database. Recently, several offline big data analytical workloads have started to explore the benefits of leveraging high-performance KV stores, which includes designing a highly efficient and scalable burst buffer layer on top of PFS such as Lustre. A Memcached-based burst buffer layer to accelerate checkpoint I/O in scientific applications running over Lustre was introduced by T. Wang et al. (2014) and Wang et al. (2015), while employing advanced features such as RDMA and replication for fault-tolerance. This has also lead to big data–centric burst buffer prototypes like Boldio, by (Shankar, Lu, and Panda, 2016), to enable the performance of local storage via distributed shared memory-based buffers, but employing a larger and slower PFS to store the actual data.

DDN IME and Cray DataWarp were similarly designed for accelerating HPC I/O using NVMe storage. These systems were mostly designed to serve as burst buffers in HPC clusters to absorb bursty I/O and flush it to PFSs in the background. In contrast, GlusterFS (Red Hat, Inc., 2021), OrangeFS (OrangeFS, 2021), Zest (Nowoczynski et al., 2008), PLFS (Bent et al., 2009), UnifyFS (Moody et al., 2017), and BurstFS (Wang et al., 2015) are HPC-oriented file systems with a specific focus on improving checkpoint I/O. Although their designs are not specifically targeted toward NVMe devices, they can leverage their higher performance. Zest uses a log-structured design with a burst buffer layer to reduce checkpoint overhead. PLFS is another file system aimed at optimizing checkpoints that follow the $N - 1$ pattern.

However, recent work by Vazhkudai et al. (2018) has shown that upto 90 percent of application runs follow the $N - N$ pattern. Other works such as UnifyCR and BurstFS present a burst buffer design using node-local storage to accelerate checkpoint I/O. Distributed file systems such as DDN IME, Cray DataWarp, GlusterFS, and OrangeFS overlay multiple software layers on local kernel file systems to access data. Thus, while these works have improved checkpoint overhead, they suffer from two basic flaws. First, these file systems still suffer from the use of POSIX file systems to access devices. Direct user-space access to devices via NVMEoF is not supported. Second, serialization of metadata operations and synchronization between clients prevents applications from fully utilizing available I/O bandwidth. IBM's Burst Buffer Shared Checkpoint File System (BSCFS) does support an NVMEoF-based data plane. However, it requires applications to be modified to use its specific non-POSIX API.

9.5 Case Studies and Performance Benefits

This section presents two early works that provide us with a futuristic perspective toward leveraging emerging NVM storage devices in an efficient manner for big data.

9.5.1 Heterogeneous Storage-Aware Big Data Middleware

Preliminary studies with PMEM (NVM such as PMEM or NVRAM) emulated over DRAM show that through the use of an NVM-aware data movement substrate (Islam et al., 2016b, Rahman et al., 2017, Li et al., 2020), one can significantly improve the communication, I/O, and application performance for big data analytics and management middleware, such as MapReduce, HDFS, Spark, and so on (see figure 9.1).

For Hadoop MapReduce, figure 9.1a presents the evaluation of PUMA workloads on an eight-node Hadoop cluster with sixty-four maps and thirty-two reducers. For

(a)

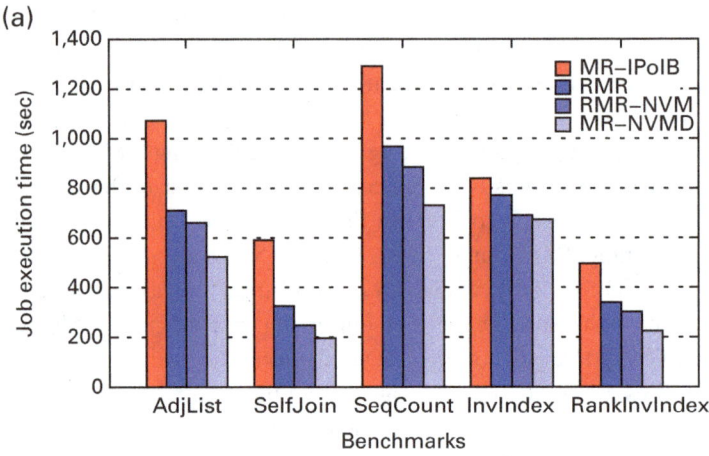

(b)

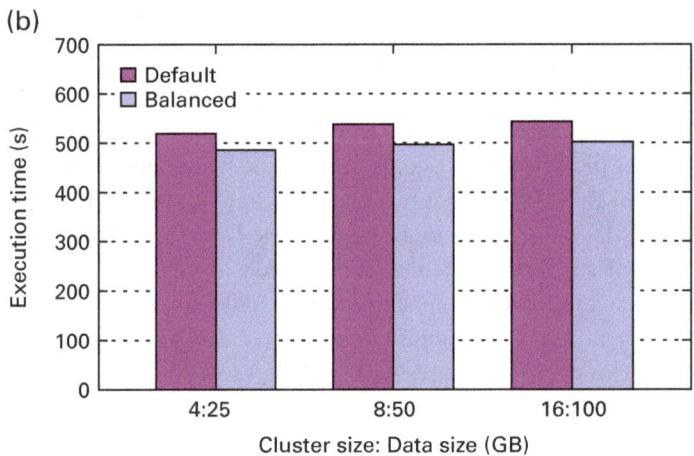

Figure 9.1
Performance benefits of heterogeneous storage-aware designs for Hadoop on SDSC Comet. (a) NVM-assisted MapReduce design. (b) Spark TeraSort over heterogeneity-aware HDFS. MR-IPoIB, Default MapReduce running with the IPoIB protocol; RMR, RDMA-based MapReduce; RMR-NVM, RDMA-based MapReduce running with NVM in a naive manner; NVMD, Non-Volatile Memory-assisted design for MapReduce and DAG execution frameworks (Rahman et al., 2017).

shuffle-intensive workloads (AdjList, SelfJoin, and RankedInvIndex), two to three times performance benefits can be observed in contrast to the default Hadoop MapReduce running IPoIB.

Performance benefits of running Spark over heterogeneity-aware HDFS designs of HHH (presented in section 9.3.2) is showcased in figure 9.1b. The experiment is performed using the Spark TeraSort with data size varying from 80 GB on four nodes to 320 GB on sixteen nodes (twenty-four concurrent tasks per node) on the SDSC Comet cluster. Each CN in this

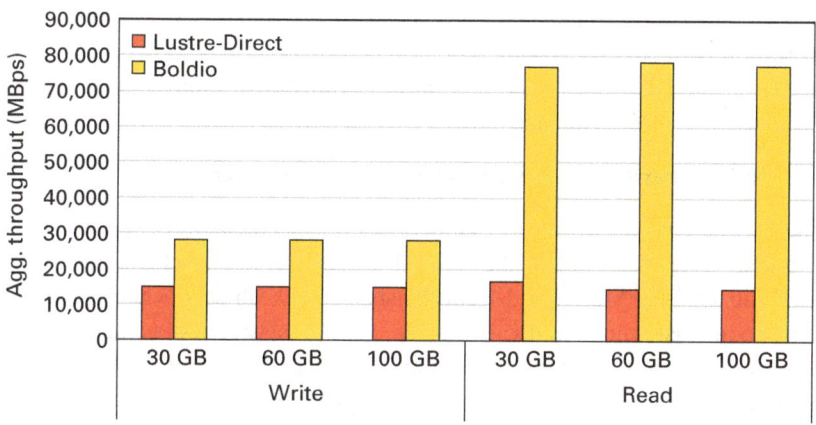

Figure 9.2
Performance benefits with RDMA-Memcached–based workloads.

cluster has two twelve-core Intel Xeon E5-2680 v3 (Haswell) processors, 128 GB DDR4 DRAM, 80 GB HDD, and 320 GB of local SATA-SSD with CentOS operating system. The network topology in this cluster is 56 Gb/s FDR InfiniBand with rack-level full bisection bandwidth and 4:1 oversubscription cross-rack bandwidth. Comet also has 7 PB of Lustre installation, that enables running the RDMA-aware HDFS design of HHH running in the HHH-L mode (see section 9.3.2.3). Specifically, figure 9.1b presents a balanced read strategy that reduces the execution time of TeraSort by up to 9–11 percent. In addition to this, the performance of NVM-aware HBase read/write operations can also be improved by 16–26 percent, over HBase over SATA-SSDs on SDSC Comet cluster, as discussed in Islam et al. (2016a).

9.5.2 Burst Buffer for Big Data I/O with RDMA and SSDs

To illustrate the performance benefits of leveraging RDMA and SSDs to accelerate big data I/O workloads, the Boldio design (Shankar, Lu, and Panda, 2016) (presented in section 9.4) is compared with Hadoop running directly over Lustre PFS. We use the production-scale cluster, namely SDSC Gordon, that is made up of 1,024 dual-socket CNs with two eight-core Intel Sandy Bridge processors and 64 GB of DRAM, connected via InfiniBand QDR links. This cluster has sixteen 300-GB Intel 710 SSDs distributed among the CNs. It consists of a Lustre setup with 1-PB capacity. We use twenty nodes on SDSC Gordon for our experiments. For Boldio, sixteen CNs on cluster A are used as YARN NodeManagers to launch Hadoop map/reduce tasks and a four-node Boldio burst buffer cluster over Lustre (replication = 2). For fair resource distribution, a twenty-node Hadoop cluster is employed for the Lustre Direct mode. From figure 9.2, it can be seen that by leveraging nonblocking RDMA-enabled KV store semantics over SSD-assisted Memcached server design, Boldio can sustain 3 and 6.7 times improvements in read and write throughputs, respectively, over default Hadoop running directly over Lustre.

It is to be noted that such software-based designs have transformed into specialized and optimized I/O hardware (e.g., DDN IME) today.

9.5.3   Boosting Hadoop Performance with NVMe-SSD in Twitter

The speed of the storage subsystem could become a dominant factor for the performance of real-world big data systems. Based on the report from Twitter and Intel (Beckett, Dave and Singer, Matt and Damle, Milind and Radhakrishnan, Rakesh and Wheeler, Barrie, 2018), a single Hadoop cluster in Twitter can have up to ten thousand nodes and about 100 PB of storage. In Beckett, Dave and Singer, Matt and Damle, Milind and Radhakrishnan, Rakesh and Wheeler, Barrie (2018), Twitter engineers find that deploying a lot of low-speed HDD devices cannot satisfy their growing needs on the performance of their Hadoop cluster. Through the collaboration with Intel, the joint team observes that the multiple data flows from different running applications in the Hadoop cluster can cause significant contentions for accessing HDDs. More specifically, the engineers find that the contentions mainly come from the HDFS data flows and temporary data flows managed by the YARN resource manager.

Based on these analyses, the engineers from Twitter and Intel choose to use a fast NVMe-SSD to store temporary data from YARN in each node. Then, they further use an intelligent caching technique on top of NVMe-SSDs to boost Hadoop performance significantly. Through all their investigations and optimizations, the team finally achieved up to 50 percent faster application runtimes and about 30 percent lower total cost of ownership. They even conclude their work with an interesting observation. They believe that the next-generation big data platforms should have more compute threads for each disk when the system can be equipped with faster NVMe-SSDs.

9.6   Summary

This chapter presents an overview of the different NVM-aware storage middleware designs that are attempting to exploit the high performance promised by the NVMe standard NVMEoF for scalable distributed storage and byte-addressable PMEM. We discuss state-of-the-art heterogeneous storage-aware architectures for I/O systems and emerging directions with NVM for data centers and beyond. As NVM-aware storage technologies are continuously evolving, many new designs are expected to be available to accelerate big data middleware and workloads on next-generation data centers and HPC systems.

10 Deep Learning over Big Data

To mine more value from the massive amount of gathered data, DLoBD is becoming one of the most efficient analyzing paradigms. An increasing number of deep learning (DL) tools and libraries are running over big data stacks, such as Apache Hadoop and Spark. By combining the advanced capabilities from deep learning libraries (e.g., Caffe, TensorFlow) and big data stacks, the DLoBD approach can enable powerful distributed deep learning and big data analytics on a unified computing cluster environment with potential performance and management benefits. However, the current-generation DLoBD stacks have multiple software layers, which may cause inefficient execution and limited scalability for running deep learning together with big data analytics. This chapter will first discuss the challenges of designing high-performance DLoBD stacks. Then we give a detailed overview of some representative DLoBD stacks. We discuss a performance characterization and analysis methodology for these stacks. We envision that the community needs a new holistic approach to design next-generation DLoBD stacks.

10.1 Overview

As the explosive growth of big data continues, this deluge of data—from large scientific facilities, advanced cyberinfrastructure, new data analysis frameworks, and more—is profoundly transforming research in all fields of science and engineering and is also forcing scientists to ask and answer new types of questions.

To mine big value from big data, deep learning has been becoming a pervasive technology forming the backbone of many modern data analytics applications, such as speech recognition, natural language processing, automatic language translation, computer vision, and so on. Many scientific applications (Kurth et al., 2017, Smola and Narayanamurthy, 2010, Chu et al., 2007) have started using deep learning techniques to handle the grand challenges faced by their communities, such as molecular dynamics, metadata classification, image processing, high-energy physics, climate research, and so on. Deep Learning techniques, together with advances in computational capabilities, have come to play a key role in big data analytics and science-driven knowledge discovery (Chen and Lin, 2014,

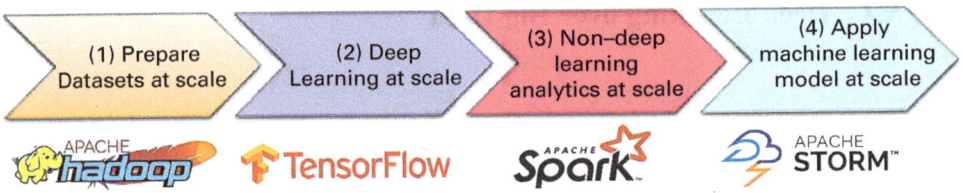

Figure 10.1
Deep learning and big data analytics pipeline. *Source*: Courtesy of Flickr (Garrigues, 2015).

Lin and Kolcz, 2012, Panda et al., 2011). Recent studies by researchers from Stanford University and Google have shown that a deep learning–based approach can be used to meet the challenges of extracting useful knowledge from big data, allowing scientists to perform breakthroughs in drug discoveries (Harris). Thus, deep learning together with big data, are being considered as the "big deals and the bases for an innovation and economic revolution" (Crego et al., 2013).

In this context, combining the power of both deep learning techniques and big data analytics capabilities is becoming one of the most highly demanded analyzing paradigms. Figure 10.1 shows a typical deep learning and big data analytics pipeline in many of the real science-driven knowledge discovery processes. The pipeline mainly includes four major steps: (1) Datasets preparation at scale, where the existing and recognized Apache Hadoop framework can be utilized to prepare an appropriate dataset format in a distributed manner. (2) Deep learning–based analysis at scale, where another recognized deep learning framework (i.e., TensorFlow) can be used to perform the high-performance deep learning analysis with the big datasets prepared in step 1. (3) Non–deep learning data analysis at scale, which represents a lot of other steps, such as indexing, querying, graph processing, non–deep learning–based machine learning, and so on, that are not running deep learning jobs. For this step, the well-known Apache Spark framework can be leveraged to perform these types of analyses. (4) Applying deep/machine learning models at scale, where the trained deep/machine learning models in steps 2 and 3 can be applied to the fresh incoming data streams.

The pipeline depicted in figure 10.1 can be realized in multiple ways. The top part of figure 10.2 shows that many of the existing systems choose to deploy two dedicated clusters (i.e., deep learning cluster and big data processing cluster) physically or logically to run deep learning jobs and other data analytics jobs independently. Then, when data sharing is needed between these two different clusters, data movement over networks is needed, which can cause performance degradation. This kind of deployment also increases the high costs due to the maintenance of two different infrastructures.

To overcome these burdens, a unified DLoBD (Lu, Shi, Javed, et al., 2017, X. Lu et al., 2018) processing paradigm is emerging in the field. An increasing number of deep learning

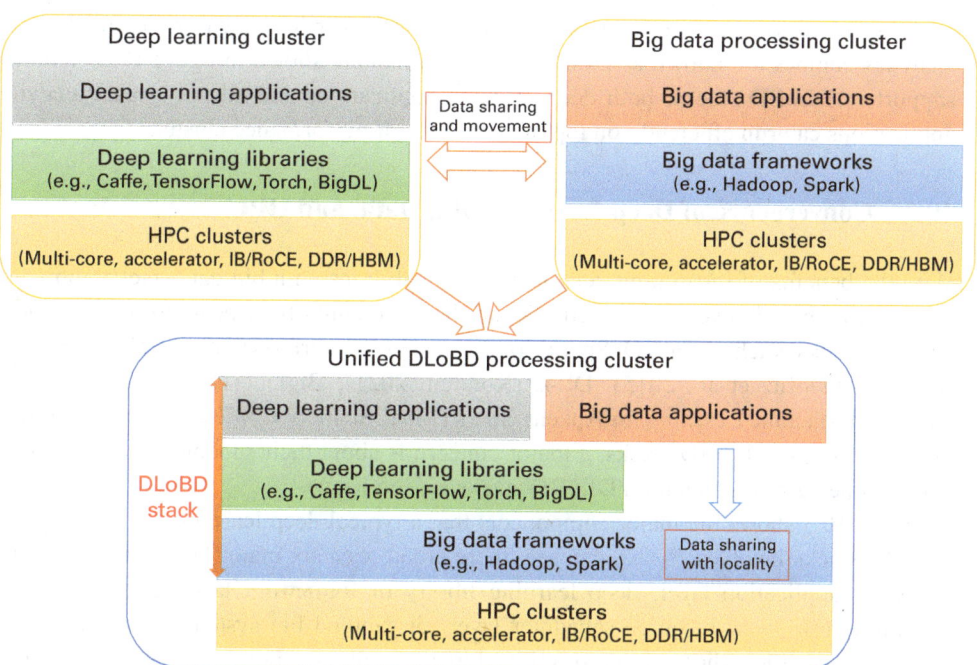

Figure 10.2
Overview of a unified DLoBD stack. IB, InfiniBand.

tools and libraries are being deployed over big data analytics engines, such as the most popular representatives: Apache Hadoop and Spark. By combining the advanced capabilities from deep learning libraries (e.g., Caffe (Jia et al., 2014), TensorFlow, and the Microsoft Cognitive Toolkit [CNTK] (Seide and Agarwal, 2016)) and big data analytics engines, the DLoBD approach can enable powerfully distributed deep learning on big data analytics clusters with at least the following three major benefits.

1. From the data analytics workflow perspective, if we run deep learning jobs on big data stacks, we can easily integrate deep learning components with other big data processing components in the whole workflow.

2. From data locality perspective, because large amounts of gathered data in companies typically are already stored or being processed in big data stacks (e.g., stored in HDFS), deep learning jobs on big data stacks can easily access the data without moving them back and forth.

3. From an infrastructure management perspective, we do not need to set up new dedicated deep learning clusters if we can run deep learning jobs directly on existing big data analytics clusters. This could significantly reduce the total cost of ownership.

Such unified software frameworks that combine both deep learning libraries and big data analytics engines are known as DLoBD stacks, which are shown in figure 10.2. With the support of DLoBD stacks, both deep learning applications and other big data analytics applications can run efficiently on top of a single high-performance cluster.

10.2   Convergence of Deep Learning, Big Data, and HPC

With the benefits of integrating deep learning capabilities with big data stacks, more and more research and development activities in the community have been proposed to build DLoBD stacks, such as CaffeOnSpark (Yahoo, 2018), TensorFlowOnSpark (Yahoo, 2021), SparkNet (Moritz et al., 2015), DL4J (Konduit, 2021), BigDL (Dai et al., 2019), and Microsoft Machine Learning for Apache Spark (MMLSpark or CNTKOnSpark) (Hamilton et al., 2018). For DLoBD stacks, a major concern is about their suboptimal performance, because the current-generation DLoBD stacks are too heavy.

Figure 10.3 shows the major components that a typical deep learning job running over DLoDB stacks involves. As we can see, there are at least six major layers: deep learning model or application layer, deep learning library or framework layer, big data analytics framework layer, resource scheduler layer, distributed file system layer, and cluster resource layer. Such a heavy layered architecture may cause inefficient execution and limited scalability for a combined deep learning and big data analytics workflow.

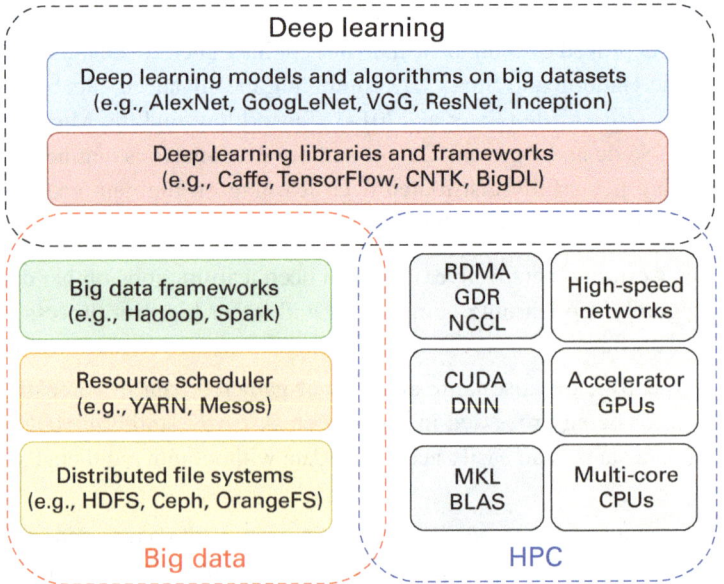

Figure 10.3
Convergence of deep learning, big data, and HPC.

The community has also observed these potential performances and scalability issues of DLoBD stacks. Thus, we clearly see a new trend in the convergence of deep learning, big data, and HPC technologies. To shown in figure 10.3, help solve the performance issues in DLoBD stacks, many high-performance technologies on advanced hardware architectures have been made available to the community, as and discussed in chapter 4. For high-speed networks, DLoBD stacks can take advantage of many new communication schemes to accelerate the performance of deep learning applications, such as RDMA protocols, GPUDirect RDMA (GDR), and NVIDIA Collective Communications Library (NCCL) for connected GPUs. From the computation perspective, GPU accelerators with the available efficient CUDA DNN (cuDNN) library and multicore CPUs with the highly optimized Intel MKL library or Basic Linear Algebra Subprograms (BLAS) libraries can also be adopted to speed up the model training performance.

As shown in figure 10.3, the emerging DLoBD stacks (such as CaffeOnSpark, Tensor-FlowOnSpark, MMLSpark, BigDL) are being designed to leverage RDMA-capable high-performance interconnects and multi-/many-core–based CPUs/GPUs. These advanced architectures and technologies can provide a lot of opportunities to redesign or codesign the DLoBD stacks.

10.3   Challenges of Designing DLoBD Stacks

There are many efforts in the field to improve the performance of each of these layers. For example, the default Caffe, TensorFlow, and CNTK can leverage the high-performance GPU accelerators with the codesigned efficient cuDNN library, while BigDL can efficiently run on Intel CPUs or Xeon Phi devices by utilizing the highly optimized Intel MKL library or BLAS libraries. Yahoo! researchers have proposed RDMA-based communication in CaffeOnSpark and TensorFlowOnSpark. Several studies (Ozery, 2018, Lu et al., 2013, 2016, Rahman et al., 2013, Islam et al., 2015) have proposed RDMA-based designs for Spark and Hadoop.

Even though these works have been proposed and well studied with their targeted workloads and environments, there exists no framework that allows users to systematically understand the cross-layer activities, analyze the impact of the advanced hardware features on DLoBD stacks, identify bottlenecks and computation-communication-I/O patterns, and further benefit DLoBD stacks and deep learning applications. We clearly see three important demands for building next-generation DLoBD frameworks. First, we need analyzing and optimization tools to systematically profile, analyze, and tune different layers in DLoBD stacks. Second, we need to design more lightweight DLoBD stacks, which can fully take advantage of the capabilities delivered by modern hardware architectures. Third, we also need to codesign a set of data-driven deep learning applications with these tools and stacks to show significant performance and scalability gains.

Solving these fundamental challenges by efficiently combining the capabilities of deep learning libraries, Big data processing engines, and HPC technologies can lead to solutions that form the foundations of global innovation and economic revolution. Such an approach has immense potential to execute deep learning and big data analytics applications on modern HPC systems with high performance, scalability, and analyzability. The current approaches to achieving these in the community are discussed in the following sections.

10.4   Distributed Deep Learning Training Basics

There are three major steps of training a deep learning model, which are forward propagation (FP), backward propagation (BP), and parameter update (PU). During the FP stage, each training process receives a batch of training data, propagates that data through the training model, and calculates the loss function. During the BP stage, each training process uses the loss value to compute the gradients of each parameter. During the PU stage, processes use the aggregated gradients to update the parameters with a particular optimizer, such as stochastic gradient descent (Zinkevich et al., 2010), Adam (Kingma and Ba, 2014), and so on. Distributed deep learning model training typically means updating the model parameters by running these three steps iteratively on a set of CPU or GPU machines in parallel, until the loss function reaches a stop condition.

On top of these distributed deep learning training steps, there are broadly two approaches for parallelizing a model training process: model parallelism and data parallelism. *Model parallelism* is when the different processing units use the same data, but the model is distributed among them. In *data parallelism*, the same model is used for every processing unit, but different parts of the data are read and processed by all processing units in parallel.

In addition, there are two major families of distributed deep learning training architectures: all-reduce and parameter server (PS). The all-reduce communication approach is originated from the HPC community, which aggregates every process's gradients in a collective manner before all the processes update their own parameters locally. The PS architecture (Li, Anderson, et al., 2014) contains two sets of processes: workers and the PS. Workers perform FP, BP, and push the gradients to the PS. The PS then aggregates the gradients from different workers and updates the parameters, which will be pulled by workers again for next-iteration training.

Nowadays, the most popular distributed training approach is data parallelism. Both all–reduce–and PS-based training architectures are being used by different deep learning systems and applications. This chapter focuses on data parallelism, which is more related to system-oriented studies.

10.5   Overview of DLoBD Stacks

This section presents a detailed overview of four popular and representative DLoBD stacks—CaffeOnSpark, TensorFlowOnSpark, MMLSpark or CNTKOnSpark, and BigDL —that support data parallelism.

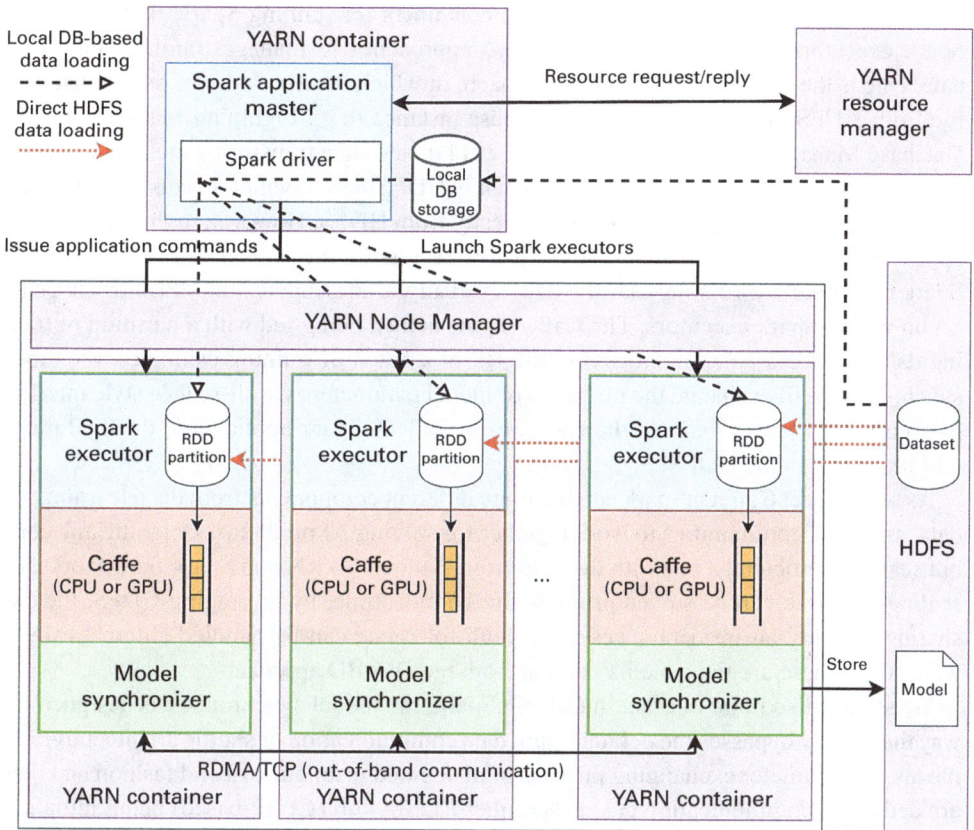

Figure 10.4
Overview of CaffeOnSpark. DB, database.

10.5.1 CaffeOnSpark Overview

CaffeOnSpark is a deep learning package designed by Yahoo! based on Apache Spark and Caffe. It inherits features from Caffe-like computing on CPU, GPU, and GPU with accelerating components (e.g., cuDNN). CaffeOnSpark enables deep learning training and testing with Caffe to be embedded inside Spark applications. Such an approach eliminates unnecessary data movement, and deep learning benefits from the high performance and scalability of Hadoop and Spark clusters. For example, the Flickr team improved image recognition accuracy significantly with CaffeOnSpark by training with the Yahoo Flickr Creative Commons 100M (Yahoo, 2021) dataset.

The system architecture of CaffeOnSpark (YARN cluster mode) is illustrated in figure 10.4. CaffeOnSpark applications are launched by standard Spark commands, and then Hadoop YARN launches a number of containers for running Spark executors. After Spark executors are running, there are two approaches to manage training and testing data. One is the local database-based approach, in which the Spark driver reads a database file from HDFS, loads it into a local database instance (e.g., Lightning Memory-Mapped Database Manager, LMDB (Chu, Howard, 2011)), and then transforms the data inside the local database into RDD. The other approach is HDFS-based, which means that the Spark executors fetch training and testing data directly from HDFS. However, in the HDFS-based approach, the raw data needs to be converted into sequence files or DataFrame format. After Spark executors are running and the data are ready, Caffe engines on GPUs or CPUs are set up within Spark executors. The Caffe engine is then being fed with a partition of training data (i.e., data parallelism). After the BP of a batch of training examples, the model synchronizer will exchange the gradients of model parameters via all-reduce style interface over either RDMA or TCP. At the end of each CaffeOnSpark application, the final model will be stored on the HDFS.

As we can see, CaffeOnSpark can integrate different components from deep learning, big data, and HPC communities to work together for solving AI problems. Default Caffe could not scale-out efficiently, but with the help from Hadoop YARN and Spark frameworks, the scaling-out issue can be solved properly. In the meantime, by leveraging HDFS, the data sharing, locality-aware data access, and fault-tolerance can be handled automatically as well. All of these are the benefits coming from the DLoBD approach.

We should also point out that in CaffeOnSpark, the model synchronizer is designed in a way that it fully bypasses the default Spark data communication or shuffle architecture. This means the parameter exchanging phase is implemented in an out-of-band fashion and there are dedicated communication channels (either RDMA- or TCP/IP-based) being initialized in the model synchronizers.

10.5.2   TensorFlowOnSpark Overview

TensorFlow has been seen as one of the most popular deep learning frameworks for both academia and industry. Vanilla TensorFlow does not provide support for training over big data stacks. SparkNet and TensorFrame (DataBricks, 2018) are some of the initial efforts toward training support, but they still leave a lot to be desired regarding the features provided. Thus, Yahoo! researchers used their experience from developing CaffeOnSpark to come up with TensorFlowOnSpark, a framework that enables execution of deep learning jobs using TensorFlow on an existing big data cluster using Spark and Hadoop YARN to distribute the training and includes support for RDMA over high-speed networks.

TensorFlowOnSpark seamlessly integrates along with other Spark components, such as SparkSQL, MLlib, and so on, in the overall Spark ecosystem, requiring minimal changes to default TensorFlow code. Figure 10.5 presents the architecture overview of Tensor-FlowOnSpark. TensorFlowOnSpark allows Spark executors to act as containers used to run TensorFlow code. It provides two different modes to ingesting data: *QueueRunner* is

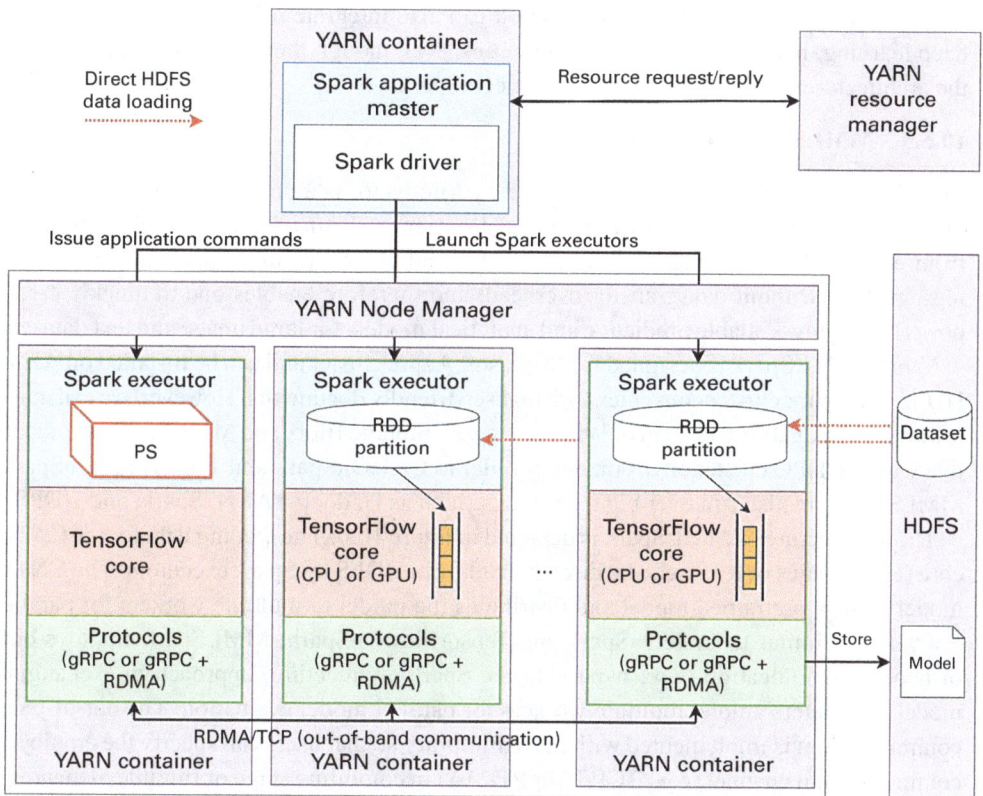

Figure 10.5
Overview of TensorFlowOnSpark.

used to read data directly from HDFS using built-in TensorFlow modules, whereas *Spark Feeding* provides the data from Spark RDDs to Spark executors, which in turn can feed it to the TensorFlow core.

Similar to CaffeOnSpark, TensorFlowOnSpark also bypasses the Spark architecture for communication (i.e., out-of-band communication) thereby achieving similar scalability as stand-alone TensorFlow jobs. The default TensorFlow can support multiple different channels for data communication, such as gRPC, gRPC+Verbs (RDMA), gRPC+MPI, gRPC+GDR, and so on. When utilizing RDMA, tensors are written directly to the memory of remote processes bypassing the kernel space. This design provides a considerable performance boost compared with the performance of the default gRPC design. One different design compared to the architecture of CaffeOnSpark is that TensorFlowOnSpark is based on the PS approach. The PS(s) will be embedded inside one or some Spark executor(s) and can talk to other tensors over different communication channels.

As we can see here, TensorFlowOnSpark can also integrate different components from deep learning, big data, and HPC communities, even though there are some differences in the architecture compared with that of CaffeOnSpark.

10.5.3   MMLSpark Overview

MMLSpark (or CNTKOnSpark), proposed by Microsoft, is a powerful toolkit for Apache Spark in accelerating deep learning and data science. It turns parallelizable algorithms from external libraries (e.g., Microsoft CNTK and OpenCV) into Spark machine learning pipelines without data transfer overhead, and therefore enables one to quickly create powerful, highly-scalable predictive and analytical models for large image and text datasets.

Vanilla MMLSpark is designed for Microsoft Azure cluster and can be installed on Azure HDInsight Spark cluster conveniently with user-friendly documents. However, we can make it work on top of HDFS instead of Windows Azure Storage Blob, and MMLSpark is compatible with the HPC cluster environment. Similar to CaffeOnSpark and TensorFlowOnSpark, MMLSpark can also run over big data stacks, such as Hadoop YARN, Spark, and HDFS.

The architecture of MMLSpark is depicted in figure 10.6. The feeding data for the CNTK core (e.g., images or texts) can be directly read from HDFS by Spark executors. The CNTK model loads a pretrained model and distributes the model to multiple workers for parallel evaluation. Similar to CaffeOnSpark and TensorFlowOnSpark, MMLSpark employs out-of-band communication (e.g., bypassing the Spark architecture) approach in exchanging model parameters among multiple workers for parallel model evaluation. The out-of-band communication is implemented with the MPI library so that users can specify the employed communication channel (e.g., TCP/IP or RDMA) in compiling stage or runtime, depending on which MPI library is chosen to run with. For example, if MMLSpark is compiled with the OpenMPI library, then the communication channel can be switched between IPoIB and RDMA through configuring the Byte Transfer Layer in the OpenMPI runtime.

10.5.4   BigDL Overview

BigDL is proposed by Intel to provide a high-performance and distributed deep learning runtime that makes efficient use of Intel processors. BigDL uses Spark to scale-out to multiple processes. It allows users to import models already trained using Caffe and Torch into the Spark framework, which is then used to pipe the models to the BigDL runtime. BigDL is written using Intel's MKL, which provides optimized support for vector primitives frequently used in deep learning applications. Therefore, it significantly outperforms most out-of-the-box deep learning frameworks on a single node.

Figure 10.7 shows the architecture of BigDL. It is also based on Parameter Server the PS that is organically designed with Spark Block Manager component. The Spark Block Manager is heavily involved in the Spark shuffle architecture. Data fed to the BigDL core are ingested by the Spark executor, which can directly load data from HDFS. PUs of the training model (i.e., major communication phase) are exchanged among BigDL cores through a

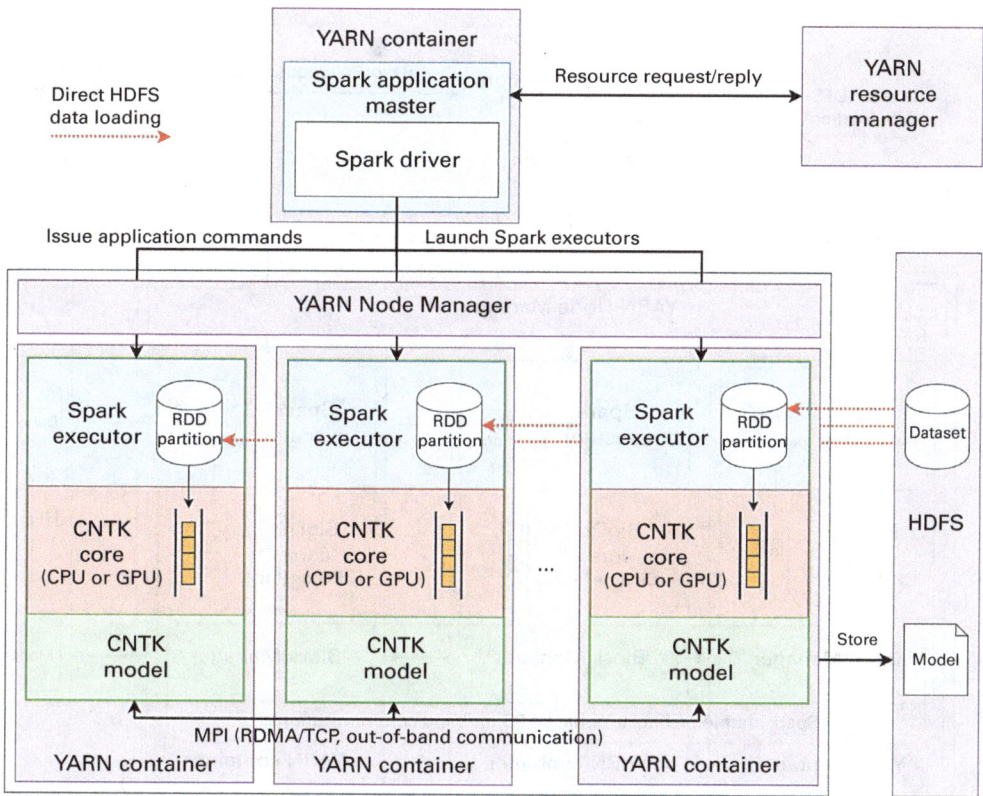

Figure 10.6
Overview of MMLSpark (CNTKOnSpark).

PS that is based on Spark shuffle architecture. We call this kind of parameter-exchanging approach *in-band communication*, which is different than the out-of-band approach (i.e., bypass Spark shuffle) applied in CaffeOnSpark, TensorFlowOnSpark, and MMLSpark. In other words, the in-band communication approach directly utilizes the Spark shuffle engine to synchronize the model.

By default, BigDL on Spark does not support RDMA-based model synchronization because the default Spark does not support RDMA-based shuffle. However, recent designs in the community (Lu et al., 2016, Ozery, 2018) can support native verbs-based high-performance RDMA shuffle in Spark, which can be used to fill this gap for BigDL. This can also be seen as the benefit of choosing the in-band communication approach. With an in-band communication approach, the BigDL framework can automatically utilize all kinds of optimizations (e.g., the RDMA-based shuffle engine), which are available in the Spark community, in the Spark core.

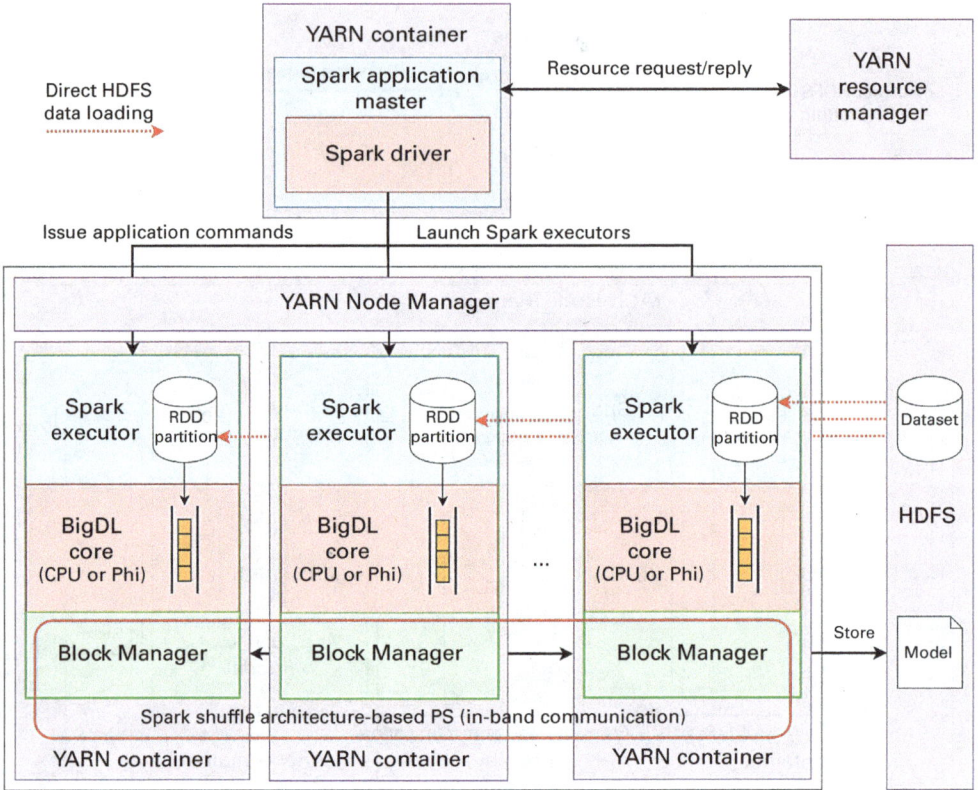

Figure 10.7
Overview of BigDL.

To summarize, these four DLoBD stacks are designed differently, and all of them can take advantage of modern HPC technologies (e.g., multicore CPUs, GPUs, RDMA) in varied ways to boost deep learning performance. In the meantime, all of them can run on top of the same big data stacks (i.e., Spark, Hadoop). These commonalities, as well as their differences, are why we chose them to represent a broad range of DLoBD stacks in this chapter.

10.6   Characterization of DLoBD Stacks

To gain a deep understanding of DLoBD stacks, a holistic characterization methodology has been proposed by Lu, Shi, Javed, et al. (2017) X. Lu et al. (2018). It is important to select representative deep learning workloads to perform benchmarking on DLoBD stacks, including popular deep learning models and open datasets. The selected models should have varied sizes of parameter spaces to cover big and small models. Similarly, the chosen

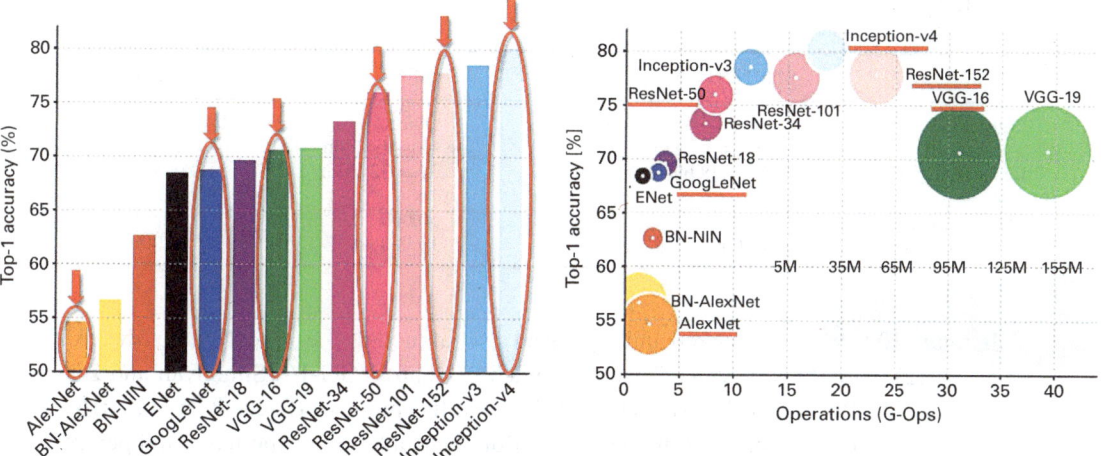

Figure 10.8
Comparison of DNNs. *Source*: Courtesy of Canziani et al. (2016). BN, Batch Nominations; ENet, efficient neural network; G-Ops, one billion (10^9) operations per second; ResNet, residual neural network; M, million; NIN, Network in Network; GoogleLeNet, a 22-layer Deep Convolutional Neural Network that's a variant of the Inception Neural Network developed by researchers at Google.

datasets can have both small and large sizes. In this way, more evaluations can be organized by covering different kinds of combinations, such as training varied-size models on both small and large datasets, which could expose different characteristics of DLoBD stacks.

10.6.1   Characterization with DNN Models

Over the recent years, more and more DNN models have been proposed in the community. Many of them have been widely adopted in AI-driven application scenarios, such as image segmentation, object detection, facial recognition, image classification, and so on. Figure 10.8 presents an interesting analysis to compare different DNNs in terms of their top-1 accuracies and sizes of model parameters. The size of the blobs in the right-hand graph is proportional to the number of model parameters. Note that the sizes of these models can span from 5 million to 155 million parameters.

As we can see in figure 10.8, different DNN models can deliver quite different accuracies while their parameter spaces are quite varied. To benchmark deep learning systems systematically, it is necessary to choose both "shallow" and "deep" models to cover a broad range of application scenarios. GoogLeNet (Szegedy et al., 2015), for example, is a twenty-two–layer model with Inception modules performing different sizes of convolutions. GoogLeNet can output seven million weights in all layers. In DLoBD stacks, these model parameters need to be exchanged among all workers. Thus, the size of model parameter space can directly influence the performance of the communication subsystem in DLoBD stacks.

Table 10.1
Image classification datasets

	MNIST	CIFAR-10	ImageNet
Category	Digit Classification	Object Classification	Object Classification
Resolution	28 × 28 black and white	32 × 32 color	256 × 256 color
Classes	10	10	1000
Training images	60,000	50,000	1.2 million
Tesing images	10,000	10,000	100,000

In figure 10.8, we have marked and suggested a set of models (i.e., AlexNet (Krizhevsky et al., 2012), GoogLeNet, VGG (Visual Geometry Group) VGG-16 (Simonyan and Zisserman, 2014), ResNet-50/152 (He et al., 2016), and Inception-v4 (Szegedy et al., 2016)) to cover different accuracies and model sizes. More benchmarking experience and performance evaluations with these models can be found in the literature (X. Lu et al., 2018, Biswas et al., 2018).

10.6.2   Characterization with Open Datasets

In addition to choosing DNN models, it is also important to choose representative open datasets to characterize DLoBD stacks. We include three popular datasets in this chapter: MNIST (LeCun et al., 2021), CIFAR-10 (Krizhevsky, 2009), and ImageNet (Deng et al., 2009). They have different categories, resolutions, classes, or scales, which are shown in table 10.1.

MNIST: MNIST consists of seventy thousand black and white handwritten digit images, which have been size-normalized and centered in a fixed-size 28 × 28 pixel box. Even though the focus of research has moved on to other much more challenging image recognition problems, the fast speed of training on the MNIST dataset means that it is still a proper problem for evaluation purposes.

CIFAR-10:   CIFAR-10 has fifty thousand training images and ten thousand test images, which are 32 × 32 pixel box RGB images in ten classes. The ten classes are airplane, automobile, bird, cat, deer, dog, frog, horse, ship, and truck. Nearly all deep learning frameworks use the CIFAR-10 dataset as one example, and there are many accuracy results reported publicly on it. Hence, the CIFAR-10 dataset is one of the most popular choices to evaluate object recognition algorithms.

ImageNet: ImageNet refers to the dataset for ImageNet Large Scale Visual Recognition Challenge (ILSVRC) 2012. In 2012, the ILSVRC competition involved a training set of 1.2 million 256 × 256 pixel box color images in a thousand categories. The ImageNet problem is one of the most challenging object recognition problems for modern computer vision research and deep learning research. Because of the long-lasting training time of complex models on the ImageNet dataset, evaluating deep learning frameworks on it becomes one of the best choices.

10.6.3 Characterization Dimensions

To characterize DLoBD stacks, it is vital to choose proper evaluation dimensions. The first important dimension is to evaluate the type of performance characteristics that different DLoBD stacks can attain on various advanced hardware platforms, because speed is one of the most important factors for choosing an appropriate DLoBD stack to train deep learning models. For instance, we typically need to evaluate the effect of powerful computing accelerators (such as NVIDIA GPU) and multicore CPUs (such as Intel Xeon processors) on DLoBD stacks. We want to find out whether DLoBD stacks can take advantage of the highly optimized libraries on GPUs (e.g., cuDNN) and CPUs (e.g., Intel MKL). For network protocols, we need to investigate the performance impact of TCP/IP and RDMA on deep learning workloads, because many deep learning models need to communicate a large number of parameters over the network during the model training process.

In addition, we need to show evaluation reports and analysis in detail based on all the evaluations. We also need to care about other metrics, such as accuracy, scalability, resource utilization, and so on. For performance, we typically need to pay attention to three major factors: end-to-end model training and testing time, consumed time to reach a certain accuracy, and epoch-level execution time. In neural network training terminology, an *epoch* is one pass of all the training examples.

Overall, the more factors and metrics are analyzed in the report, the more characteristics of deep learning stacks and associated applications can be observed. These useful characteristics can guide the designing of efficient next-generation DLoBD stacks.

10.7 Case Studies and Performance Benefits

This section presents several case studies and the performance benefits of popular deep learning frameworks over big data infrastructures.

10.7.1 A Case Study with CaffeOnSpark

To show an example of characterizing the performance and accuracy of DLoBD stacks running over IPoIB and RDMA protocols, we choose to train AlexNet on the ImageNet dataset with CafeeOnSpark. In the experiment, we report the training time of reaching 70 percent accuracy. The cluster used in the experiment comprises twenty nodes connected via Mellanox single-port InfiniBand EDR (100 Gb/s) HCA. Each node is equipped with two Intel Broadwell (E5-2680-V4) fourteen-core processors, 128 GB of RAM, 120 GB of local HDD, and two NVIDIA Tesla K80 GPUs. As shown in figure 10.9, for CaffeOnSpark,

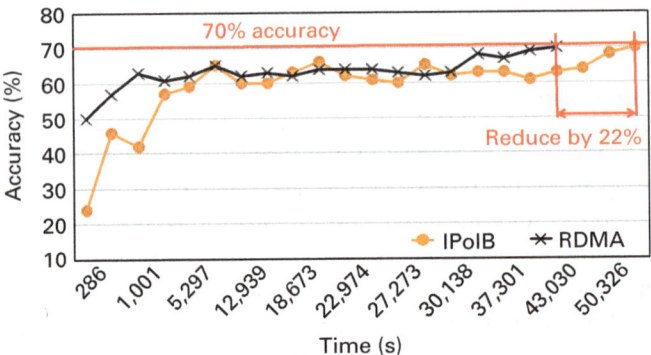

Figure 10.9
Performance and accuracy comparison of training AlexNet on ImageNet with CaffeOnSpark running over IPoIB and RDMA. *Source*: Courtesy of X. Lu et al. (2018).

replacing IPoIB with RDMA reduces the overall processing time of reaching 70 percent accuracy by 22 percent for training AlexNet on ImageNet.

Other results of experiments on CaffeOnSpark (X. Lu et al., 2018) show that CaffeOn-Spark has a communication overhead at a larger scale (e.g., sixteen nodes) for training DNN models over high-performance networks. The characterization report indicates that CaffeOnSpark benefits from the high performance of RMDA compared with that of IPoIB once the communication overhead becomes significant. More characterizations can be done with other advanced computation and communication techniques, such as cuDNN, NCCL, GPUDirect RDMA, and so on.

10.7.2 TensorFlow versus TensorFlowOnSpark

TensorFlow is a widely used deep learning framework in the market currently. Here, we try to characterize the performance of TensorFlow and TensorFlowOnSpark. As discussed in section 10.5.2, TensorFlowOnSpark enables direct tensor communication among different processes such as workers and PSs. With this architecture, TensorFlowOnSpark can achieve scalability and performance similar to that of stand-alone TensorFlow.

Figure 10.10 shows the time spent in different phases of training in TensorFlowOnSpark and stand-alone TensorFlow, respectively. TensorFlowOnSpark spends up to 15.5 percent of its time in the Apache Hadoop YARN scheduler layer, and up to 18.1 percent of its execution time in the Spark job execution layer for this particular test and settings. From figure 10.10, we can clearly see some performance overhead incurred by both the YARN scheduler and Spark execution layer in the TensorFlowOnSpark architecture. Because the data size is small, we do not see much overhead from the HDFS layer. In contrast, the stand-alone TensorFlow installation does not suffer from this kind of extra overhead. However, note that the overhead observed in figure 10.10 can be amortized in long-running deep

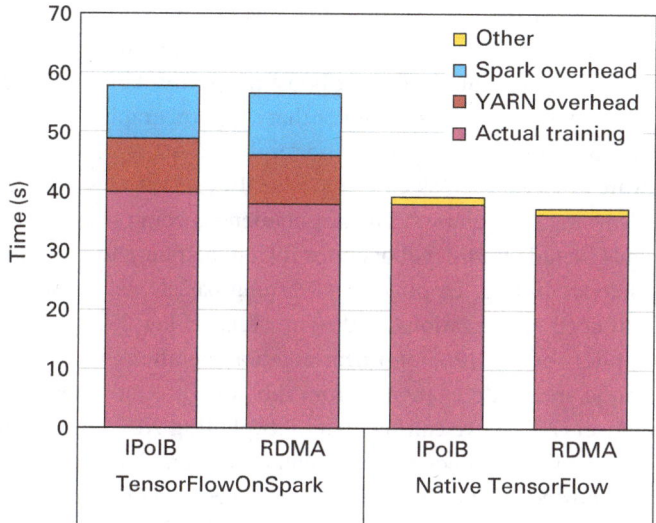

Figure 10.10
Performance analysis of TensorFlowOnSpark and stand-alone TensorFlow (lower is better). The numbers were taken by training the SoftMax Regression model over the MNIST dataset on a four-node cluster, which includes one PS and three workers. *Source*: Courtesy of X. Lu et al. (2018).

learning jobs over big datasets and eventually becomes negligible. These are the common cases for many deep learning training processes, because many deep learning models may be trained for minutes, hours, or even days. Overall, we can observe that the actual training time remains similar for both TensorFlowOnSpark and stand-alone TensorFlow. In the meantime, TensorFlowOnSpark can obtain benefits from running deep learning jobs on big data infrastructures as we will discuss in chapter 10.1. Similar results are observed in Bitfusion (2017), which are reported by Bitfusion researchers as well.

10.8   Discussions on Optimizations for Deep Learning Workloads

Due to the popularity of deep learning technologies, many performance studies and optimization approaches for TensorFlow, PyTorch, BigDL, and other deep learning frameworks have been proposed in the community in recent years. These approaches can be broadly classified as accelerations in the following deep learning categories: computation, communication, and I/O.

10.8.1   Accelerations on Deep Learning Computation

The community has proposed many advanced runtime libraries for deep learning workloads, such as cuDNN, Intel MKL, and BLAS libraries (see discussion in chapter 4). In addition, many advanced domain-specific compilers, such as TVM (T. Chen et al., 2018), TensorFlow Accelerated Linear Algebra (XLA) (XLA, 2021), and Astra (Sivathanu et al., 2019), are becoming available in the community to further boost the performance of deep learning workloads. TVM (T. Chen et al., 2018) is a compiler that supports graph-level and operator-level optimizations that can provide performance portability to many deep learning workloads on diverse hardware devices, such as CPU, mobile GPU, server-class GPUs, and FPGA-based generic deep learning accelerators. The key challenges being resolved by TVM for deep learning workloads include high-level operator fusion, mapping to arbitrary hardware primitives, memory latency hiding, employing a learning-based cost modeling approach for rapid exploration of code optimizations, and so on. TensorFlow XLA (XLA, 2021) is a domain-specific compiler for accelerating linear algebra operations in Tensor-Flow models with potentially no source code changes. From the post by Kanwar Pankaj and Brandt Peter and Zhou Zongwei (2020), we can see that by exploiting XLA, the training throughput of the BERT (Bidirectional Encoder Representations from Transformers) model (Chen et al., 2020) can be boosted by approximately seven times on a Google Cloud virtual machine with eight Volta V100 GPUs attached (each with 16 GB of GPU memory). Astra (Sivathanu et al., 2019) is a compilation and execution framework for optimizing the execution of a deep learning training job. Astra exploits domain knowledge about deep learning jobs (i.e., unique repetitiveness and predictability) to perform online exploration of the optimization state space while making progress on the training job.

10.8.2   Accelerations on Deep Learning Communication

There are many studies in the community for accelerating deep learning communication. As we discussed in chapter 4, deep learning systems and applications can take advantage of many new communication schemes to accelerate the performance of communication, such as RDMA protocols over InfiniBand or RoCE high-speed Ethernet, GPUDirect RDMA, NCCL, and so on. These technologies can be applied to both all-reduce–and PS-based training architectures, as discussed in section 10.4. To further improve the performance of the all-reduce training architecture, some recent studies propose to leverage the hierarchical topology of the cluster to minimize the traffic at bottleneck links. For example, BlueConnect (Cho et al., 2019) proposes to decompose a single all-reduce operation into a group of parallel reduce-scatter and all-gather operations, which can exploit the trade-off between latency and bandwidth and adapt to a variety of network configurations. Blink (Wang et al., 2020) is an example for exploiting intranode topology on systems equipped with multiple GPUs. Blink is a collective communication library that leverages heterogeneous communication channels (such as NVLinks and PCIe links) for hybrid and faster data transfers. BytePS (Jiang et al., 2020) proposes a unified communication framework for accelerating distributed training in heterogeneous GPU/CPU clusters. The developers of BytePS demonstrate that existing all-reduce and PS training approaches become two special cases of BytePS. Javed et al. (2019) propose to use the Roofline Trajectory model to identify optimization opportunities on both computation and communication for deep learning training workloads. This modeling-based approach can help scientists to optimize their deep learning model training performance, especially for those who may not have enough knowledge of low-level systems.

10.8.3 Accelerations on Deep Learning I/O

Many of the previously mentioned studies have focused on optimizing the computation and communication of deep learning training workloads. In many scenarios, deep learning workloads still suffer from the low performance of data I/O subsystems. Pumma et al. (2019), have performed a large-scale I/O analysis for deep learning workloads on a 9,216-core system. They find that I/O can take up to 90 percent of the total training time in deep learning training workloads, which is unimaginable. The authors further propose optimizations for the vanilla LMDB to improve the I/O performance of deep learning training workloads. In Elangovan, Aparna (2020), Amazon researchers also share their experience of optimizing the I/O speed for GPU-based deep learning training applications in Amazon SageMaker. Amazon researchers also present their performance optimization case studies with training deep learning workloads on Amazon SageMaker using Amazon FSx for Lustre and Amazon EFS file systems (Kastuar, Vidhi and Ochandarena, Will and Saxena, Tushar, 2019).

10.9   Summary

DLoBD is becoming one important analyzing paradigm for data-driven AI. This chapter presented a detailed architectural overview of four representative DLoBD stacks— CaffeOnSpark, TensorFlowOnSpark, MMLSpark or CNTKOnSpark, and BigDL. We discussed the opportunities of the convergence among deep learning, big data, and HPC technologies, and we discussed the associated challenges still facing the community these days. We further presented a performance characterization approach for benchmarking DLoBD stacks. We provided two case studies on CaffeOnSpark and TensorFlowOnSpark with our benchmarking methodology. Finally, we discussed the trends of recent studies for accelerations on deep learning computation, communication, and I/O perspectives.

For future research avenues, we envision that the community needs a more holistic and lightweight approach to design next-generation high-performance DLoBD stacks through proposing new abstractions and mechanisms, because the current-generation DLoBD stacks are mainly built by composing software layers or components from different communities, which may cause unnecessary overheads.

In addition, we think the community needs useful and efficient performance characterization tools to allow users to systematically perform the performance evaluation experiments with DLoBD stacks and help users understand the cross-layer activities. Such kinds, of toolkits will be very helpful to analyze the impact of the advanced hardware features on DLoBD stacks, identify performance and scalability bottlenecks in DLoBD stacks, and expose computation-communication-I/O patterns, as well as to further benefit deep learning applications.

Last but not the least, we also need to codesign more data-driven deep learning applications with the envisioned tools and stacks to demonstrate the potential performance and scalability gains.

11 Designs with Cloud Technologies

HPC and cloud computing have been converging to create a new field, known as "HPC cloud." Cloud computing technologies are able to provide fast deployment and flexible management for HPC and big data workloads. However, due to the internal virtualization overhead, it is a challenging task to achieve near-native performance for HPC and big data applications running in the clouds. This chapter first presents an overview of HPC cloud technologies, including virtual machine (VM), container, single root (SR)-IOV, and OpenStack. Then, we discuss several representative research efforts in the community about how to satisfy the demands of efficiently running HPC and big data applications in the cloud. For each of these aspects, we discuss the current state-of-the-art designs and how they achieve the desired goals.

11.1 Overview

Cloud computing has become an essential platform for enabling modern applications in the fields of data analytics, deep learning, web services, and even HPC. Recently, the fields of HPC and cloud computing have been converging to create a new field, HPC cloud. The advent of cloud computing has revolutionized the HPC and big data fields. Cloud computing provides flexibility, fast deployment, reliability, and security while enabling significant cost savings for HPC, big data, and AI applications. This has led HPC, big data, and AI applications and frameworks to be adapted to the cloud so as to exploit these benefits.

With this increase in the usability and popularity of "HPC in the cloud," several efforts have been directed toward enabling HPC cloud environments to not compromise on performance. As a result, many of the performance-related shortcomings of cloud computing have been addressed. For instance, hardware-assisted IOV techniques are being extensively leveraged over the low-performance software-based approaches. In addition, many cloud providers furnish modern networking technologies such as InfiniBand and RoCE. Such features are making the use of the cloud for HPC workloads more accessible and attractive to application researchers today.

The popularity of accelerators and NVMe-based storage devices in the HPC and big data communities has made its way to the cloud. Recent technological advancements are also making GPU and NVMe devices affordable for cloud providers. These include cloud infrastructure giants, such as Amazon EC2, Microsoft Azure, Alibaba Cloud, and Google Cloud which, that provide instances supporting access to such hardware resources. For example, Amazon EC2 I3 instances provide NVMe-SSDs, while P2 and G2 instances provide GPGPUs. Moreover, GPGPUs are being heavily used in machine learning and deep learning applications these days to process vast amounts of matrix and vector manipulations. NVMe-based SSDs are in huge demand for big data analytics owing to their extremely low latency and good random read/write performance. In addition, their inherent parallelism in request processing makes them ideal to be used in virtualized environments, where sharing of resources is a common and given scenario.

Most cloud providers currently offer only pass-through device access or software-based virtualization. The use of these virtualization and resource-sharing approaches in cloud environments breaks away from the performance-centric view of the HPC community. The biggest concern is to achieve near-native performance and scalability while still reaping the benefits of virtualization. Significant effort has been poured into solving this challenge over the past few years.

The recently introduced SR-IOV (PCI-SIG, 2021) approach provides hardware-based IOV by directly allowing VMs or containers access to the PCIe device. The SR-IOV enabled virtualization technology has significantly improved the application performance of using advanced devices (such as high-speed networks, NVMe-SSDs, GPUs) in clouds. Even though these advanced devices have become popular in clouds, not many solutions exist that can efficiently make use of these devices in virtualized environments. In addition to this, performance is not the only metric of interest, but QoS and availability (e.g., live migration support) are essential features offered by cloud providers. Thus, it is imperative to provide solutions that can support these features to benefit cloud-based applications.

This chapter provides a comprehensive overview of HPC cloud technologies and discusses the state-of-the-art designs which can efficiently run HPC and Big Data applications in the cloud. Then, we discuss research efforts in the community about how to satisfy the demands of efficiently running HPC and big data applications in the cloud. For each of these aspects, we discuss the current state-of-the-art solutions and how they achieve the desired goals.

11.2   Overview of High-Performance Cloud Technologies

In this section, we provide an overview of common high-performance cloud technologies.

11.2.1   Virtual Machine

A VM is an emulated computer system providing the same functionalities as a physical computer. VMs run atop a piece of computer software known as the hypervisor (or VM monitor). Popular hypervisors include Xen, Kernel-based Virtual Machine (KVM), and

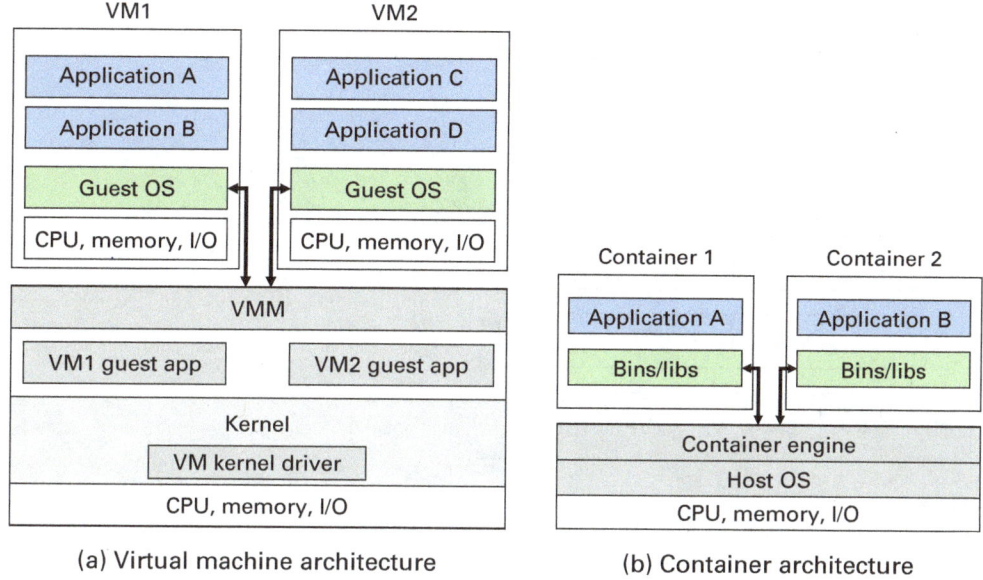

(a) Virtual machine architecture   (b) Container architecture

Figure 11.1
Overview of virtualization techniques. (a) VM architecture. (b) Container architecture. libs, libraries; OS, operating system.

VMWare. The hypervisor is responsible for virtualizing hardware resources such as CPUs, disks, and network cards. Multiple instances of a variety of operating systems may share the virtualized hardware resources. This enables multiple isolated and secure servers to run on the same physical hardware and effectively share available resources. Figure 11.1a shows the VM architecture. The hypervisor typically runs as an application on the host system with an associated kernel driver for hardware virtualization. The guest operating system relies on the hypervisor for interacting with hardware and allows applications to be run unmodified within the VM.

11.2.2 Container

Container-based virtualization is a lightweight alternative to the hypervisor-based virtualization. The host kernel allows the execution of several isolated user-space instances running a different software stack (e.g., system libraries, services, applications). Container-based virtualization provides self-contained execution environments, effectively isolating applications that rely on the same kernel in the Linux operating system, but container-based virtualization does not introduce a layer of virtual hardware. There are two core mature Linux technologies to build containers. First, namespace isolation isolates a group of processes at various levels: networking, file system, users, process identifiers, and so on. Second, cgroups (control groups) combines processes and limits their resources usage. Several container-based solutions have been developed, such as Docker, LXC, and Google's lmctfy.

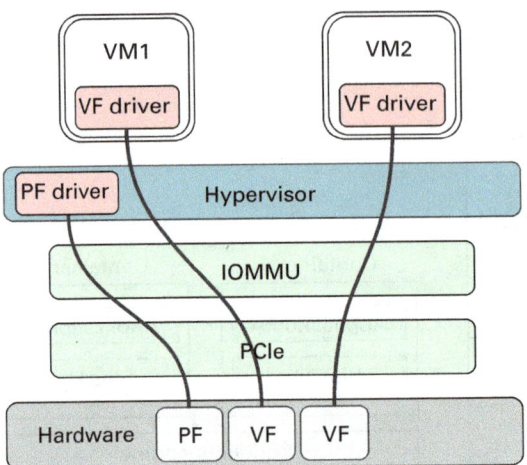

Figure 11.2
SR-IOV architecture. IOMMU, input-output memory management unit.

Figure 11.1b shows the general container architecture. As seen from the figure, containers share the host operating system and kernel, and this is a stark contrast from VMs.

11.2.3   SR-IOV

SR-IOV (PCI-SIG, 2021) is a standard for PCIe that specifies the native IOV capabilities in PCIe adapters. By using SR-IOV, a single physical device or physical function (PF), can be presented as multiple virtual devices or virtual functions (VFs). As shown in figure 11.2, a single VM can be assigned a virtual device through PCI pass-through, which allows direct access to the VF from each VM. SR-IOV is a hardware-based approach for implementing IOV. Thus, the drivers of the PFs can also be used for VFs, and their performance is generally higher than that of the traditional software-based IOV methods.

11.2.4   OpenStack

OpenStack is a cloud management system that aggregates computation, storage, and networking resources and provides this infrastructure as a service to end-users. A graphical interface accessible through the web is available to users for conveniently provisioning and interacting with resources. Beyond standard infrastructure-as-a-service functionality, additional components provide orchestration, fault management, and service management among other services to ensure high availability of user applications. Some of the main components of OpenStack and their functionalities include Nova, which manages computation resources; Neutron, which manages networking services; Swift, which provides object storage; Cinder, which provides block storage; Heat, which enables complex resource orchestration; and Keystone, which serves as the authentication agent.

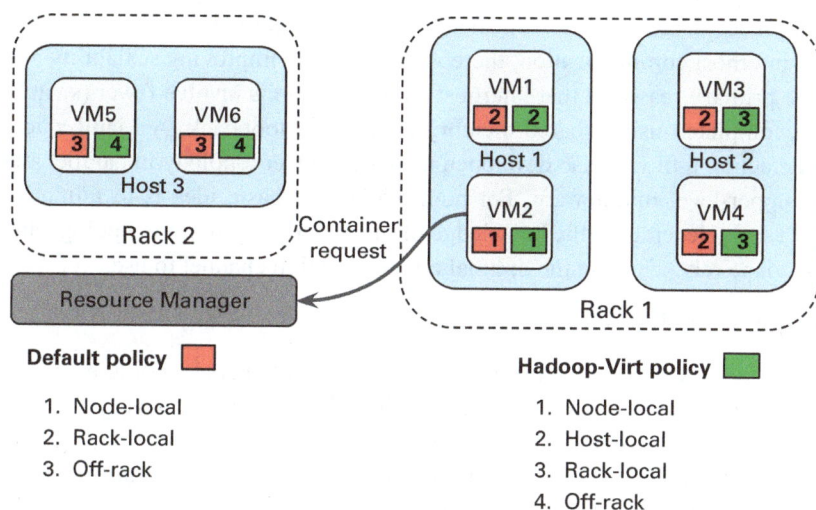

Figure 11.3
Topology-aware resource allocation in Hadoop-Virt.

11.3   State-of-the-Art Designs

In this section, we present several designs that tackle different aspects related to efficiently running HPC and big data applications in the cloud.

11.3.1   Locality- and Topology-Aware Designs

Networking performance is extremely important in the context of HPC stacks. Hardware-based network virtualization techniques such as SR-IOV have been implemented by major network vendors and can be used to reduce the overheads of network virtualization. Despite the availability of SR-IOV, the networking overheads are unacceptable for many HPC applications. To achieve near-native performance, several works have proposed locality- and topology-aware designs. The main idea is to minimize communication costs for VMs that are located on the same physical machine. For instance, Zhang et al. (2014) proposed the use of shared memory channels (such as inner-VM shared memory [IVSHMEM]) for communicating between colocated VMs as a way to optimize MPI libraries in virtualized environments. Their work introduces a locality detector to identify colocated VMs and determine the appropriate channel to use for communication with peers.

Along the same lines, Gugnani et al. (2016) proposed Hadoop-Virt, a set of topology- and locality-aware designs for optimizing communication patterns in big data stacks (such as Hadoop). Their work proposed a job scheduling framework to maximize communication between colocated VMs. Figure 11.3 shows the container allocation policy in Hadoop-Virt. The authors modified the map task placement and container allocation policies to give priority to host-local tasks. In this manner, the overall communication pattern will be dominated by interhost communication, thereby significantly improving scalability and performance. The primary reason is that interhost communications involve fewer (switch) hops and can be completed using shared memory or network loopback. A MapReduce-based topology detection tool was also developed to detect cluster topology in parallel and expose it to the upper-level middleware. For both works, the basic idea is to minimize communication costs by leveraging the knowledge of the underlying runtime topology and intelligently and adaptively selecting the optimal communication channel to use.

11.3.2 Supporting Live Migration

Live migration provides the ability to move running VMs or containers to a different physical host. It is a critical feature in any virtualized environment because it is very effective in load-balancing and resource consolidation. Google researchers (Ruprecht et al., 2018) present performance results of VM live migration across different workloads and measure the impact of various performance optimizations at a scale of performing one million migrations per month at Google. They list the various reasons for live migration of VMs such as data center maintenance and hardware upgrades, kernel and VM monitor (VMM) upgrades, and bin packing and load-balancing. Among these factors, the major cause for live migrations at Google is disruptive software updates.

The study by Ruprecht et al. (2018) explains how live migration at Google works. First, they modified the Borg (Verma et al., 2015) cluster manager, where each VM runs in its own container as a Borg task. Under Borg, tasks can be restarted at any time. So, all the configuration and state information are available to Borg and allows Borg to mark jobs that are migratable and to perform live migration. Second, memory migration uses precopy. When the write rate exceeds network throughput, blackouts (where VM is paused) took a long time. During memory migration, the system does not send unpopulated pages to the target. This is achieved by passing this information from the host kernel to the user-space migration logic via a system call. To support VMs with large dirty sets and high write rates, Ruprecht et al. (2018) added memory postcopy to the memory migration algorithm. Post-copy uses background fetching (at target) to continue migrating memory from the source. The network component migration is optimized by activating the new configuration at the target during the blackout. If any packets are received by the source before the configuration propagates through the network, those packets are forwarded to the target. The storage migration (local SSD) is done via write-mirroring. Here, the system hooks into the stream of guest I/O to disk and notes all the write and discard operations. If any of these operations affect the already migrated regions, they are mirrored to the target before the guest is notified of their completion.

As mentioned, in a virtualized cloud environment, software updates to an operational hypervisor can be highly disruptive, because VMs running on a hypervisor must be either migrated away or shut down. Doddamani et al. (2019) the authors proposed a new approach, called HyperFresh, that is a live hypervisor replacement method using nested virtualization and is less disruptive than live migration of VM. A lightweight shim layer, called hyperplexor, runs beneath the hypervisor and instantiates a new replacement hypervisor as a guest on top of hyperplexor. Then all states of the VMs are transferred from the old hypervisor to the replacement hypervisor via intrahost live VM migration. To avoid significant memory copying overhead, the hyperplexor relocates the ownership of the VM's memory pages from the old to the replacement hypervisor. As most of the remapping happens outside the critical path of the VM state transfer, the replacement time is not related to the size of the VM being migrated. To mitigate the performance overhead introduced by nested virtualization during normal execution of the VMs (when no hypervisor replacement is occurring) the authors propose a series of optimizations. They showed that the hypervisor replacement time remains constant for busy and idle VMs under HyperFresh because the time comprises only the time taken for virtual CPU and I/O device state information. HyperFresh has a shorter downtime and less performance impact than intrahost live migration does because pages are not transferred.

With hardware-based virtualization technologies like SR-IOV being increasingly adopted by cloud providers, it is essential to provide solutions that can enable the live migration of VMs in an application-oblivious manner. Several solutions have been proposed to provide this feature. One interesting solution by J. Zhang et al. (2017) presents a hypervisor- and device driver–independent design through the use of an external migration controller. Such a solution is more practical because it can be deployed in any cloud environment and prevents the vendor-locking problem. Their design uses a shared memory device to communicate between the hypervisor and VM and uses a stage-driven approach that allows applications to suspend all I/O activities until the migration is complete. They show how codesigning with frameworks such as MPI is possible to achieve application-oblivious migration.

11.3.3 Accelerator Virtualization

A rising demand exists for GPU virtualization with good performance, full features, and sharing capability. Kun et al. Tian et al. (2014) propose gVirt, a product-level GPU virtualization implementation with full GPU virtualization running native graphics driver in the guest and mediated pass-through that achieves both good performance and scalability and also secure isolation among guests. Dong et al. (2009) show that the GPU could be directly passed through to a specific VM, thus enabling easier access to virtualized GPU resources. NVIDIA GRID vGPU (NVIDIA, 2021i) enables multiple VMs to have simultaneous, direct access to a single physical GPU, using the same NVIDIA graphics drivers that are deployed on nonvirtualized operating systems. GPU virtualization could also be achieved through API remoting techniques. API remoting forward graphics commands from the guest operating system to the host. Gupta et al. (2009) propose GViM, which provides effective sharing of available GPU resources among VMs by exploiting the fact that all accelerator kernel invocations are routed through the management domain. Shi et al. (2009) presented a very early CUDA-oriented GPU virtualization solution. It uses API interception to capture CUDA calls on the guest operating system with a wrapper library and to redirect the calls to the host operating system where a stub service was running. GPUvm (Suzuki et al., 2014) is a GPU virtualization solution on a NVIDIA card. It implements both para- and full virtualization. However, full virtualization exhibits a considerable overhead for memory-mapped I/O handling. Compared to the native, the performance of optimized para-virtualization is two to three times slower.

In container-based cloud environments, we see increasing adoption of GPUs. Therefore, it is desirable to first enable sharing of hardware among distributed clients. Container technologies have been designed such that sharing of hardware is virtually the same as multiple applications sharing hardware in a bare-metal setting. There has also been work to virtualize GPU devices through remote access over the network. For instance, rCUDA (Duato et al., 2010) is a library that provides remote access to GPUs by using a client-server model. The client library allows unmodified binaries to run on remote GPUs, by translating local GPU requests to remote requests. The biggest advantage of remote access is the ability to share the GPU among multiple clients. The real challenge, however, is to enable efficient communication between application processes operating on GPU memory. As we discussed in chapter 4, several state-of-the-art GPU communication schemes exist, such as cuda-Memcpy, cudaIPC, GDR, and GDRCOPY. The performance characteristics of the schemes in a virtualized context differ widely from the native context. This is because the latency of different communication schemes depends a great deal on the locations of the source and destination GPU buffers. The problem is complicated by the fact that global topology information is unknown to applications. Zhang et al. (2019) present a solution to this problem. They propose a locality-aware detection module based on shared interprocess communications as well as a communication coordinator for selecting the appropriate communication channel to attain the lowest latency.

11.3.4   NVMe-Based Storage Virtualization

The performance of NVMe devices in virtualized environments is a source of concern. The NVMe standard provides SR-IOV for NVMe device virtualization. With SR-IOV, the NVMe device can be presented to each VM as a separate physical device. The VM can directly access the device without hypervisor intervention. Some studies (Jose et al., 2013, Liu, 2010, Song et al., 2014) have shown that SR-IOV performance is close to the native context. Cloud providers should provide support for SR-IOV–based NVMe virtualization to allow for maximum performance. The NVMe standard offers the ability to create namespaces as a way to achieve logical isolation. Namespaces are a set of logical blocks in the hardware. Each namespace contains a distinct continuous set of logical blocks and can be addressed using a namespace ID. This design does not provide any guarantee of performance isolation because the namespaces are logical and the physical blocks used to store the data are common for all namespaces. Separate namespaces offer complete isolation only in terms of security. In a cloud environment, each VM can be assigned a separate namespace for security isolation; however, performance isolation is still a matter of concern. Recent work by Song et al. (2014) has shown that modifying the Flash Translation Layer to map namespaces to separate physical blocks can show significant improvement in I/O performance in virtualized environments.

Apart from using SR-IOV, another method for storage virtualization is to enable remote sharing through the use of the NVMEoF standard (NVMe Express, 2016). This standard allows hosts to access a remote SSD using fast RDMA communication. Consequently, several hosts may remotely connect to the SSD and share the available storage capacity and bandwidth. As mentioned earlier, separate namespaces can be created for each client to achieve logical isolation. Although remote access adds some latency overhead, it allows the device to be shared among several clients that can reside on different physical machines. Compared to SR-IOV, NVMEoF provides more flexibility. Furthermore, it has also been shown that NVMeoF only adds 10 percent overhead at the application level compared with the overhead for local access (Guz et al.). Therefore, NVMeoF is emerging as a more critical storage virtualization technique.

In the community, researchers have also proposed to accelerate NVMe-SSD virtualization through using FPGAs. For example, FVM (Kwon et al., 2020) provides hardware-assisted storage virtualization by using FPGA. Along with running virtual device emulation on an FPGA, FVM also has a hardware-based device-control mechanism to allow FPGA to directly manage physical storage devices (hardware-level NVMe interface). Though being implemented in hardware, FVM allows programmable VM management (migration, caching, replication, and so on) such as a software-based storage virtualization solution. FVM uses FPGA's on-chip memory for SQ-CQ pairs and doorbell registers that can be directly accessed by VMs and NVMe devices through PCIe. Through the SR-IOV implementation in FVM, VMs can enter a virtualization layer without any arbitration support from the host software. Through these designs, FVM shows more promising performance and manageability than other solutions.

Overall, all of the above-mentioned NVMe-based storage virtualization solutions have the potential to significantly improve the I/O performance of many big data, AI, and HPC applications on next-generation cloud computing platforms.

11.3.5 Cost- and Power-Efficient Designs

The cloud was designed with the goal of providing a cost-effective platform to users. Considering the scale of modern data centers and the amount of power and associate costs required to run them, there is a need for designs that can effectively reduce the overall cost and power requirements. Along these lines, Kingfisher (Sharma et al., 2011) is an elastic and cost-aware cloud system that minimizes cost by selecting the optimal virtual machine configuration and reducing the time to transition to new configurations. Convolbo and Chou (2016) propose a cost-aware DAG scheduling algorithm for complex parallel scientific applications. They provide optimal as well as heuristic solutions to the DAG resource scheduling problem for reducing both makespan and resource utilization. Beloglazov and Buyya (2010) propose an energy-efficient VM allocation policy. The idea is to perform VM consolidation using live migration and turn off idle machines to save power. Relocation of VMs is performed based on a heuristic algorithm that ensures that service-level agreements (SLAs) are not violated and resource utilization is maximized. Along the same lines, Goudarzi et al. (2012) presented a convex optimization and dynamic programming algorithm for resource allocation to deal with the energy and client satisfaction trade-off. Cloud systems are penalized for requests not meeting prespecified limits. The algorithm tries to obtain probabilistic bounds on SLA violations. Through simulation, the authors demonstrate the effectiveness of their work compared to previous solutions.

11.3.6 QoS-Aware Designs

In modern cloud environments, users expect a guarantee of service for their applications. Cloud providers typically negotiate SLAs with users as a way to provide these guarantees. In such a scenario, cloud middleware and runtime should be able to provide mechanisms to satisfy these guarantees as QoS. From an enduser's perspective, the SLA provisioning should be completely transparent. So, the service guarantee mechanisms should be completely application-oblivious.

In the context of storage devices (such as NVMe-SSDs), the goal is to provide guarantees of I/O bandwidth. The NVMe standard provides provisions for hardware request arbitration such as weighted round robin (WRR). Figure 11.4 shows an overview of this scheme as proposed in the standard. Broadly, there are three priority classes and each is assigned a weight. Multiple submission queues can be associated with a single priority class. The NVMe controller processes requests from submission queues based on their priority and weight. Processing for queues in the same priority is done in a round-robin manner. Gugnani et al. (2018) use the hardware-provided arbitration mechanisms that are part of NVMe-SSDs as a way of providing applications with a guarantee of I/O bandwidth. Furthermore, they show that the existing WRR arbitration policy is insufficient in providing fair bandwidth QoS, and deficit round robin (DRR) performs much better in this context. The authors also show how DRR can be efficiently implemented in hardware. Reflex (Klimovic et al., 2017) is a system that provides fast remote access to flash NVMe-SSDs in a multi-tenant cloud environment. Reflex provides tail latency and throughput service guarantees by using a software-based QoS algorithm. The main idea behind the algorithm is token-based processing using a cost model derived using the performance characteristics of the SSD. In this manner, each tenant is allocated a fixed amount of tokens for a time period based on its SLA. Tenant requests are submitted based on the cost model and available tokens, resulting in both latency and throughput QoS. Libra (Shue and Freedman) is an I/O scheduling framework that aims to provide applications with throughput guarantees. Libra

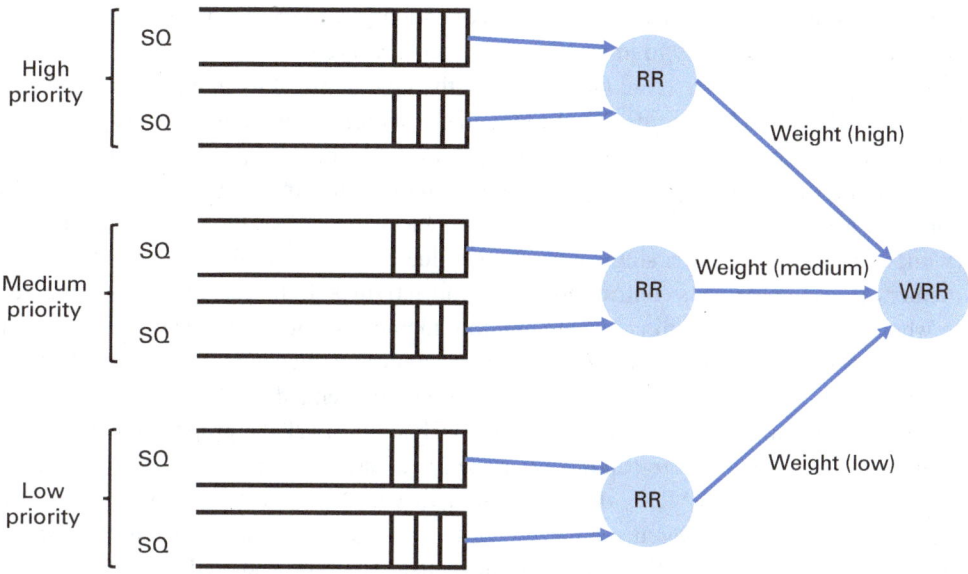

Figure 11.4
NVMe hardware arbitration overview.

tracks application I/O consumption down to low-level I/O operations and uses a cost model based on virtual I/O operations, which results in an I/O size agnostic solution.

11.3.7 Elastic Memory Management for Big Data Applications

Big data processing and management frameworks, such as Hadoop and Spark, are typically implemented in managed languages (e.g., Java) where garbage collection (GC) causes memory and CPU overhead. Also, if JVM heap size is not configured properly, the application fails with an out-of-memory (OOM) error. Accurate estimation of memory before application execution is not possible. Thus, memory is overprovisioned for data-intensive applications and leads to underutilized cluster resources and job queuing delays. This has become an important problem when we run big data applications on cloud computing environments because the resources in the cloud need to be efficiently shared with multiple tenants.

Chen et al. (2019) present Pufferfish for addressing these issues to decrease performance degradation, OOM errors, and underutilization of cluster resources. To solve the problem of OOM, Pufferfish limits the memory usage of a task in operation system containers while setting a large heap size. This leads to disk swapping instead of running out of memory. Now, disk swapping degrades system performance, hence, Pufferfish temporarily suspends the swapping of containers by capping its CPU resource such that thrashing is throttled but the task is alive. Pufferfish alleviates the problem of static cluster resource allocation by dynamically adjusting the memory allocation of containers on the fly. In this way, Pufferfish achieves memory elasticity to improve cluster memory usage.

By setting the heap size (e.g., 64 GB) larger than the physical memory limit of the container (4 GB), Pufferfish avoids OOM errors. The memory overcommitment supported by the operating system allows JVMs to run with a much larger heap size than the physical memory limit by swapping. To suppress this memory thrashing, the CPU allocation of a container is reduced to 1 percent (by experiment) via cgroup subsystem and pin task threads (work and GC threads) to a single core. This reduces the rate at which the JVM accesses memory so as to reduce its contribution to paging activities. Pufferfish identifies containers in which processes are running out of container memory using the following heuristics: if the sum of the memory usage and swap usage is larger than the memory limit and there are swapping activities detected, the container should be suspended.

Pufferfish is built on top of Apache YARN. Chen et al. (2019) propose a new type of container, FLEX, whose size can be dynamically adjusted during execution by Pufferfish. A regular container is YARN's default container, whose size is static and determined at request time. Another difference between the two is that the container size and JVM size are overprovisioned in YARN, while only heap size is overprovisioned in Pufferfish. The key component of Pufferfish includes one node memory manager per NodeManger, one container monitor per container, and a scheduling plugin in ResourceManager. The node memory manager is responsible for monitoring node memory usage and instructing the container monitor to resize a container. The container monitor is a per-container daemon responsible for memory monitoring and adjusting. The scheduling plugin causes the existing cluster schedulers to be memory-aware. Experiments with Spark and MapReduce on real-world traces (chosen from HiBench) show Pufferfish improves cluster memory utilization by 2.7 times and the median job runtime by 5.5 times compared to a memory overprovisioning solution. The overhead caused by Pufferfish is less than 10 percent across various applications and can be amortized for long-running applications.

11.3.8   Efficient Storage Management for Containers

As discussed before, container technology, like Docker, has been heavily used in many cloud computing environments. Docker-enabled hosts require a copy of an image on local storage to create and run containers. Docker provides a set of interfaces, implemented by the Docker daemon, which allows users to conveniently package an application and its dependencies in images. This design introduces tight coupling between the daemon and the node's local storage. This leads to problems such as wastage of storage space, wastage of network bandwidth, and high application start-up time (in the case of large-scale applications where each node has to pull the same image).

Zheng et al. (2018) propose Wharf, an architecture for container runtime that provides a way of storing images in a shared file system. Wharf also enables distributed Docker daemons to collaboratively retrieve and store container images in shared storage and create containers from shared images. This significantly reduces network and storage overheads by holding only one copy of an image in central storage for all daemons. The authors start by outlining a naïve solution where the distributed storage is partitioned and each daemon receives its own partition where it can store all of its state and image data. This approach still overutilizes storage space as each daemon has to pull an image separately and store it in its partition. The Docker graph driver defines a way of creating a union mount and performing copy-on-write (COW). The core idea of Wharf is to split the graph driver contents into global and local states and synchronize accesses to the global states. The global state contains data that need to be shared across daemons, that is, image and layer data, and static metadata such as layer hierarchies and image manifests. This ensures that all image-related information is stored once and not duplicated across daemons. The global state is stored in the distributed file system to be accessible by every daemon.

The Wharf architecture consists of Wharf daemons, which run on individual Docker hosts and manage their own local state; image management interface, which allows Wharf daemons to access the shared global state; and a shared store, which hosts all of the global state. Wharf exploits the structure of Docker images to reduce synchronization overhead. Images consist of several layers that can be downloaded in parallel. Wharf implements a fine-grain layer lock to coordinate access to the shared image store. This allows daemons to pull different images and different layers of the same image in parallel and therefore increase network utilization and avoid excessive blocking. Also, a layer lock ensures that only a consistent layer state can be seen by each daemon and prevents the entire cluster from failing when a single daemon fails. In terms of pull latency and network overhead, Wharf reduces image pulling by a factor of up to twelve times compared to Docker on local storage and introduces an overhead of 2.6 percent during container execution due to remote accesses.

To improve the storage efficiency and deployment speed of containers, Docker stores images in many layers and enables each layer to be read-only and sharable between multiple containers. The storage driver of Docker containers is used to provide a unified view for multiple image layers and support COW for read-only files. Existing storage drivers fall either into file-based or block-based mechanisms. Guo et al. (2019) observe that the I/O performance and cache efficiency of these drivers have the following limitations. First, the file-based COW needs to copy the entire file before the update. This leads to a coarse-grained COW that incurs a large write overhead and degrades I/O performance. Second, for block-based drivers, when multiple containers read data from the same file, a large number of redundant I/O requests are introduced. Third, block-based drivers generate many copies of a block in page cache due to its inefficiency in sharing cached data. Similarly, file-based drivers introduce duplicate data after performing COW on shared files. This degrades cache efficiency.

To solve these inefficiencies, Guo et al. (2019) propose a new approach called HP-Mapper, which supports fine-grained COW operations, intercepts redundant I/Os, and efficiently manages duplicate cache data. HP-Mapper consists of three modules. First, the address mapper employs a two-level mapping strategy to support two different block sizes in a logical volume and adopts on-demand block allocation for different write requests. Second, the I/O interceptor detects redundant I/O requests to read data from the page cache instead of disks. It achieves this by recording the physical block number (PBN) of a read request in a hash table and checking new requests by looking up their PBN in the hash table. The third component is the cache manager that removes redundant data in the page cache by considering page redundancy and hotness. The cache manager adaptively adjusts the eviction strategy to balance cache hit ratio and cache redundancy. The CPU overhead of HP-Mapper is around 4 percent as it locates all duplicate pages by scanning metadata maintained in the hash table and fetches their characteristics from page flags maintained by the Linux kernel. Experiments show that fine-grained COW operations reduce latency by up to 99.8 percent compared with file-based drivers; the I/O interceptor is able to intercept 92.6 percent of redundant read requests; and the cache management mechanism is able to reduce cache usage by 65–75 percent.

To resolve these inefficiencies, Guo et al. (2019) propose a new approach called HP-Mapper, which supports fine-grained COW operations, intercepts redundant I/Os, and efficiently manage duplicate cache data. HP-Mapper consists of three modules. First, the address mapper employs a two-level mapping strategy to support two different block sizes in logical volume and adopts on-demand block allocation for different write requests. Second, the I/O interceptor detects redundant I/O requests to read data from the page cache instead of disks. It achieves this by recording the PBN of a read request in a hash table and checking new requests by looking up their PBN in a hash table. Third, the cache manager removes redundant data in the page cache by considering page redundancy and hotness. The cache manager adaptively adjusts the eviction strategy to balance cache hit ratio and cache redundancy. The CPU overhead of HP-Mapper is around 4 percent as it locates all duplicate pages by scanning metadata maintained in the hash table and fetches their characteristics from page flags maintained by Linux kernel. Experiments show that fine-grained COW operations reduce latency by up to 99.8 percent compared with file-based drivers; the I/O interceptor is able to intercept 92.6 percent of redundant read requests; and the cache management mechanism is able to reduce cache usage by 65–75 percent.

11.3.9   Accelerated Object Storage Systems

Unstructured data in the cloud are typically stored in object storage systems, such as Amazon S3 (Amazon, 2021b), OpenStack Swift (OpenStack, 2021b), and so on. OpenStack Swift is an object store that is scalable, redundant, and distributed. It is typically used by users for uploading/downloading unstructured/semistructured data, such as software, simulation input files, experimental results, large datasets, VM images, and configuration files. OpenStack Swift is able to easily scale by adding more servers to the object store. If a server or disk dies, its contents are replicated onto new nodes. This creates an easy-to-manage and low-cost, but highly redundant place for file storage. One of the major limitations with state-of-the-art object stores such as OpenStack Swift is the lack of ability to write or read directly to or from them with desired performance. For instance, uploading and downloading objects from Swift incurs significant network and I/O overhead, especially for large objects. The default Swift system is written in Python with network communication implemented using TCP sockets-based communication. As we already know, Python performance is lower than that of other common languages, such as C. In addition, TCP communication has several known performance bottlenecks, such as context switch and extra buffer copies for each message communication.

As discussed in chapter 4, modern high-performance interconnects such as InfiniBand and RoCE offer advanced features such as RDMA and provide high bandwidth and low-latency communication. InfiniBand has been widely used in modern HPC clusters, while InfiniBand and RoCE networks are also being deployed in many cloud computing environments. In this context, Gugnani et al. (2017) propose an enhanced design for Swift, called Swift-X, which contains two new designs to improve the performance and scalability of Swift. Swift-X redesigns the network communication and I/O modules in Swift based on RDMA to provide the fastest possible object transfer. In addition, Swift-X proposed both client-oblivious design and metadata server-based design for Swift to further improve the performance under different object store usage scenarios in the cloud. The authors also discussed how to use efficient hashing algorithms to accelerate object verification in Swift. This study has demonstrated that there are a lot of opportunities for cloud computing components to be optimized or redesigned for big data applications.

11.4 Case Studies and Performance Benefits

To illustrate the benefits of the different state-of-the-art designs discussed in the previous sections, we present two case studies for locality- and topology-aware designs in Hadoop-Virt and QoS-aware designs.

11.4.1 Benefits with Locality- and Topology-Aware Designs in Hadoop-Virt

Apache Hadoop has been deployed in virtualized cloud environments because of the flexibility and elasticity offered by cloud-based systems. Hadoop supports topology awareness through topology-aware designs in all of its major components. However, automatically detecting the underlying network topology in a scalable and efficient manner for Hadoop components was still missing for the big data community. Moreover, the topology-aware designs in Hadoop were not optimized for virtualized cloud environments. To address these issues, as discussed in section 11.3.1, a new library called Hadoop-Virt (Gugnani et al., 2016) was proposed in the literature based on RDMA-Hadoop, which provides an automatic topology-detection module and virtualization-aware designs in Hadoop to fully take advantage of virtualized environments.

To demonstrate the performance benefits of Hadoop-Virt, the following experiments were taken on a test bed that consists of nine physical nodes on the Chameleon cloud (Chameleon, 2021),where each node has a twenty-four–core, 2.3-GHz Intel Xeon E5-2670 (Haswell) processor with 128 GB of main memory and is equipped with Mellanox ConnectX-3 FDR (56 Gb/s) HCAs and PCIe Gen-3 interfaces. The InfiniBand network provides native SR-IOV support. KVM in Linux was used as the VMM. For consistency, the same operating system and software versions were used for the VMs as well.

Traditionally, Hadoop is deployed with each slave node running one NodeManager and one DataNode. However, in virtualized environments, it may be beneficial to separate the data (HDFS) and computation (YARN) components of Hadoop, so they can be scaled independently, and any change in one component does not affect the other. Thus, for deploying Hadoop in a Distributed Mode, NodeManagers and DataNodes are deployed in separate nodes. In the following tests, thirty-two slave nodes ran one NodeManager each and thirty-two additional slave nodes ran one DataNode each. The Distributed Mode thus allows users to scale HDFS and YARN independently allowing for more control and flexibility.

In these tests, sixty-five VMs in total with eight VMs each on eight bare-metal nodes (totaling sixty-four VMs) are used as Hadoop slave nodes, and one VM on one bare-metal node is used as the Hadoop master node. Seventy percent of the RAM disk is used for data storage. HDFS block size is kept to 256 MB. The NameNode runs in the master node of the Hadoop cluster, and the benchmark is run in the NameNode. Each NodeManager is configured to assign a minimum of 4 GB of memory per container.

Benchmarking tests with Wordcount, PageRank, Sort, and Self-Join are performed and the results are shown in figure 11.5. From these graphs, we can clearly see up to 35.5, 34.7, 52.6, and 55.7 percent improvement with Hadoop-Virt as compared to RDMA-Hadoop for Wordcount, PageRank, Sort, and Self-Join, respectively. Overall, we see much more improvement for the Distributed Mode as compared to the default way of deploying Hadoop on nodes. This is because, in the Default Mode, most of the map tasks and container allocations are node-local. For the Distributed Mode, there are no node-local map tasks or container allocations because the NodeManagers and DataNodes are running on separate nodes. Thus, in this case, RDMA-Hadoop makes all rack-local allocations, whereas Hadoop-Virt makes all host-local allocations, reducing interhost network traffic and leading to large performance improvements. More details about these experiments can be found in Gugnani et al. (2016).

11.4.2   Benefits with QoS-Aware Designs for NVMe-SSD

QoS support for end applications should be provided in a transparent manner on modern cloud computing platforms. In the context of NVMe storage, as discussed in section 11.3.6, the design goal in Gugnani et al. (2018) is to use the hardware-provided arbitration mechanisms as a way of providing applications with a guarantee of I/O bandwidth. Because these schemes should be application-oblivious, the standardized SPDK runtime is enhanced to

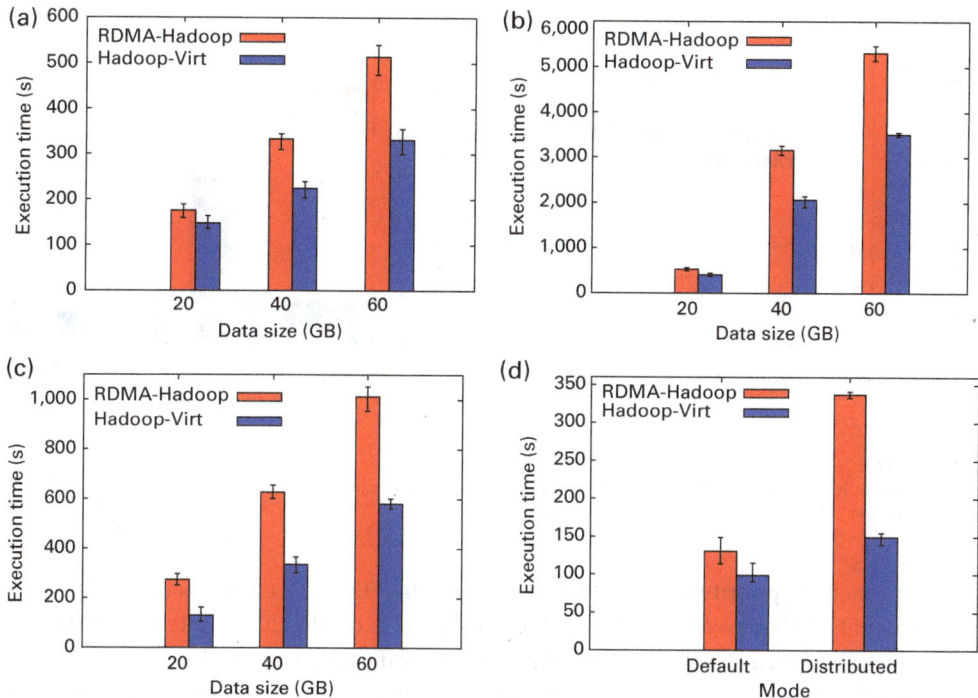

Figure 11.5
Performance benefits of Hadoop-Virt on HPC clouds. Execution times for (a) WordCount, (b) PageRank, (c) Sort, and (d) Self-Join (30 GB).

allow QoS support for NVMe-SSDs. The SPDK runtime has support for the NVMe WRR scheme. However, this is left to the discretion of the user. To use the WRR scheme, the user has to explicitly enable it using an SPDK function and set the priority for each submission queue created. This existing mechanism does not satisfy the application-oblivious requirements. Gugnani et al. (2018) thus proposed a new priority mapping design in SPDK that does not require application changes to modify priority. To this end, an approach was proposed to use the Linux I/O priority framework as a means to transfer the priority class information from the application to the SPDK runtime, similar to the approach proposed by Joshi et al. (2017).

In the following experiments, because the current-generation SSDs rarely support hardware-based arbitration, a QEMU-based (Quick EMUlator). emulator approach was used. A modified version of QEMU was built for emulating open-channel SSDs. To show the benefits of a QoS-aware SPDK runtime, five application scenarios are simulated and the bandwidth achieved by each job over time is measured. For all scenarios, the priority class weights are set to (32, 16, 8) for WRR and (128k, 64k, 32k) for DRR, ensuring that the weight ratios are the same. For the first scenario, one high-priority job with 4k requests

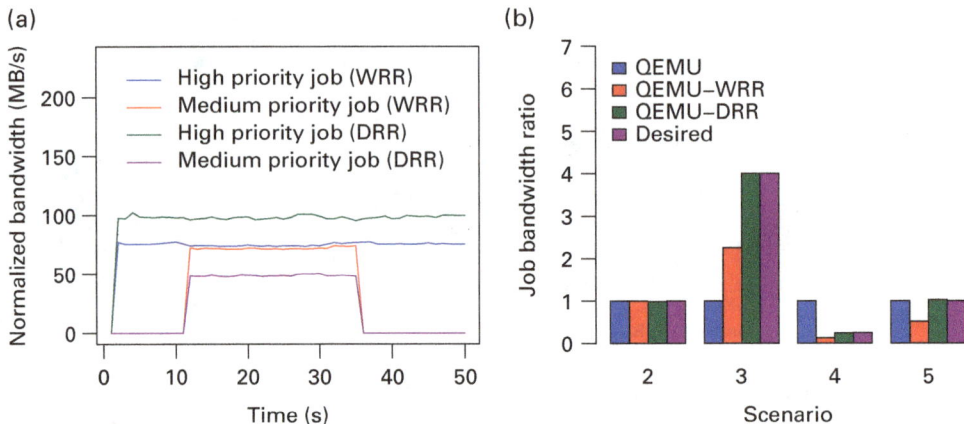

Figure 11.6
Evaluation with synthetic application scenarios. (a) Bandwidth over time with scenario 1. (b) Job bandwidth ratio for scenarios 2–5.

and one medium priority job with 8k requests are run. Figure 11.6a shows the results of this experiment. With the WRR, both jobs receive the same bandwidth despite having different priorities. Each request is given equal priority regardless of its size, leading to a skewed bandwidth distribution. However, DRR is able to achieve near-perfect bandwidth distribution as per priority weights. Note that the bandwidth over time is relatively stable pointing to robust hardware emulation.

The remaining scenarios all have two simultaneous jobs submitting back-to-back requests. Scenario 2 has two high-priority jobs, both with 4k requests. Scenario 3 has a high-priority job with 4k requests and a low priority job with 8k requests. Scenario 4 is the same as scenario 3 with the priorities exchanged. Scenario 5 has two high-priority jobs, one submitting 4k and 8k requests and the other 8k and 16k requests. The total average bandwidth for both jobs in each scenario and the bandwidth ratio are measured. This analysis is presented in figure 11.6b. In all scenarios, the difference in request sizes leads WRR to incorrectly favor the job with a larger request size, while DRR is able to achieve close to the expected ratio. The only case where WRR provides the desired ratio is scenario 2, where the request sizes for both jobs are the same. Default QEMU just provides equal throughput distribution, not providing any service guarantees whatsoever. This analysis clearly demonstrated the superiority of DRR over WRR in achieving bandwidth guarantees.

Although these results are based on NVMe emulation, similar behavior with actual hardware is expected. There are two reasons for this expectation. One, both WRR and DRR have been shown to be easy to implement in hardware (Shreedhar and Varghese, 1996, Katevenis et al., 1991) and provide accurate bandwidth ratios. Two, using separate hardware queues for each application along with lockless request submission and completion paths ensures that the hardware performance characteristics are reflected in the application performance. These results significantly demonstrate that DRR should either become part of the NVMe standard or be accepted by vendors as a good implementation choice for vendor-specific arbitration schemes.

11.4.3 Benefits of SSD-Based Storage for Big Data Applications on Clouds

NVMe-based SSD devices and SSD-based block storage services are widely available on modern cloud computing platforms, such as Amazon EC2, Google Cloud, Microsoft Azure Cloud, Oracle Cloud, Alibaba Cloud, and so on. The availability of fast and powerful storage hardware devices or services has the potential to significantly accelerate big data processing workloads. In the community, there are many studies to demonstrate promising solutions by exploiting the performance benefits of SSDs for big data applications. For example, Lee and Fox (2019) have used multiple built-in Hadoop benchmark tools, including WordCount, PageRank, K-Means, TeraSort, and DFSIO, to represent a wide range of big data applications running on top of Hadoop systems by varying workload volumes and configurations. With these benchmark tools, the authors have been able to systematically evaluate various big data workloads on multiple modern clouds (e.g., Amazon EC2, Google Cloud, Microsoft Azure Cloud, and Oracle Cloud) that can provide NVMe-SSD devices or SSD-based storage services. The authors have compared the performance of Hadoop workloads on these different cloud platforms, understood the difference between SSD-based block storage and local NVMe-SSDs for I/O intensive workloads, and provided the analysis of cost-efficiency potentially reducing storage costs.

Finally, Lee and Fox (2019) found that high-performance storage services can make a significant impact on HDFS-based applications with a large number of virtual CPU cores. The "just a bunch of disks" style of storage is able to achieve six times better performance with additional volume counts and CPUs per server than large numbers of low-end servers can. With these observations, the authors envision that data-intensive applications could gain better scalability and efficiency on high-capacity servers.

11.5 Summary

In this chapter, we first presented an overview of the primary HPC cloud technologies. Following this, we discussed state-of-the-art solutions in the community to provide desired aspects to HPC and big data applications in a cloud environment. Overall, significant work has been done by the community to alleviate the dreaded native and virtualized performance gap. Similarly, a lot of work has been done to exploit the cloud features in HPC and big data frameworks and applications. Nevertheless, a lot remains to be done. In particular, there is significant scope for providing QoS-aware big data processing and management middleware and improving I/O performance by exploiting technologies such as NVMEoF.

12 Frontier Research on High-Performance Big Data Computing

As Moore's law is ending, we are going to enter a new computing era, in which there will be highly increased demands on parallel and distributed computing architectures and systems for solving the big data challenges. In this new era, we are facing serious challenges in attaining high performance, which is not only coming from exploiting more parallelisms from modern and next-generation multicore processors and accelerators, but also from exploiting more potential parallelisms in low-level I/O devices and networking resources. There are many active research studies along these directions in the computing field. In this final chapter, we discuss the ongoing and possibly high-reward future research avenues for achieving high-performance big data computing.

12.1 Heterogeneity-Aware Big Data Processing and Management Systems

Modern HPC clusters and data centers are supporting heterogeneous computing resources, such as GPU and FPGA devices for the acceleration of computation. Current-generation schemes being used to provide efficient designs for big data processing and management systems using GPU and FPGA devices are available in the field.

12.1.1 Big Data Processing Frameworks on Heterogeneous Clusters

Big data processing frameworks such as Hadoop and Spark can take advantage of accelerator-equipped systems for accelerating particular big data applications that are more compute-bound and less shuffle- and I/O-intensive. Much work has been done toward achieving a high-performance Hadoop or Spark framework with accelerators (see chapter 8). Similar approaches have been proposed for Spark with GPU as well to improve the performance of deep learning applications (see chapter 10). Due to the popularity and massive computing power of modern accelerators, there are still a lot of potentials to propose new schemes to improve the existing designs in heterogeneity-aware big data processing frameworks on accelerator-based platforms.

HeteroSpark (Li, Luo, et al., 2015) is a middleware framework that enables acceleration of Spark workloads using GPUs. HeteroSpark places GPUs in Spark's worker nodes and employs JVMs to exchange data among CPUs and GPUs. Function calls from Spark applications are intercepted by HeteroSpark and invoke GPU kernels for the acceleration of chosen functions. During CPU-GPU communication, the CPU worker serializes the data partition and sends it to the remote method on GPU-JVM through the Remote Method Invocation communication interface. As discussed in section 12.1.2, such type of data movement may incur expensive overhead and compromise the system's fault-tolerance ability. HeteroSpark allows the use of existing GPU libraries by integrating them with GPU-JVM via JNI. One challenge of using HeteroSpark is that it is not designed as a general-purpose system and Spark applications still need to be ported on HeteroSpark as pointed out in the Spark-GPU work by Yuan et al. (2016).

To provide a more generic approach of using GPUs in Spark, the design of Spark-GPU (Yuan et al., 2016) provides a CPU-GPU hybrid data analytics system. Spark-GPU uses GPUs to accelerate the performance of data analytics applications based on specific computing demands for Spark. Many types of data analytics applications have data parallelism via RDDs and are computation-intensive. Thus, the nature of the task makes GPU an ideal choice to accelerate Spark's performance. Previous work redesigned the system based on GPU's characteristics without considering the performance of CPU operations. Spark-GPU uses heuristic rules to offload SQL queries to GPUs and provides block processing ability for GPUs to get the good performance of data analytics applications.

Spark-GPU overcomes the challenges to run applications on both GPU and CPU by modifying several components. First, Spark uses a one-element-at-a-time iterator model to execute applications that can underutilize GPU resources. Spark-GPU extends Spark's iterator model to support block processing on GPUs via GPU-RDD. GPU-RDD buffers all its data in either row format or column format in native memory. To utilize the parallelism of GPU, the block interface provides the ability to apply an operation on GPU-RDD to all the buffered data at the same time. Spark runs on top of JVM, whereas Spark-GPU utilizes native memory instead of Java heap memory to buffer data in GPU-RDDs. One advantage of this design is that it saves data copying operation inside Java heap memory as data in native memory can be directly transferred to GPUs to process. Second, Spark-GPU extends Spark's SQL module to offload SQL queries to GPUs. It supports five important GPU query operators—GPU scan, GPU broadcast join, GPU hash join, GPU aggregation, and GPU sort—that can be used as the building blocks for a wide range of SQL queries. Third, Spark-GPU extends Spark's cluster manager and the task scheduler to manage GPUs in the cluster. The fundamental factor that determines the benefit of GPU sharing is the amount of work that can be executed in parallel when running tasks on GPUs. Spark-GPU enables GPU sharing in the cluster. It abstracts each GPU into multiple logical GPUs. Each logical GPU can run one GPU task. Because each GPU task needs a CPU core to initiate the task, a total number of concurrent GPU tasks cannot exceed the total number of CPU cores on that node. The benefits of GPU sharing come from overlapping the current GPU kernel execution with PCIe data transfer for subsequent GPU tasks.

In addition to supporting GPUs in Spark, the community has also proposed similar support in Hadoop, which is another popular big data processing framework. He et al. (2015) propose Hadoop+, which is a heterogeneity model to predict the shared resource contention between the simultaneously running tasks of a MapReduce application when heterogeneous computing resources (e.g., CPUs and GPUs) are allocated. In a heterogeneous cluster, both the performance and cost for a Hadoop MapReduce application would vary with consumed resources. Hadoop+ takes the shared resource contention into consideration and provides a heterogeneity model to guide users to select their desired resource utilization for the purpose of improving performance or cost-efficiency or to guide the GPU allocation across multiple simultaneously running MapReduce applications. Hadoop+ also enables user-provided CUDA/OpenCL functions to be integrated as plug-ins into Hadoop. Using the heterogeneity model, the optimal or cost-effective resource configuration for an application is determined.

As we can see from these existing designs for supporting GPUs in both Spark and Hadoop, there are many challenges to enable GPUs in the complex Spark and Hadoop data processing pipelines. Even though several of the components get enhanced through the above-mentioned designs to bring GPUs into the pipelines, there are still many open challenges to be resolved, which can be seen as valuable future research directions. For instance, so far many of these existing designs need to do significant changes or build new components to enable GPUs in Spark and Hadoop. A question here is whether the built-in Spark libraries could easily utilize GPUs without significant changes. One way might be to parallelize all existing Spark RDD operations on GPUs such that applications built on top of RDDs can automatically utilize GPUs. However, this approach is quite complex and cannot guarantee high performance, because it may perform even worse than the original Spark. As discussed in Yuan et al. (2016), the Spark RDD abstraction does not have semantic information about the applications, which may cause unnecessary data copying operations between Java heap memory and native memory and uncoalesced GPU memory accesses. Considering the data movement and access overhead when we try to enable GPUs in big data processing frameworks, other types of possible overheads also need to be carefully considered, such as task scheduling for heterogeneous devices, contention management, performance modeling, fault-tolerance, cost-efficiency, and so on. We believe these topics are high-reward future research avenues to explore for achieving high-performance big data computing on heterogeneous platforms.

12.1.2   KV Stores on GPUs and FPGAs

Modern KV stores, such as Memcached, have almost been exclusively deployed on CPU-based servers, which are most prevalent in current HPC clusters and modern data center environments. Consider a KV pair request at the server, such as Set(K, V) or Get(K), there is hardly any intensive computation involved. The essence is a large number of random local memory accesses and moving data between memory and network for every request, implying low instruction count and need for high memory-level parallelism for scaling-out. Thus, traditional Memcached server designs are severely limited by the CPU-memory gap.

In contrast to CPUs, offload-based accelerators such as CUDA-enabled GPUs are capable of offering much higher data accessing throughputs than CPUs are due to two features: higher memory bandwidth and SIMD capabilities. On the other hand, FPGAs enable implementing fully pipelined dataflow architectures that can fully exploit fine-grained task- and instruction-level parallelism to achieve high data processing rates. With these advanced features on GPUs and FPGAs, there have been many research works dedicated to enabling in-memory KV storage systems, such as Memcached, to leverage GPUs (e.g., MemcachedGPU [Hetherington et al., 2015]) and FPGAs (e.g., KVS Accelerator with Xilinx FPGAs [Blott et al., 2013]) in both HPC and data center environments. For each of these two devices, a critical analysis of different designs in literature reveals two generic schemes: device-centric GPU/FPGA approach and hybrid CPU+(GPU/FPGA) approach.

Both GPU- and FPGA-based solutions provide efficient server designs that are essential for scaling-out KV store clusters. Both device-centric and CPU-GPU/FPGA hybrid approaches outperform their respective CPU counterparts (Hetherington et al., 2015, K. Zhang, Wang, et al., 2015). However, these designs also introduce bottlenecks or limitations atypical to traditional CPU-based designs. The bottlenecks on these emerging designs also open up new research opportunities as high-reward future research avenues.

In the hybrid designs, we can see that an obvious bottleneck is moving data between CPU and GPU for large KV pairs. This typically makes such hybrid designs impractical for improving the overall performance of end applications over heterogeneity-aware KV stores. The GPU community has been working on proposing new schemes for optimizing the data movement performance for CPU-to-GPU and GPU-to-GPU paths, such as page-locked (or pinned) memory, unified memory (Harris Mark, 2017), GDR (NVIDIA, 2021d), GDRCOPY (NVIDIA, 2021e), and so on. While approaches such as unified memory provide a single memory address space accessible from any CPU/GPU processor in a system, the on-demand paging scheme in the unified memory mechanism is still much slower than the prefetching scheme or the static scheme. However, the CPU-GPU data movement overheads with these approaches can potentially be amortized by codesigning KV store operation-specific kernels to overlap with other computations.

In GPU-based designs, we typically also see another clear performance bottleneck that is the high overhead of kernel launching on GPU. To alleviate this bottleneck, GPU-based designs usually require us to buffer requests in large enough batches that can be offloaded to the GPU. Small batches are impractical as they are dominated by kernel launch overheads. But employing large batches comes at the cost of client-side latency, because the clients have to wait longer for their requests to be processed at the server on receipt. While server throughput is gained, end-to-end performance is hurt. Approaches such as persistent kernels on GPUs are being explored to lower the kernel launch overheads (Gupta et al., 2012) that can potentially make smaller batches more practical.

FPGAs are predominantly programmed by using hardware description languages such as Verilog and VHSIC. These typically require expertise that is not familiar to most software engineers. Apart from the language-related challenges, rare but probable pipeline stalls, which could occur when two or more operations (e.g., read-after-write) on the same hash index reside within a small and critical part of the pipeline, can cause further usage and programming complexities on FPGAs. Many FPGA-only approaches efficiently implement a subset of KVS commands on an FPGA, that is, Set and Get, on top of the User Datagram Protocol (UDP) to fit the design in the target FPGA. But for medium and large message requests, such as Sets, reliable TCP is more preferable. These complicate the FPGA-based implementations, making it difficult for an FPGA-only solution to be complete and practical. One way of alleviating these issues is to use a hybrid CPU+FPGA approach, similar to the design proposed by (Lavasani et al., 2014). Latency-critical commands such as Get over UDP are in the "fast path" and update commands such as Set/Delete that need reliability to employ TCP are directed to the slower CPU path. This separation for conflicting command paths can simplify FPGA's usage.

Integrated FPGA designs can benefit from the optimized hash table and memory allocator designs that leverage parallelism available in FPGA and minimize the number of PCIe-DMA requests. This could be enabled by maximizing the use of the on-board DRAM buffer available on programmable NICs by implementing a hardware-based load dispatcher and caching component in FPGA to fully utilize on-board DRAM bandwidth and capacity. Efficient data placement could be enabled by placing frequently accessed data structures, such as a hash table index and a cache for hot KV pairs, close to the FPGA. Less frequent (warm or cold) KV pairs can be maintained in the DRAM, thus reducing data movement overhead.

12.2 Big Data Processing and Management for Hybrid Storage Systems

With the deployment of PMEM and NVMe-SSDs on modern and next-generation HEC clusters, the traditional data access approaches (e.g., POSIX I/O, MPI-I/O) for these advanced hybrid storage systems may incur a lot of performance bottlenecks for applications. Traditional file systems were designed considering the performance characteristics of archaic hardware, such as spinning disks and Ethernet networks, and now obsolete standards, such as SATA and SAS. With emerging PMEM and networking technologies such as NVRAM, NVMe-SSD, and RDMA networks becoming mainstream, new bottlenecks are emerging in existing storage system designs for emerging big data processing and management applications. We list the primary bottlenecks that exist and describe why they exist. We rely on extensive prior research (Min et al., 2016, Zadok et al., 2017, Boito et al., 2018, Next Platform, 2017) into the shortcomings of legacy storage systems as well as our own research to complete this analysis.

12.2.1   Synchronous Global Namespace

POSIX semantics require a single global namespace for all files that matches with traditional application requirements to be able to access files anywhere the file system is mounted. This requirement is particularly useful in distributed file systems where applications can run on different sets on CNs each time without worrying about access to previously written data. However, synchronous access to the global namespace creates significant performance issues. This is primarily because the global namespace must be able to handle concurrent requests while maintaining consistency and durability. This requires coordination across a large number of nodes, leading to poor scalability. Furthermore, several applications do not require a global namespace. For example, it is common for HPC applications that checkpoint internal state for fault-tolerance to write a single file per process (Vazhkudai et al., 2018). Another example is the MapReduce framework, where intermediate data generated by a map task are not needed by any other process. This highlights the need for rethinking the requirement of a global namespace.

12.2.2   Kernel Overhead

Traditionally POSIX file systems were designed to be part of the kernel itself. Building the file system layer into the kernel allowed them to leverage the security, memory management, DMA, and interrupt support provided by the kernel. Despite these advantages, kernel file systems have inherent performance bottlenecks. Each file system call must go through the kernel and additionally requires strict permission checking. The user-space to kernel context switch as well as access control checks add significant latency overhead, which is detrimental to the performance of the system. This problem is exacerbated when we consider the ultralow latency of modern storage devices. Furthermore, MPI-I/O inherits these overheads because most MPI-I/O implementations (including the popular ROMIO [Thakur et al.]) are built on kernel file systems.

12.2.3   File Handles

Both the POSIX and MPI-I/O standards require the creation of file handles (or descriptors) to create and manipulate files and associated data. In other words, they can be said to be stateful. This requirement limits the scalability of the storage runtime, particularly for highly parallel workloads because the runtime must keep track of all open file handles and implement some mechanism of dealing with stale handles. Usually, this requires the use of complex data structures, the state of which must be updated frequently. As millions or even billions of clients try to access files from the same file system, this becomes a major performance bottleneck. The cost of opening a file increases linearly with the number of concurrent clients (Next Platform, 2017). This is even before data can be read or written to the file.

12.2.4 Lack of Asynchronous API

POSIX uses a synchronous one-at-a-time API that is inefficient for cloud services (Zadok et al., 2017). Cloud services typically manage a large number of small-sized objects. The use of synchronous POSIX API is particularly bad in this scenario and leads to high access latency. To access objects, a KV-oriented interface (e.g., get, set functions) or the ability to send compound operations (e.g., open, read, and close) as a single request in an asynchronous manner will be useful for application programmers.

12.2.5 Poor Multicore Implementation

Achieving concurrency in file systems is a must on modern clusters with dense multicore CPUs. Unfortunately, kernel file systems and MPI-I/O implementations rarely scale beyond a few cores (Min et al., 2016, Boito et al., 2018). There are a few reasons why this happens: (1) coarse-grained locking is used to maintain consistency; (2) consistency mechanisms commonly employed by file systems (such as journaling and COW) are not scalable; and (3) files cannot be concurrently updated, even if no dependencies exist between the updates. Prior research (Min et al., 2016) has investigated the multicore scalability of popular Linux kernel file systems on an eighty-core machine and concluded that no file system provides good scalability for all workload types. Moreover, scalability is not portable; some file systems have peculiar scalability characteristics for different operations. The multicore design and implementation of file systems need to be seriously reconsidered.

12.2.6 Strong Consistency Requirements

POSIX file writes (`write()` or `fsync()`) are expected to be atomic operations and the write-read or write-write reordering on the same file segment is not allowed (Lockwood, 2017). In replicated distributed storage systems, providing atomicity is hard and requires synchronous writes to all replicas. This can lead to poor scalability, particularly if the number of replicas is large or if one of the replicas is overloaded. Moreover, many applications do not require such strong guarantees of consistency. These include emerging web 2.0 and 3.0 applications (e.g., e-commerce shopping cart, online password checking). Clearly, existing POSIX semantics are unnecessarily too restrictive and prohibit applications from getting the best possible performance.

12.2.7 Superfluous Caching

POSIX and MPI-I/O implementations often rely on client-side caches (e.g., the operating system page cache) to improve latency. This approach was beneficial for file systems backed by spinning disks or SATA-SSDs because their latency was unpredictable and high. The cache was useful in hiding the latency of slow media. However, new media based on 3DXP-like technology has very low latency, making the benefits of having a cache between the application and storage media virtually nonexistent. In some cases having a cache for these new devices may even negatively affect performance because of superfluous copies. Besides, in situations where fault-tolerance is immediately desired, there are no benefits to caching data. Recent research (Min et al., 2016) has also shown that the scalability of the Linux page cache is poor. Even in scenarios with high I/O locality, which conventional wisdom states is the golden rule to improve I/O performance, the cache is unable to scale.

Despite significant research into developing storage systems for new hardware devices such as NVRAM or NVMe-SSD, the proposed solutions are still unable to extract the best possible performance out of the hardware by resolving all or any of the above-mentioned performance bottlenecks. To the best of our knowledge, there is no single system that can utilize all of these advanced devices in an integrated manner (NVRAM NVMe-SSD, GPU, and CPU). In particular, the interactions between GPU and NVRAM, and GPU and NVMe-SSD have not been well studied. We discuss some possible solutions that can achieve tighter integration between these devices through a cross-layer holistic system design approach.

12.2.8 Multirail and Pipelined Design

UMR-EC (Shi et al.) showed how heterogeneous hardware in modern clusters can be used to perform erasure coding in parallel using a multirail design. Lessons can be learned from this work and the same multirail principle can be used as a guide to the design of distributed storage systems. The computational capabilities of new devices and their asynchronous processing nature make it feasible to design a multirail, asynchronous, and pipelined end-to-end I/O path. For instance, consider the functionalities that a storage server in a typically distributed file system might perform. For incoming requests, it might encode and store data in SSDs or NVRAM. For outgoing requests, it might use hashing to locate the object, decode it, and send it out. Both of these pipelines can be optimized by making full use of all available hardware. Furthermore, the operations of each hardware unit can be pipelined. So, while data are being read from the SSD and transferred to the GPU, the GPU can operate on the previous chunk, and the NIC can send out the chunk already processed by the GPU.

12.2.9 Extending GPU Memory

GPU memory size is a critical factor that limits the size of tasks that can be offloaded to the GPU. This is particularly important for deep learning workloads where the size of the model is often limited by the GPU-RAM. Recent studies such as GPUswap (Kehne et al.), NVMMU (J. Zhang et al., 2015), and Dragon (Markthub et al.) enable out-of-core GPU training. GPUswap uses the GPU's ability to access DRAM to enable transparent swapping of pages. NVMMU proposes a custom NVMe driver to directly issue NVMe commands from the GPU using GDR. Dragon extends GPU memory using node-local NVMe storage. It modifies the GPU driver to handle page faults by loading in data from the SSD. However, these technologies do need significant changes in to the drivers and/or kernels, which may incur some performance overhead. If we could achieve a similar effect in user-level runtimes, such as out-of-core deep learning training with MPI (Awan et al., 2018), then there may be more opportunities to codesign applications that can obtain better performance benefits.

12.3 Efficient and Coherent Communication and Computation in Network for Big Data Systems

As discussed in chapter 5, for next-generation HPC and data center clusters, the network speed and capabilities will need to be improved significantly. With the deployment of heterogeneous computing and storage devices in the same system, there are many research opportunities to design efficient and coherent communication and computation schemes in high-speed networks for accelerating big data processing.

12.3.1 Efficient Data Movement

Data movement speed between different devices within a node as well as across nodes is critical to the performance of end applications. Various technologies have been proposed for efficient data communication between devices. For example, NVMEoF allows fast remote access to NVMe-SSD using RDMA, GDR allows fast remote access to GPU memory, CUDA Unified Memory can automatically move data between CPU and GPU memory, and RDMA by itself can be used to transfer data between local and remote memory (DRAM or NVRAM). In addition, server-side NVMEoF processing can be offloaded to SmartNICs. All of these data paths create incredible complexity for developers who only want to deal with a single abstraction for data movement. C-GDR (Zhang et al., 2019) proposes optimized locality-aware communication between GPUs for MPI in a cloud environment. We believe that similar schemes can be developed for big data systems and runtimes as well. The potential designs should then deal with the complexities and subtleties of data movement. Ideally, a locality detector can be designed to determine the data source and destination. Then, an adaptive protocol can be designed atop this detector that optimizes latency based on the available communication channels. Similarly, a high-level interface for accelerating data computation and communication tasks that abstract away the hardware details can be developed. cuDF (RAPIDS, 2021) is a GPU data frame library that provides data management features that can be used to design complex workflows. Currently, cuDF is only for the management of data on the GPU. We believe that a library that provides such communication and computation primitives for distributed heterogeneous devices to system developers will make it much easier to design efficient big data processing and management runtimes. The design of such a library is, therefore, a good avenue for future research.

12.3.2 Coherent Communication and Computation in Network

Modern high-speed networks can support computation offloading mechanisms, which provide the capabilities to perform small computation kernels on SmartNICs or switches. For instance, the HCAs of Mellanox ConnectX-4/5 and later SmartNICs offer an erasure coding

(EC) calculation engine that allows applications to offload EC tasks on the NIC (Mellanox, 2016). Such an offloading scheme provides a new design methodology that brings benefits such as low CPU utilization and more overlapping opportunities. More importantly, the computation offloading scheme on modern SmartNICs provides the possibility of offloading both calculations (e.g., EC) and communications to the NIC.

As discussed in Shi and Lu (2020), there are two kinds of schemes to take advantage of the computation offloading capability on SmartNICs (e.g., Mellanox InfiniBand): incoherent calculation and networking and coherent calculation and networking. The incoherent calculation and networking scheme means the systems have to take care of calculations and data transmissions separately. Taking EC encoding, for example, an initiator node first encodes data chunks on the NIC and then sends data chunks and generated parity chunks to other receiver nodes *(encode-then-send)*. On the other hand, with the coherent calculation and networking scheme (or coherent communication and computation in network), the systems can offload both computations and communications simultaneously to the network. Taking the EC flow as the example, with coherent communication and computation capability, the initiator node only needs to post one network operation (i.e., post_encode_and_send). This operation can fully offload the EC computation and communication stages to the NIC *(encode-and-send)*. In contrast to the encode-then-send scheme, the encode-and-send scheme has the potential to deliver higher performance, because it ideally could save several (or a few) post_send and DMA operations in the pipeline (Shi and Lu, 2019). We believe coherent computation and communication in network schemes are the start-of-the-art or promising approaches for efficiently leveraging next-generation high-speed networks.

However, the current-generation design for coherent computation and communication in the network may not take advantage of the optimized parallel paradigms. As Shi and Lu (2019) discussed in their recent work, the default EC offloading paradigm in modern SmartNICs follows a bipartite graph-based parallel approach (called BiEC) that could not achieve an optimal overlapping pipeline and performance. To address the shortcomings of the default BiEC scheme, the authors proposed a novel scheme (called TriEC) based on a tripartite graph-based EC pipelining. They demonstrated that the TriEC scheme can achieve better performance and overlapping than the default BiEC scheme on modern HPC clusters for big data workloads. We believe a similar approach can be adopted to design next-generation capabilities of coherent communication and computation in network for more types of computation kernels, not just with EC.

12.4 Summary

In this chapter, we discussed several aspects of the frontier research on high-performance big data computing. We mainly focused on discussing the promising designs and research directions for next-generation high-performance big data computing systems, runtimes, and applications on advanced computing, communication, and storage hardware devices. It is impossible to cover all possible aspects along this research line, however, we have demonstrated several important future research avenues in this field. We believe that with the continuous research effort from the community, more and more advanced and optimized designs and schemes will be proposed and adopted in next-generation HPC and data center clusters to build up more powerful and efficient high-performance big data computing platforms.

References

Abadi., Daniel J, Adam Marcus, Samuel R. Madden, and Kate Hollenbach. 2009. SW-Store: A Vertically Partitioned DBMS for Semantic Web Data Management. *VLDB Journal* 18 (2): 385–406.

Abadi, Martín, Ashish Agarwal, Paul Barham, Eugene Brevdo, Zhifeng Chen, Craig Citro, Gregory S. Corrado, Andy Davis, Jeffrey Dean, Matthieu Devin, et al. 2016. TensorFlow: Large-scale Machine Learning on Heterogeneous Distributed Systems. arXiv/*CoRR* abs/1603.04467. http://arxiv.org/abs/1603.04467.

Abadi, Martín, Paul Barham, Jianmin Chen, Zhifeng Chen, Andy Davis, Jeffrey Dean, Matthieu Devin, Sanjay Ghemawat, Geoffrey Irving, Michael Isard, et al. 2016. TensorFlow: A System for Large-scale Machine Learning. *OSDI* 16: 265–283.

Aerospike, Inc. 2021. Key-Value Operations. https://www.aerospike.com/docs/guide/kvs.html.

Ahmad, Faraz, Seyong Lee, Mithuna Thottethodi, and T. N. Vijaykumar. 2012. PUMA: Purdue MapReduce Benchmarks Suite. ttps://engineering.purdue.edu/~puma/puma.pdf

Aker, Brian. 2011. libMemcached. https://libmemcached.org/libMemcached.html.

Akinaga, Hiroyuki, and Hisashi Shima. 2010. Resistive Random Access Memory (ReRAM) Based on Metal Oxides. *Proceedings of the IEEE* 98: 2237–2251.

Alarcon, Nefi. 2019. GPU-accelerated Spark XGBoost—A Major Milestone on the Road to Large-Scale AI. https://developer.nvidia.com/blog/gpu-accelerated-sparkxgboost/.

Albutiu, Martina-Cezara, Alfons Kemper, and Thomas Neumann. 2012. Massively Parallel Sort-Merge Joins in Main Memory Multi-Core Database Systems. *Proceedings of the VLDB Endowment* 5 (10): 1064–1075.

Algo-Logic Systems. 2020. Low Latency KVS on Xilinx Alveo U200—Algo-Logic Systems Inc. https://web2020.algo-logic.com/kvs/Alveo.

Al-Kiswany, Samer, Abdullah Gharaibeh, and Matei Ripeanu. 2013. GPUs as Storage System Accelerators. *IEEE Transactions on Parallel and Distributed Systems* 24 (8): 1556–1566.

Allen, Grant, and Mike Owens. 2010. *The Definitive Guide to SQLite*. 2nd ed. Berkeley, CA: Apress.

Alok, Gupta. 2020. Architecture Apocalypse Dream Architecture for Deep Learning Inference and Compute—VERSAL AI Core. https://www.xilinx.com/support/documentation/white_papers/EW2020-Deep-Learning-Inference-AICore.pdf.

Alverson, Bob, Edwin Froese, Larry Kaplan, and Duncan Roweth. 2012. Cray® XC^{TM} Series Network-Cray. https://www.alcf.anl.gov/files/CrayXCNetwork.pdf.

Amazon, AWS. 2019. Best Practices Design Patterns: Optimizing Amazon S3 Performance. https://d1.awsstatic.com/whitepapers/AmazonS3BestPractices.pdf.

Amazon, AWS. 2021a. Amazon Elastic Block Store (EBS). https://aws.amazon.com/ebs/.

Amazon, AWS. 2021b. Amazon S3 REST API Introduction. https://docs.aws.amazon.com/AmazonS3/latest/API/Welcome.html.

Amazon, AWS. 2021c. Boto3 documentation. https://boto3.amazonaws.com/v1/documentation/api/latest/index.html.

Amazon, AWS. 2021d. Elastic Fabric Adapter. https://aws.amazon.com/cn/hpc/efa/.

Amazon, AWS. 2021e. Using High-level (S3) Commands with the AWS CLI. https://docs.aws.amazon.com/cli/latest/userguide/cli-services-s3-commands.html.

AMD. 2021a. AMD. https://www.amd.com.

AMD. 2021b. AMD Instinct™ MI Series Accelerators. https://www.amd.com/en/graphics/servers-radeon-instinct-mi.

AMD. 2021c. AMD Instinct™ MI100 Accelerator. https://www.amd.com/en/products/server-accelerators/instinct-mi100.

AMD. 2021d. AMD ROCm. https://rocmdocs.amd.com/en/latest/.

AMD. 2021e. AMD "Zen 3" Core Architecture. https://www.amd.com/en/technologies/zen-core-3.

AMPLab, UC Berkeley. 2021. Big Data Benchmark. https://amplab.cs.berkeley.edu/benchmark/.

Apache Flink. 2020. Accelerating Your Workload with GPU and Other External Resources. https://flink.apache.org/news/2020/08/06/external-resource.html.

Apache Software Foundation. 2016. Class MultithreadedMapper. https://hadoop.apache.org/docs/r2.6.5/api/org/apache/hadoop/mapreduce/lib/map/MultithreadedMapper.html.

Apache Software Foundation. 2019. Apache Crail. http://crail.incubator.apache.org/.

Apache Software Foundation. 2020a. HDFS Erasure Coding. https://hadoop.apache.org/docs/r3.3.0/hadoop-project-dist/hadoop-hdfs/HDFSErasureCoding.html.

Apache Software Foundation. 2020b. Memory Storage Support in HDFS. https://hadoop.apache.org/docs/r3.3.0/hadoop-project-dist/hadoop-hdfs/MemoryStorage.html.

Apache Software Foundation. 2020c. MapReduce Tutorial. https://hadoop.apache.org/docs/current/hadoop-mapreduce-client/hadoop-mapreduce-client-core/MapReduceTutorial.html.

Apache Software Foundation. 2021a. AffinityFunction (Ignite 2.10.0)—Apache Ignite. https://ignite.apache.org/releases/latest/javadoc/org/apache/ignite/cache/affinity/AffinityFunction.html.

Apache Software Foundation. 2021b. Apache Arrow. https://arrow.apache.org/.

Apache Software Foundation. 2021c. Apache Cassandra. https://cassandra.apache.org/.

Apache Software Foundation. 2021d. Apache Flink®—Stateful Computations over Data Streams. https://flink.apache.org.

Apache Software Foundation. 2021e. Apache Hadoop. https://hadoop.apache.org/.

Apache Software Foundation. 2021f. Welcome to Apache HBase™. https://hbase.apache.org/.

Apache Software Foundation. 2021g. Apache Heron (Incubating). https://incubator.apache.org/clutch/heron.html.

Apache Software Foundation. 2021h. Apache Hive™. https://hive.apache.org/.

Apache Software Foundation. 2021i. Apache Impala Is the Open Source, Native Analytic Database for Apache Hadoop. https://impala.apache.org.

Apache Software Foundation. 2021j. Apache Kafka. https://kafka.apache.org/.

Apache Software Foundation. 2021k. Apache Spark™ Is a Unified Analytics Engine for Large-scale Data Processing. https://spark.apache.org/.

Apache Software Foundation. 2021l. Apache Storm. http://storm.apache.org.

Apache Software Foundation. 2021m. Apache Zepplin. https://zeppelin.apache.org/.

Apache Software Foundation. 2021n. Apache Zepplin. https://zeppelin.apache.org.

Apache Software Foundation. 2021o. Apache ZooKeeper™. https://zookeeper.apache.org/.

Apache Software Foundation. 2021p. Apache Ignite: Distributed Database for High-performance Computing with In-memory Speed. https://ignite.apache.org/.

Apache Software Foundation. 2021q. Welcome To Apache Giraph! https://giraph.apache.org/.

Apache Software Foundation. 2021r. Spark SQL and DataFrames—Apache Spark. https://spark.apache.org/sql/.

Apache Software Foundation. 2021s. Spark SQL Is Apache Spark's Module for Working with Structured Data. https://spark.apache.org/sql/.

Apache Software Foundation. 2021t. Spark Streaming Makes It Easy to Build Scalable Fault-tolerant Streaming Applications. https://spark.apache.org/streaming/.

Apache Software Foundation. 2021u. Working with SQL. https://ignite.apache.org/docs/latest/SQL/sql-introduction.

Arafa, Mohamed, Bahaa Fahim, Sailesh Kottapalli, Akhilesh Kumar, Lily P. Looi, Sreenivas Mandava, Andy Rudoff, Ian M. Steiner, Bob Valentine, Geetha Vedaraman, et al. 2019. Cascade Lake: Next Generation Intel Xeon Scalable Processor. *IEEE Micro* 39 (2): 29–36.

Argonne National Lab. 2021. Aurora Argonne Leadership Computing Facility. https://www.alcf.anl.gov/aurora.

ARM. 2021a. Arm Neoverse V1 Platform: Unleashing a New Performance Tier for Arm-based Computing. https://community.arm.com/developer/ip-products/processors/b/processors-ip-blog/posts/neoverse-v1-platform-a-new-performance-tier-for-arm.

ARM. 2021b. High Performance Computing. https://www.arm.com/solutions/infrastructure/high-performance-computing.

Armstrong, Timothy G., Vamsi Ponnekanti, Dhruba Borthakur, and Mark Callaghan. 2013. LinkBench: A Database Benchmark Based on the Facebook Social Graph. *Proceedings of the 2013 ACM SIGMOD International Conference on Management of Data*, 1185–1196.

Arulraj, Joy, Andrew Pavlo, and Subramanya R. Dulloor. 2015. Let's Talk about Storage and Recovery Methods for Non-Volatile Memory Database Systems. *Proceedings of the 2015 ACM SIGMOD International Conference on Management of Data*, 707–722.

ASCAC Subcommittee on Exascale Computing, The. 2010. The Opportunities and Challenges of Exascale Computing. https://science.osti.gov/-/media/ascr/ascac/pdf/reports/Exascale_subcommittee_report.pdf

Atikoglu, Berk, Yuehai Xu, Eitan Frachtenberg, Song Jiang, and Mike Paleczny. 2012. Workload Analysis of a Large-scale Key-Value Store. *SIGMETRICS Performance Evaluation Review* 40 (1): 53–64.

Awan, A. A., C. Chu, H. Subramoni, X. Lu, and D. K. Panda. 2018. OC-DNN: Exploiting Advanced Unified Memory Capabilities in CUDA 9 and Volta GPUs for Out-of-Core DNN Training. *2018 IEEE 25th International Conference on High Performance Computing (HIPC)*, 143–152.

Bader, David A., John R. Gilbert, Jeremy Kepner, and Kamesh Madduri. 2021. HPC Graph Analysis. http://www.graphanalysis.org/index.html.

Baidu Research. 2021. DeepBench. https://github.com/baidu-research/DeepBench.

Bakkum, Peter, and Kevin Skadron. 2010. Accelerating SQL Database Operations on a GPU with CUDA. *Proceedings of the 3rd Workshop on General-purpose Computation on Graphics Processing Units*, 94–103.

Balakrishna, Vijay. 2016. Delivering on NoSQL Database Performance Requirements with NVMe SSDs (Samsung). *Proceedings of the 2016 Flash Memory Summit*. https://www.samsung.com/us/labs/pdfs/2016-fms-nosql-v1.pdf

Barthels, Claude, Simon Loesing, Gustavo Alonso, and Donald Kossmann. 2015. Rack-scale In-memory Join Processing Using RDMA. *Proceedings of the 2015 ACM SIGMOD International Conference on Management of Data*, 1463–1475.

Barthels, Claude, Ingo Müller, Timo Schneider, Gustavo Alonso, and Torsten Hoefler. 2017. Distributed Join Algorithms on Thousands of Cores. *Proceedings of the VLDB Endowment* 10 (5): 517–528.

Beckett, Dave, Matt Singer, Milind Damle, Rakesh Radhakrishnan, and Barrie Wheeler. 2018. Boosting Hadoop Performance and Cost Efficiency with Caching, Fast SSDs, and More Compute. https://www.intel.com/content/dam/www/central-libraries/us/en/documents/csp-twitter-white-paper.pdf.

Behrens, Tobias, Viktor Rosenfeld, Jonas Traub, Sebastian Breß, and Volker Markl. 2018. Efficient SIMD Vectorization for Hashing in OpenCL. In *the 21th International Conference on Extending Database Technology (EDBT)*. 489–492. https://openproceedings.org/2018/conf/edbt/paper-330.pdf

Beloglazov, Anton, and Rajkumar Buyya. 2010. Energy Efficient Allocation of Virtual Machines in Cloud Data Centers. *2010 10th IEEE/ACM International Conference on Cluster, Cloud and Grid Computing*, 577–578.

Bengio, Yoshua, Réjean Ducharme, Pascal Vincent, and Christian Jauvin. 2003. A Neural Probabilistic Language Model. *Journal of Machine Learning Research* 3 (Feb): 1137–1155.

Bent, John, Garth Gibson, Gary Grider, Ben McClelland, Paul Nowoczynski, James Nunez, Milo Polte, and Meghan Wingate. 2009. PLFS: A Checkpoint Filesystem for Parallel Applications. *Proceedings of the Conference on High Performance Computing Networking, Storage and Analysis, (SC '09). Association for Computing Machinery, New York, NY, USA, Article 21*, 1–12. DOI:https://doi.org/10.1145/1654059.1654081

Besta, Maciej, Dimitri Stanojevic, Johannes De Fine Licht, Tal Ben-Nun, and Torsten Hoefler. 2019. Graph Processing on FPGAs: Taxonomy, Survey, Challenges. *arXiv preprint. arXiv:1903.06697.* http://unixer.de/publications/img/graphs-fpgas-survey.pdf

Birrittella, Mark S., Mark Debbage, Ram Huggahalli, James Kunz, Tom Lovett, Todd Rimmer, Keith D. Underwood, and Robert C. Zak. 2015. Intel Omni-path Architecture: Enabling Scalable, High Performance Fabrics. In *Proceedings of the 2015 IEEE 23rd Annual Symposium on High-Performance Interconnects (HOTI '15).* 1–9. DOI:https://doi.org/10.1109/HOTI.2015.22

Bisong, Ekaba. 2019. Google Colaboratory. *Building Machine Learning and Deep Learning Modules on Google Cloud Platform*, 59–64. Berkeley, CA: Apress.

Bisson, Tim, Ke Chen, Changho Choi, Vijay Balakrishnan, and Yang-suk Kee. 2018. Crail-KV: A High-performance Distributed Key-Value Store Leveraging Native KV-SSDs over NVMe-oF. *2018 IEEE 37th International Performance Computing and Communications Conference (IPCCC)*, 1–8.

Biswas, R., X. Lu, and D. K. Panda. 2018. Accelerating TensorFlow with Adaptive RDMA-based GRPC. In *2018 IEEE 25th International Conference on High Performance Computing (HIPC)*, 2–11. https://doi.org/10.1109/HiPC.2018.00010.

Bitfusion. 2017. Deep Learning Frameworks with Spark and GPUs. https://on-demand.gputechconf.com/gtc/2017/presentation/s7737-subbu-rama-deep-learning-with-spark%20.pdf.

Blazegraph. 2021. Welcome to Blazegraph. https://blazegraph.com/.

BlazingSQL. 2021. blazingSQL—Open Source SQL in Python. https://blazingsql.com/.

Blott, Michaela, Kimon Karras, Ling Liu, Kees Vissers, Jeremia Bär, and Zsolt István. 2013. Achieving 10Gbps Line-rate Key-Value Stores with FPGAs. *5th USENIX Workshop on Hot Topics in Cloud Computing (HotCloud 13).* San Jose, CA: USENIX Association. https://www.usenix.org/conference/hotcloud13/workshop-program/presentations/blott.

Boito, Francieli Zanon, Eduardo C. Inacio, Jean Luca Bez, Philippe O. A. Navaux, Mario A. R. Dantas, and Yves Denneulin. 2018. A Checkpoint of Research on Parallel I/O for High-Performance Computing. *ACM Computing Surveys (CSUR)* 51 (2): 23.

Bostock, Mike. 2020. D3.js. https://d3js.org/.

Braam, Peter J., and Rumi Zahir. 2002. Lustre: A Scalable, High-Performance File System. *Cluster File Systems, Inc.* https://cse.buffalo.edu/faculty/tkosar/cse710/papers/lustre-whitepaper.pdf

Brytlyt. 2021. BrytlytDB. https://www.brytlyt.com/what-we-do/brytlytdb/.

Buluç, Aydin, Tim Mattson, Scott McMillan, José Moreira, and Carl Yang. 2017a. Design of the GraphBLAS API for C. *2017 IEEE International Parallel and Distributed Processing Symposium Workshops (IPDPSW)*, 643–652.

Buluç, Aydın, Timothy Mattson, Scott McMillan, José Moreira, and Carl Yang. 2019. The Graph-BLAS C API Specification. http://people.eecs.berkeley.edu/~aydin/GraphBLAS_API_C_v13.pdf

Canziani, Alfredo, Adam Paszke, and Eugenio Culurciello. 2016. An Analysis of Deep Neural Network Models for Practical Applications. *arXiv/CoRR* abs/1605.07678. http://arxiv.org/abs/1605.07678.

Cao, Wei, Zhenjun Liu, Peng Wang, Sen Chen, Caifeng Zhu, Song Zheng, Yuhui Wang, and Guoqing Ma. 2018. Polarfs: An Ultra-low Latency and Failure Resilient Distributed File System for Shared Storage Cloud Database. *Proceedings of the VLDB Endowment* 11 (12): 1849–1862.

Carbone, Paris, Asterios Katsifodimos, Stephan Ewen, Volker Markl, Seif Haridi, and Kostas Tzoumas. 2015. Apache Flink: Stream and Batch Processing in a Single Engine. *Bulletin of the IEEE Computer Society Technical Committee on Data Engineering* 38 (4). 28–38. http://sites.computer.org/debull/A15dec/p28.pdf.

Caulfield, Adrian M., Todor I. Mollov, Louis Alex Eisner, Arup De, Joel Coburn, and Steven Swanson. 2012. Providing Safe, User Space Access to Fast, Solid State Disks. *ACM SIGPLAN Notices* 47 (4): 387–400.

Ceph. 2021. Ceph Delivers Object, Block, and File Storage in a Single, Unified System. https://ceph.com/en/discover/technology/.

Cerebras. 2021a. Cerebras. https://cerebras.net.

Cerebras. 2021b. Cerebras Systems: Achieving Industry Best AI Performance through a Systems Approach. https://cerebras.net/wp-content/uploads/2021/04/Cerebras-CS-2-Whitepaper.pdf.

CGCL-codes. 2017. TensorFlow RDMA. https://github.com/CGCL-codes/Tensorflow-RDMA.

Chabot, C. 2009. Demystifying Visual Analytics. *IEEE Computer Graphics and Applications* 29 (2): 84–87. https://doi.org/10.1109/MCG.2009.23.

Chameleon. 2021. A Configurable Experimental Environment for Large-scale Edge to Cloud Research. https://www.chameleoncloud.org/.

Chandramouli, Badrish, Guna Prasaad, Donald Kossmann, Justin Levandoski, James Hunter, and Mike Barnett. 2018. Faster: A Concurrent Key-Value Store with In-place updates. *Proceedings of the 2018 International Conference on Management of Data*, 275–290.

Chen, Chen, Xianzhi Du, Le Hou, Jaeyoun Kim, Jing Li, Yeqing Li, Abdullah Rashwan, Fan Yang, and Hongkun Yu. 2020. TensorFlow official model garden. https://github.com/tensorflow/models/tree/master/official.

Chen, Tianqi, Mu Li, Yutian Li, Min Lin, Naiyan Wang, Minjie Wang, Tianjun Xiao, Bing Xu, Chiyuan Zhang, and Zheng Zhang. 2015. Mxnet: A Flexible and Efficient Machine Learning Library for Heterogeneous Distributed Systems. *arXiv* preprint. arXiv:1512.01274.

Chen, Tianqi, Thierry Moreau, Ziheng Jiang, Lianmin Zheng, Eddie Yan, Haichen Shen, Meghan Cowan, Leyuan Wang, Yuwei Hu, Luis Ceze, et al. 2018. TVM: An Automated End-to-End Optimizing Compiler for Deep Learning. *13th USENIX Symposium on Operating Systems Design and Implementation (OSDI 18)*, 578–594. https://www.usenix.org/conference/osdi18/presentation/chen.

Chen, Wei, Aidi Pi, Shaoqi Wang, and Xiaobo Zhou. 2019. Pufferfish: Container-driven Elastic Memory Management for Data-intensive Applications. *SoCC '19: Proceedings of the ACM Symposium on Cloud Computing*, 259–271. https://doi.org/10.1145/3357223.3362730.

Chen, Xinyu, Hongshi Tan, Yao Chen, Bingsheng He, Weng-Fai Wong, and Deming Chen. 2021. ThunderGP: HLS-based Graph Processing Framework on FPGAs. *The 2021 ACM/SIGDA International Symposium on Field-programmable Gate Arrays*, 69–80.

Chen, Xue-wen, and Xiaotong Lin. 2014. Big Data Deep Learning: Challenges and Perspectives. *Access, IEEE* 2: 514–525. https://doi.org/10.1109/ACCESS.2014.2325029.

Chen, Youmin, Jiwu Shu, Jiaxin Ou, and Youyou Lu. 2018. HiNFS: A Persistent Memory File System with Both Buffering and Direct-access. *ACM Transactions on Storage (TOS)* 14 (1): 1–30.

Chen, Yu-Ting, Jason Cong, Zhenman Fang, Jie Lei, and Peng Wei. 2016. When Spark Meets FPGAs: A Case Study for Next-generation {DNA} Sequencing Acceleration. *8th USENIX Workshop on Hot Topics in Cloud Computing (HotCloud)*. USENIX Association. https://www.usenix.org/system/files/conference/hotcloud16/hotcloud16_chen.pdf.

Cheng, Wang. 2019. APUS: Fast and Scalable Paxos on RDMA. https://github.com/hku-systems/apus.

Ching, Avery, Sergey Edunov, Maja Kabiljo, Dionysios Logothetis, and Sambavi Muthukrishnan. 2015. One Trillion Edges: Graph Processing at Facebook-Scale. *Proceedings of the VLDB Endowment* 8 (12): 1804–1815.

Chintapalli, Sanket, Derek Dagit, Bobby Evans, Reza Farivar, Thomas Graves, Mark Holderbaugh, Zhuo Liu, Kyle Nusbaum, Kishorkumar Patil, Boyang Jerry Peng, and Paul Poulosky. 2016. Benchmarking Streaming Computation Engines: Storm, Flink and Spark Streaming. *2016 IEEE International Parallel and Distributed Processing Symposium Workshops (IPDPSW)*, 1789–1792. https://doi.org/10.1109/IPDPSW.2016.138.

Chintapalli, Sanket, Derek Dagit, Bobby Evans, Reza Farivar, Thomas Graves, Mark Holderbaugh, Zhuo Liu, Kyle Nusbaum, Kishorkumar Patil, Boyang Jerry Peng, and Paul Poulosky. 2021. Yahoo Streaming Benchmarks. https://github.com/yahoo/streaming-benchmarks.

Cho, Kyunghyun, Bart Van Merriënboer, Caglar Gulcehre, Dzmitry Bahdanau, Fethi Bougares, Holger Schwenk, and Yoshua Bengio. 2014. Learning Phrase Representations Using RNN Encoder-Decoder for Statistical Machine Translation. arXiv preprint. arXiv:1406.1078.

Cho, Minsik, Ulrich Finkler, David Kung, and Hillery Hunter. 2019. BlueConnect: Decomposing All-reduce for Deep Learning on Heterogeneous network Hierarchy. *Proceedings of Machine Learning and Systems* 1: 241–251. https://proceedings.mlsys.org/paper/2019/file/9b8619251a19057cff70779273e95aa6-Paper.pdf.

Chu, Chengtao, Sang K. Kim, Yi an Lin, Yuanyuan Yu, Gary Bradski, Kunle Olukotun, and Andrew Y. Ng. 2007. Map-Reduce for Machine Learning on Multicore. In *Advances in Neural Information Processing Systems 19: Proceedings of the 2006 Conference.* eds. B. Schölkopf, J. C. Platt, and T. Hoffman, 281–288. Cambridge, MA: MIT Press. https://proceedings.neurips.cc/paper/2006/file/77ee3bc58ce560b86c2b59363281e914-Paper.pdf.

Chu, Ching-Hsiang, Sreeram Potluri, Anshuman Goswami, Manjunath Venkata, Neena Inam, and Chris J. Newburn. 2018. Designing High-performance In-memory Key-Value Operations with Persistent GPU Kernels and OpenSHMEM. In *Workshop on OpenSHMEM and Related Technologies (OpenSHMEM).* 148–164. Springer International Publishing. https://link.springer.com/chapter/10.1007/978-3-030-04918-8_10.

Chu, Howard. 2011. MDB: A Memory-Mapped Database and Backend for OpenLDAP. http://www.lmdb.tech/media/20111010LDAPCon%20MDB.pdf.

CloudSuite Team. 2021. CloudSuite: A Benchmark Suite for Cloud Services. https://www.cloudsuite.ch/.

Collobert, Ronan, Cl'ement Farabet, Koray Kavukcuoglu, and Soumith Chintala. 2021. Torch—Scientific Computing for LuaJIT. http://torch.ch/.

Collobert, Ronan, Samy Bengio, and Johnny Marithoz. 2002. Torch: A Modular Machine Learning Software Library. IDIAP Research Report. http://publications.idiap.ch/downloads/reports/2002/rr02-46.pdf.

Condit, Jeremy, Edmund B. Nightingale, Christopher Frost, Engin Ipek, Benjamin Lee, Doug Burger, and Derrick Coetzee. 2009. Better I/O through Byte-addressable, Persistent Memory. *Proceedings of the ACM SIGOPS 22nd Symposium on Operating Systems Principles*, 133–146.

Convolbo, Moïse W., and Jerry Chou. 2016. Cost-aware DAG Scheduling Algorithms for Minimizing Execution Cost on Cloud Resources. *Journal of Supercomputing* 72 (3): 985–1012.

Cooper, B. F., A. Silberstein, E. Tam, R. Ramakrishnan, and R. Sears. 2010. Benchmarking Cloud Serving Systems with YCSB. In *The Proceedings of the ACM Symposium on Cloud Computing (SoCC '10).* Association for Computing Machinery, New York, NY, USA, 143–154. DOI:https://doi.org/10.1145/1807128.1807152

Cornelis. 2020. Cornelis Networks. https://www.cornelisnetworks.com/.

Cray. 2021. HPE. 2021. XC^{TM} Series DataWarp^{TM} User Guide. https://www.hpe.com/psnow/doc/a00114122en_us.

Crego, E., G. Munoz, and F. Islam. 2013. Big Data and Deep Learning: Big Deals or Big Delusions? http://www.huffingtonpost.com/george-munoz-frank-islam-and-ed-crego/big-data-and-deep-learnin_b_3325352.html.

CSCS, Swiss National Supercomputing Centre. 2021. Piz Daint. https://www.cscs.ch/computers/piz-daint/.

Dagum, Leonardo, and Ramesh Menon. 1998. OpenMP: An Industry Standard API for Shared-Memory Programming. *IEEE Computational Science and Engineering* 5 (1): 46–55.

Dai, Jason (Jinquan), Yiheng Wang, Xin Qiu, Ding Ding, Yao Zhang, Yanzhang Wang, Xianyan Jia, Li (Cherry) Zhang, Yan Wan, Zhichao Li, et al. 2019. BigDL: A Distributed Deep Learning Framework for Big Data. *Proceedings of the ACM Symposium on Cloud Computing. SoCC '19*, 50–60. https://doi.org/10.1145/3357223.3362707.

Dalessandro, Dennis, Ananth Devulapalli, and Pete Wyckoff. 2005. Design and Implementation of the iWarp Protocol in Software. In *The IASTED International Conference on Parallel and Distributed Computing and Systems (PDCS).* 471–476.

Dandu, Satish Varma, Chinmay Chandak, Jarod Maupin, and Jeremy Dyer. 2020. cuStreamz: More Event Stream Processing for Less with NVIDIA GPUs and RAPIDS Software. https://medium.com/rapids-ai/gpu-accelerated-streamprocessing-with-rapids-f2b725696a61.

DataBricks. 2018. TensorFrames. https://github.com/databricks/tensorframes.

Databricks. 2021. Collaborative Notebooks: Collaborative Data Science with Familiar Languages and Tools. https://databricks.com/product/collaborative-notebooks.

Datadog. 2015. Monitor Cassandra with Datadog. https://www.datadoghq.com/blog/monitoring-cassandra-with-datadog/.

Datadog. 2021. DATADOG—Unified Monitoring for the cloud age. https://www.datadoghq.com/unified-monitoring/.

DataMPI Team. 2021. DataMPI: Extending MPI for Big Data with Key-Value based Communication. http://datampi.org/.

Davis, Timothy A. 2019. Algorithm 1000: SuiteSparse:GraphBLAS: Graph Algorithms in the Language of Sparse Linear Algebra. *ACM Transactions on Mathematical Software* 45 (4): article 44.

DDN. 2021. Infinite Memory Engine: Break Free from the Challenges and Inefficiencies Caused by I/O Bottlenecks. https://www.ddn.com/products/ime-flash-native-data-cache/.

Dean, Jeffrey, and Sanjay Ghemawat. 2008. MapReduce: Simplified Data Processing on Large Clusters. *Communications of the ACM* 51 (1): 107–113.

Deep500 Team. 2021. Deep500: An HPC Deep Learning Benchmark and Competition. https://www.deep500.org/.

Deepnote. 2021. Deepnote. https://deepnote.com/.

DeepSpeech Team. 2021. Project DeepSpeech. https://github.com/mozilla/DeepSpeech.

Deng, Jia, Wei Dong, Richard Socher, Li-Jia Li, Kai Li, and Fei-Fei Li. 2009. ImageNet: A Large-Scale Hierarchical Image Database. In *2009 IEEE conference on computer vision and pattern recognition* (pp. 248–255). https://ieeexplore.ieee.org/document/5206848.

Department of Energy, U.S. 2011. Terabits Networks for Extreme Scale Science. DOE Workshop Report. https://science.osti.gov/-/media/ascr/pdf/program-documents/docs/Terabit_networks_workshop_report.pdf.

Difallah, Djellel Eddine, Andrew Pavlo, Carlo Curino, and Philippe Cudre-Mauroux. 2013. OLTP-Bench: An Extensible Testbed for Benchmarking Relational Databases. *Proceedings of the VLDB Endowment* 7 (4): 277–288.

Doddamani, Spoorti, Piush Sinha, Hui Lu, Tsu-Hsiang K. Cheng, Hardik H. Bagdi, and Kartik Gopalan. 2019. Fast and Live Hypervisor Replacement. *VEE 2019: Proceedings of the 15th ACM SIGPLAN/SIGOPS International Conference on Virtual Execution Environments* 45–58. https://doi.org/10.1145/3313808.3313821.

Domo. 2021. Data Never Sleeps 8.0. https://www.domo.com/learn/data-never-sleeps-8.

Dong, Yaozu, Jinquan Dai, Zhiteng Huang, Haibing Guan, Kevin Tian, and Yunhong Jiang. 2009. Towards High-quality I/O Virtualization. *Proceedings of SYSTOR 2009: The Israeli Experimental Systems Conference. (SYSTOR '09). Association for Computing Machinery, New York, NY, USA, Article 12, 1–8. DOI:https://doi.org/10.1145/1534530.1534547.*

Dormando. 2021. What is Memcached? http://memcached.org/.

Douglas, Chet. 2015. RDMA with PMEM Software Mechanisms for Enabling Access to Remote Persistent Memory. http://www.snia.org/sites/default/files/SDC15_presentations/persistant_mem/ChetDouglas_RDMA_with_PM.pdf.

Doweck, J., W. Kao, A. K. Lu, J. Mandelblat, A. Rahatekar, L. Rappoport, E. Rotem, A. Yasin, and A. Yoaz. 2017. Inside 6th-Generation Intel Core: New Microarchitecture Code-Named Skylake. *IEEE Micro* 37 (2): 52–62.

Dragojević, A., D. Narayanan, M. Castro, and O. Hodson. 2014. FaRM: Fast Remote Memory. *11th USENIX Symposium on Networked Systems Design and Implementation (NSDI 14)*, 401–414.

Duato, José, Antonio J. Pena, Federico Silla, Rafael Mayo, and Enrique S Quintana-Ortí. 2010. rCUDA: Reducing the Number of GPU-based Accelerators in High Performance Clusters. *2010 International Conference on High Performance Computing and Simulation*, 224–231.

Dulloor, Subramanya R., Sanjay Kumar, Anil Keshavamurthy, Philip Lantz, Dheeraj Reddy, Rajesh Sankaran, and Jeff Jackson. 2014. System Software for Persistent Memory. *Proceedings of the Ninth European Conference on Computer Systems*, 15.

Dynatrace. 2021a. Apache Spark monitoring. https://www.dynatrace.com/technologies/apache-spark-monitoring/.

Dynatrace. 2021b. Dynatrace. https://www.dynatrace.com/.

Dynatrace. 2021c. Hadoop Performance Monitoring. https://www.dynatrace.com/technologies/hadoop-monitoring/.

Dysart, Timothy, Peter Kogge, Martin Deneroff, Eric Bovell, Preston Briggs, Jay Brockman, Kenneth Jacobsen, Yujen Juan, Shannon Kuntz, Richard Lethin, et al. 2016. Highly Scalable Near Memory Processing with Migrating Threads on the Emu System Architecture. *2016 6th Workshop on Irregular Applications: Architecture and Algorithms (IA3)*, 2–9. https://doi.org/10.1109/IA3.2016.007.

E8 Storage. 2021. E8 Storage E8-D24 Rack Scale Flash, Centralized NVMe Solution. https://nvmexpress.org /portfolio-items/e8-storage-e8-d24-rack-scale-flash-centralized-nvme-solution/.

Eisenman, Assaf, Darryl Gardner, Islam AbdelRahman, Jens Axboe, Siying Dong, Kim Hazelwood, Chris Petersen, Asaf Cidon, and Sachin Katti. 2018. Reducing DRAM Footprint with NVM in Facebook. *Eurosys'18*, 42.

Elangovan, Aparna. 2020. Optimizing I/O for GPU Performance Tuning of Deep Learning Training in Amazon SageMaker. https://aws.amazon.com/blogs/machine-learning/optimizing-i-o-for-gpu-performance-tuning-of -deep-learning-training-in-amazon-sagemaker/.

Elasticsearch B.V. 2021. Elasticsearch: The Heart of the Free and Open Elastic Stack. https://www.elastic.co /elasticsearch/.

Emani, Murali, Venkatram Vishwanath, Corey Adams, Michael E. Papka, Rick Stevens, Laura Florescu, Sumti Jairath, William Liu, Tejas Nama, and Arvind Sujeeth. 2021. Accelerating Scientific Applications with SambaNova Reconfigurable Dataflow Architecture. *Computing in Science Engineering* 23 (2): 114–119.

Engel, Jörn, and Robert Mertens. 2005. LogFS—Finally a Scalable Flash File System. *Proceedings of the 12th International Linux System Technology Conference*. https://citeseerx.ist.psu.edu/viewdoc/download?doi=10.1.1. 83.1239&rep=rep1&type=pdf.

Facebook. 2018. RocksDB. https://rocksdb.org/.

Facebook AI Team. 2021. Facebook AI Performance Evaluation Platform. https://github.com/facebook/FAI-PEP.

Fan, Bin, David G. Andersen, and Michael Kaminsky. 2013. MemC3: Compact and Concurrent MemCache with Dumber Caching and Smarter Hashing. *10th USENIX Symposium on Networked Systems Design and Implementation (NSDI 13)*, 371–384.

Fan, Ziqi, Fenggang Wu, Jim Diehl, David H. C. Du, and Doug Voigt. 2018. CDBB: An NVRAM-based Burst Buffer Coordination System for Parallel File Systems. *Proceedings of the High Performance Computing Symposium*, 1.

FASTDATA.io. 2021. PlasmaENGINE®. https://fastdata.io/plasma-engine/.

Fent, Philipp, Alexander van Renen, Andreas Kipf, Viktor Leis, Thomas Neumann, and Alfons Kemper. 2020. Low-Latency Communication for Fast DBMS Using RDMA and Shared Memory. *2020 IEEE 36th International Conference on Data Engineering (ICDE)*, 1477–1488. https://doi.org/10.1109/ICDE48307.2020.00131.

Fielding, Roy Thomas. 2000. Chapter 5: Representational State Transfer (REST). *Architectural Styles and the Design of Network-based Software Architectures*. PhD dissertation, University of California, Irvine.

Fikes, Andrew. 2010. Storage Architecture and Challenges. *Google Faculty Summit*, 535.

Foley, D., and J. Danskin. 2017. Ultra-Performance Pascal GPU and NVLink Interconnect. *IEEE Micro* 37 (02): 7–17.

Fujitsu. 2021. FUJITSU Processor A64FX. https://www.fujitsu.com/global/products/computing/servers/super computer/a64fx/.

Garrigues, Pierre. 2015. How Deep Learning Powers Flickr. *RE.WORK Deep Learning Summit 2015*. http://bit .ly/1KIDfof.

Ghasemi, E. and Chow, P., 2016, July. Accelerating Apache Spark Big Data Analysis with FPGAs. *2016 International IEEE Conferences on Ubiquitous Intelligence & Computing, Advanced and Trusted Computing, Scalable Computing and Communications, Cloud and Big Data Computing, Internet of People, and Smart World Congress (UIC/ATC/ScalCom/CBDCom/IoP/SmartWorld)*. (pp. 737–744). IEEE.

Ghemawat, Sanjay, Howard Gobioff, and Shun-Tak Leung. 2003. The Google File System. *Proceedings of the 19th ACM Symposium on Operating Systems Principles*, 20–43.

Goldenberg, Dror, Michael Kagan, Ran Ravid, and Michael S. Tsirkin. 2005. Transparently Achieving Superior Socket Performance Using Zero Copy Socket Direct Protocol over 20 Gb/s InfiniBand Links. *2005 IEEE International Conference on Cluster Computing (Cluster)*, 1–10.

Gonzalez, Joseph E., Reynold S. Xin, Ankur Dave, Daniel Crankshaw, Michael J. Franklin, and Ion Stoica. 2014. Graphx: Graph Processing in a Distributed Dataflow Framework. *OSDI*, 599–613.

Google. 2010. Our New Search Index: Caffeine. https://googleblog.blogspot.com/2010/06/our-new-search-index -caffeine.html.

Google. 2021a. Cloud TPU. https://cloud.google.com/tpu/.

Google. 2021b. TensorFlow. https://www.tensorflow.org/.

Gottschling, Paul. 2019. Monitor Apache Hive with Datadog. https://www.datadoghq.com/blog/hive/.

Goudarzi, Hadi, Mohammad Ghasemazar, and Massoud Pedram. 2012. SLA-based Optimization of Power and Migration Cost in Cloud Computing. *12th IEEE/ACM International Symposium on Cluster, Cloud and Grid Computing, CCGrid 2012*. https://doi.org/10.1109/CCGrid.2012.112.

Govindaraju, N. K., B. He, Q. Luo, and W. Fang. 2011. Mars: Accelerating MapReduce with Graphics Processors. *IEEE Transactions on Parallel and Distributed Systems* 22 (04): 608–620. https://doi.org/10.1109/TPDS.2010. 158.

Govindaraju, Naga K., Brandon Lloyd, Wei Wang, Ming Lin, and Dinesh Manocha. 2004. Fast Computation of Database Operations Using Graphics Processors. *Proceedings of the 2004 ACM SIGMOD International Conference on Management of Data. SIGMOD '04*, 215–226. https://doi.org/10.1145/1007568.1007594.

Graham, Richard L., Devendar Bureddy, Pak Lui, Hal Rosenstock, Gilad Shainer, Gil Bloch, Dror Goldenerg, Mike Dubman, Sasha Kotchubievsky, Vladimir Koushnir, et al. 2016. Scalable Hierarchical Aggregation Protocol (SHArP): A Hardware Architecture for Efficient Data Reduction. *Proceedings of the First Workshop on Optimization of Communication in HPC*, 1–10.

Graph500. http://www.graph500.org.

Graphcore. 2021. https://www.graphcore.ai.

Greeneitch, Nathan G., Jing Xu, and Shailendrsingh Kishore Sobhee. 2019. Getting Started with Intel® Optimization for PyTorch. https://software.intel.com/content/www/us/en/develop/articles/getting-started-with -inteloptimization-of-pytorch.html.

Groupon. 2017. Sparklint. https://github.com/groupon/sparklint.

gRPC Authors. 2021. gRPC: A High Performance, Open Source Universal RPC Framework. https://grpc.io/

Gubner, Tim, and Peter A. Boncz. 2017. Exploring Query Compilation Strategies for JIT, Vectorization and SIMD. *Eighth International Workshop on Accelerating Analytics and Data Management Systems Using Modern Processor and Storage Architectures (ADMS)*. Vol. 2. http://adms-conf.org/2017/camera-ready/exploring-query-execution. pdf.

Gugnani, Shashank, Xiaoyi Lu, and Dhabaleswar K. Panda. 2016. Designing Virtualization-aware and Automatic Topology Detection Schemes for Accelerating Hadoop on SR-IOV-Enabled Clouds. *2016 IEEE International Conference on Cloud Computing Technology and Science (CloudCom)*, 152–159.

Gugnani, Shashank, Xiaoyi Lu, and Dhabaleswar K. (DK) Panda. 2017. Swift-X: Accelerating OpenStack Swift with RDMA for Building an Efficient HPC Cloud. *Proceedings of the 17th IEEE/ACM International Symposium on Cluster, Cloud and Grid Computing. CCGRID '17*, 238–247. https://doi.org/10.1109/CCGRID.2017 .103.

Gugnani, Shashank, Xiaoyi Lu, and Dhabaleswar K. Panda. 2018. Analyzing, Modeling, and Provisioning QoS for NVMe SSDs. *2018 IEEE/ACM 11th International Conference on Utility and Cloud Computing (UCC)*, 247–256.

Guo, Fan, Yongkun Li, Min Lv, Yinlong Xu, and John C. S. Lui. 2019. Hp-Mapper: A High Performance Storage Driver for Docker Containers. *Proceedings of the ACM Symposium on Cloud Computing*, 325–336. https://doi. org/10.1145/3357223.3362718.

Gupta, K., J. A. Stuart, and J. D. Owens. 2012. A Study of Persistent Threads Style GPU Programming for GPGPU Workloads. *2012 Innovative Parallel Computing (InPar)*, 1–14.

Gupta, Vishakha, Ada Gavrilovska, Karsten Schwan, Harshvardhan Kharche, Niraj Tolia, Vanish Talwar, and Parthasarathy Ranganathan. 2009. GViM: GPU-accelerated Virtual Machines. *Proceedings of the 3rd ACM Workshop on System-level Virtualization for High Performance Computing (HPCVirt '09). Association for Computing Machinery, New York, NY, USA*, 17–24. DOI:https://doi.org/10.1145/1519138.1519141

Gurajada, Sairam, Stephan Seufert, Iris Miliaraki, and Martin Theobald. 2014. TriAD: A Distributed Shared-nothing RDF Engine Based on Asynchronous Message Passing. *Proceedings of the 2014 ACM SIGMOD International Conference on Management of Data. SIGMOD '14*, 289–300.

Guz, Zvika, Harry Huan Li, Anahita Shayesteh, and Vijay Balakrishnan. 2017. NVMe-over-Fabrics Performance Characterization and the Path to Low-Overhead Flash Disaggregation. *Proceedings of the 10th ACM International Systems and Storage Conference, SYSTOR '17*, 16.

Habana Gaudi. 2019. Gaudi^TM Training Platform White Paper. https://habana.ai/wp-content/uploads/2019/06 /Habana-Gaudi-Training-Platform-whitepaper.pdf.

Habana Goya. 2019. Goya^TM Inference Platform White Paper. https://habana.ai/wp-content/uploads/pdf/habana_ labs_goya_whitepaper.pdf.

Hamilton, Mark, Sudarshan Raghunathan, Akshaya Annavajhala, Danil Kirsanov, Eduardo de Leon, Eli Barzilay, Ilya Matiach, Joe Davison, Maureen Busch, Miruna Oprescu, et al. 2018. Flexible and Scalable Deep Learning with MMLSpark. *arXiv* preprint. arXiv:1804.04031.

Handy, Jim. 2015. Understanding the Intel/Micron 3D XPoint Memory. In 2015 Storage Developer Conference *(SDC)* [Presentation]. 68. https://www.snia.org/sites/default/files/SDC15_presentations/persistant_mem /JimHandy_Understanding_the-Intel.pdf.

Harris, Derrick. 2015. Google, Stanford Say Big Data is Key to Deep Learning for Drug Discovery. https://gigaom. com/2015/03/02/google-stanford-say-big-data-is-key-to-deep-learning-for-drug-discovery.

Harris, Mark. 2017. Unified Memory for CUDA Beginners. https://developer.nvidia.com/blog/unified-memory -cuda-beginners/.

Harzog, Bernd. 2019. Modern Applications Require Modern APM Solutions: A SolarWinds APM Suite White-paper. https://www.solarwinds.com/-/media/solarwinds/swresources/datasheet/modern-apps-require-moderm -apm-solutions_apm-experts_whitepaper.ashx.

He, Kaiming, Xiangyu Zhang, Shaoqing Ren, and Jian Sun. 2016. Deep Residual Learning for Image Recognition. *Proceedings of the IEEE Conference on Computer Vision and Pattern Recognition*, 770–778.

He, Wenting, Huimin Cui, Binbin Lu, Jiacheng Zhao, Shengmei Li, Gong Ruan, Jingling Xue, Xiaobing Feng, Wensen Yang, and Youliang Yan. 2015. Hadoop+ Modeling and Evaluating the Heterogeneity for Map-Reduce Applications in Heterogeneous Clusters. *Proceedings of the 29th ACM International Conference on Supercomputing*, 143–153.

Henseler, Dave, Benjamin Landsteiner, Doug Petesch, Cornell Wright, and Nicholas J. Wright. 2016. Architecture and Design of Cray DataWarp. *Cray User Group CUG*. https://cug.org/proceedings/cug2016_proceedings /includes/files/pap105s2-file1.pdf.

Herodotou, Herodotos, Harold Lim, Gang Luo, Nedyalko Borisov, Liang Dong, Fatma Bilgen Cetin, and Shivnath Babu. 2011. Starfish: A Self-tuning System for Big Data Analytics. *Proceedings of the Fifth Biennial Conference on Innovative Data Systems Research (CIDR)* 11: 261–272. www.cidrdb.org.

HeteroDB. 2021. PG-Strom. http://heterodb.github.io/pg-strom/.

Hetherington, Tayler H., Mike O'Connor, and Tor M. Aamodt. 2015. Memcachedgpu: Scaling-up Scale-out Key-Value Stores. *Socc '15: Proceedings of the Sixth ACM Symposium on Cloud Computing*, 43–57. https://doi.org /10.1145/2806777.2806836.

Hetherington, T. H., T. G. Rogers, L. Hsu, M. O'Connor, and T. M. Aamodt. 2012. Characterizing and Evaluating a Key-Value Store Application on Heterogeneous CPU-GPU Systems. *2012 IEEE International Symposium on Performance Analysis of Systems Software*, 88–98.

Hochreiter, Sepp, and Jürgen Schmidhuber. 1997. Long Short-term Memory. *Neural Computation* 9 (8): 1735–1780.

Howard, Andrew G., Menglong Zhu, Bo Chen, Dmitry Kalenichenko, Weijun Wang, Tobias Weyand, Marco Andreetto, and Hartwig Adam. 2017. Mobilenets: Efficient Convolutional Neural Networks for Mobile Vision Applications. *arXiv/CoRR* abs/1704.04861. http://arxiv.org/abs/1704.04861.

Huang, J., K. Schwan, and M. K. Qureshi. 2014. NVRAM-aware Logging in Transaction Systems. *Proceedings of the VLDB Endowment* 8 (4) *(December 2014)*, 389–400. https://doi.org/10.14778/2735496.2735502.

Huang, Ting-Chang, and Da-Wei Chang. 2016. TridentFS: A Hybrid File System for Non-volatile RAM, Flash Memory and Magnetic Disk. *Software: Practice and Experience* 46 (3): 291–318.

Huang, Yihe, Matej Pavlovic, Virendra Marathe, Margo Seltzer, Tim Harris, and Steve Byan. 2018. Closing the Performance Gap between Volatile and Persistent Key-Value Stores Using Cross-referencing Logs. *2018 USENIX Annual Technical Conference (USENIX ATC 18)*, 967–979.

Hyper. 2021. HyPer—A Hybrid OLTP&OLAP High Performance DBMS. https://hyper-db.de/.

Iandola, Forrest N., Song Han, Matthew W. Moskewicz, Khalid Ashraf, William J. Dally, and Kurt Keutzer. 2016. SqueezeNet: AlexNet-level Accuracy with 50x Fewer Parameters and 0.5 MB Model Size. *arXiv* preprint. arXiv:1602.07360.

IBM. 2018. ibmgraphblas. https://github.com/IBM/ibmgraphblas.

IBM. 2020. IBM Reveals Next-generation IBM POWER10 Processor. https://newsroom.ibm.com/2020-08-17 -IBM-Reveals-Next-Generation-IBM-POWER10-Processor.

IBM. 2021. IBM General Parallel File System (GPFS) Product Documentation. https://www.ibm.com/support /knowledgecenter/SSFKCN/gpfs_content.html.

IBMSparkGPU. 2016. SparkGPU. https://github.com/IBMSparkGPU/SparkGPU.

IBTA, Inifiniband Trade Association. 2021a. Infiniband Trade Association. https://www.infinibandta.org/.

IBTA, Inifiniband Trade Association. 2021b. RoCE Is RDMA over Converged Ethernet. http://www.roceinitiative .org/.

Infiniband Trade Association. 2010. Supplement to Infiniband Architecture Specification Volume 1, Release 1.2. 1: Annex A16: RDMA over Converged Ethernet (RoCE) Apr.

insideHPC. 2021. Microchip Technology Inc.: Introducing First PCI Express 5.0 Switches. https://insidehpc.com /2021/02/microchip-technology-inc-introducing-first-pci-express-5-0-switches/.

Intel. 2012. Intel® Data Direct I/O Technology (Intel® DDIO): A Primer. Technical report, Intel - Technical brief. https://www.intel.com/content/dam/www/public/us/en/documents/technology-briefs/data-direct-i-o -technology-brief.pdf.

Intel. 2015a. Linux-pmfs/pmfs: Persistent Memory File System. https://github.com/linux-pmfs/pmfs.

Intel. 2015b. Performance Benchmarking for PCIe and NVMe Enterprise Solid-State Drives. White paper. https://www.intel.com/content/dam/www/public/us/en/documents/white-papers/performance-pcie-nvme-enter prise-ssds-white-paper.pdf.

Intel. 2017. Intel SPDK. https://www.spdk.io/.

Intel. 2019a. Cascade Lake. https://www.intel.com/content/www/us/en/design/products-and-solutions/processors -and-chipsets/cascade-lake/2nd-gen-intel-xeon-scalable-processors.html.

Intel. 2019b. Intel® Distribution of Caffe. https://github.com/intel/caffe.

Intel. 2019c. Intel Unveils New GPU Architecture with High-performance Computing and AI Acceleration, and oneAPI Software Stack with Unified and Scalable Abstraction for Heterogeneous Architectures. https://newsroom .intel.com/news-releases/intel-unveils-new-gpu-architecture-optimized-for-hpc-ai-oneapi/.

Intel. 2019d. Next-generation Intel Xeon Scalable Processors to Deliver Breakthrough Platform Performance with up to 56 Processor Cores. https://newsroom.intel.com/news/next-generation-intel-xeon-scalable-processors -deliver-breakthrough-platform-performance-56-processor-cores/.

Intel. 2019e. PMDK: Persistent Memory Development Kit. https://github.com/pmem/pmdk/.

Intel. 2020a. HiBench Suite: The BigData Micro Benchmark Suite. https://github.com/intel-hadoop/HiBench.

Intel. 2020b. Intel Announces Its Next Generation Memory and Storage Products. https://newsroom.intel.com /news/next-generation-memory-storage-products/#gs.w7bhtu.

Intel. 2020c. Intel Unpacks Architectural Innovations and Reveals New Transistor Technology at Architecture Day 2020. https://newsroom.intel.com/wp-content/uploads/sites/11/2020/08/intel-2020-architecture-day-fact -sheet.pdf.

Intel. 2021a. 24. Hash Library. Data Plane Development Kit 21.05.0 documentation. https://doc.dpdk.org/guides /prog_guide/hash_lib.html.

Intel. 2021b. AHCI Specification for Serial ATA. https://www.intel.com/content/www/us/en/io/serial-ata/ahci .html.

Intel. 2021c. Intel® Advanced Vector Extensions 512 (Intel® AVX-512). https://www.intel.com/content/www /us/en/architecture-and-technology/avx-512-overview.html.

Intel. 2021d. Intel AI Hardware. https://www.intel.ai/intel-nervana-neural-network-processors-nnp-redefine-ai -silicon/.

Intel. 2021e. The Intel Intrinsics Guide. https://software.intel.com/sites/landingpage/IntrinsicsGuide/.

Intel. 2021f. Intel® oneAPI Math Kernel Library. https://software.intel.com/content/www/us/en/develop/tools /oneapi/components/onemkl.html.

Intel. 2021g. Intel® Xeon® Processors. https://www.intel.com/content/www/us/en/products/details/processors /xeon.html.

Intel. 2021h. oneAPI Deep Neural Network Library (oneDNN). https://github.com/oneapi-src/oneDNN.

Intel. 2021i. Restricted Transactional Memory Overview. Intel® C++ Compiler Classic Developer Guide and Reference. https://software.intel.com/en-us/cpp-compiler-developer-guide-and-reference-restricted-transactional -memory-overview.

Intel. 2021j. Storage Performance Development Kit. https://github.com/spdk/spdk.

Interface AffinityFunction. 2021a. Apache Ignite (Ignite 2.10.0)—Apache Ignite. https://ignite.apache.org/releases /latest/javadoc/org/apache/ignite/cache/affinity/AffinityFunction.html. Apache Ignite. 2021b. Working with SQL. https://ignite.apache.org/docs/latest/SQL/sql-introduction.

Islam, N. S., M. W. Rahman, J. Jose, R. Rajachandrasekar, H. Wang, H. Subramoni, C. Murthy, and D. K. Panda. 2012. High Performance RDMA-based Design of HDFS over InfiniBand. *SC '12: Proceedings of the International Conference on High Performance Computing, Networking, Storage and Analysis*, 2012, pp. 1–12, doi: 10.1109/SC.2012.65.

Islam, Nusrat S., Xiaoyi Lu, Md. W. Rahman, and D. K. Panda. 2013. Can Parallel Replication Benefit Hadoop Distributed File System for High Performance Interconnects? 2013 *IEEE 21st Annual Symposium on High-performance Interconnects*, 75–78, doi: 10.1109/HOTI.2013.24.

Islam, Nusrat S., Xiaoyi Lu, Md. Wasi-ur Rahman, and Dhabaleswar K. (DK) Panda. 2014. SOR-HDFS: A SEDA-based Approach to Maximize Overlapping in RDMA-enhanced HDFS. *Proceedings of the 23rd International Symposium on High-performance Parallel and Distributed Computing. HDC '14*, 261–264.

Islam, Nusrat Sharmin, Xiaoyi Lu, Md. Wasi-ur Rahman, Jithin Jose, and Dhabaleswar K. (DK) Panda. 2012. A Micro-benchmark Suite for Evaluating HDFS Operations on Modern Clusters. In *WBDB 2012: Specifying Big Data Benchmarks*, 129–147. Lecture Notes in Computer Science 8163. New York: Springer.

Islam, Nusrat Sharmin, Xiaoyi Lu, Md. Wasi-ur Rahman, Dipti Shankar, and Dhabaleswar K. Panda. 2015. Triple-H: A Hybrid Approach to Accelerate HDFS on HPC Clusters with Heterogeneous Storage Architecture. *2015 15th IEEE/ACM International Symposium on Cluster, Cloud and Grid Computing (CCGRID)*, 101–110.

Islam, Nusrat Sharmin, Md. Wasi-ur Rahman, Xiaoyi Lu, and Dhabaleswar K. (DK) Panda. 2016a. Efficient Data Access Strategies for Hadoop and Spark on HPC Cluster with Heterogeneous Storage. *2016 IEEE International Conference on Big Data (Big Data)*, 223–232.

Islam, Nusrat Sharmin, Md. Wasi-ur Rahman, Xiaoyi Lu, and Dhabaleswar K. Panda. 2016b. High Performance Design for HDFS with Byte-addressability of NVM and RDMA. *Ics '16: Proceedings of the 2016 International Conference on Supercomputing*, 8–1814. http://doi.acm.org/10.1145/2925426.2926290.

ISO/IEC. 2016. ISO/IEC 9075-1:2016: Information Technology—Database Languages—SQL—Part 1: Framework (SQL/Framework). https://www.iso.org/standard/63555.html.

Izraelevitz, Joseph, Jian Yang, Lu Zhang, Juno Kim, Xiao Liu, Amirsaman Memaripour, Yun Joon Soh, Zixuan Wang, Yi Xu, Subramanya R. Dulloor, et al. 2019. Basic Performance Measurements of the Intel Optane DC Persistent Memory Module. *arXiv/CoRR* abs/1903.05714.

Jacob, Leverich. 2021. Mutilate: A High-Performance Memcached Load Generator. https://github.com/leverich /mutilate.

Javed, M. Haseeb, Khaled Z. Ibrahim, and Xiaoyi Lu. 2019. Performance Analysis of Deep Learning Workloads Using Roofline Trajectories. *CCF Transactions on High Performance Computing* 1 (3): 224–239. https://doi.org /10.1007/s42514-019-00018-4.

Javed, M. H., X. Lu, and D. K. Panda. 2018. Cutting the Tail: Designing High Performance Message Brokers to Reduce Tail Latencies in Stream Processing. *2018 IEEE International Conference on Cluster Computing (CLUSTER)*, 223–233. 10.1109/CLUSTER.2018.00040.

Jia, Yangqing, Evan Shelhamer, Jeff Donahue, Sergey Karayev, Jonathan Long, Ross Girshick, Sergio Guadarrama, and Trevor Darrell. 2014. Caffe: Convolutional Architecture for Fast Feature Embedding. *Proceedings of the 22nd ACM International Conference on Multimedia*, 675–678.

Jiang, Yimin, Yibo Zhu, Chang Lan, Bairen Yi, Yong Cui, and Chuanxiong Guo. 2020. A Unified Architecture for Accelerating Distributed DNN Training in Heterogeneous GPU/CPU Clusters. *14th USENIX Symposium on Operating Systems Design and Implementation (OSDI 20)*, 463–479. https://www.usenix.org/conference/osdi20/presentation/jiang.

Jiang, Zihan, Wanling Gao, Lei Wang, Xingwang Xiong, Yuchen Zhang, Xu Wen, Chunjie Luo, Hainan Ye, Yunquan Zhang, Shengzhong Feng, et al. 2019. HPC AI500: A Benchmark Suite for HPC AI Systems. *International Symposium on Benchmarking, Measuring and Optimization* (pp. 10–22). Springer, Cham. http://www.benchcouncil.org/HPCAI500/index.html.

Jose, J., H. Subramoni, K. Kandalla, M. Wasi ur Rahman, H. Wang, S. Narravula, and D. K. Panda. 2012. Scalable Memcached Design for InfiniBand Clusters Using Hybrid Transports. *12th IEEE/ACM International Symposium on Cluster, Cloud and Grid Computing (CCGRID '12)*, 236–243.

Jose, J., H. Subramoni, M. Luo, M. Zhang, J. Huang, Md. Wasi-ur Rahman, N. S. Islam, X. Ouyang, H. Wang, S. Sur, and D. K. Panda. 2011. Memcached Design on High Performance RDMA Capable Interconnects. *Proceedings of the 2011 International Conference on Parallel Processing. International Conference on Parallel Processing*, 2011, pp. 743–752, doi: 10.1109/ICPP.2011.37.

Jose, Jithin, Mingzhe Li, Xiaoyi Lu, Krishna Chaitanya Kandalla, Mark Daniel Arnold, and Dhabaleswar K. Panda. 2013. SR-IOV Support for Virtualization on InfiniBand Clusters: Early Experience. *Proceedings of IEEE/ACM International Symposium on Cluster, Cloud and Grid Computing (CCGRID)*, 385–392.

Joshi, Kanchan, Kaushal Yadav, and Praval Choudhary. Enabling NVMe WRR Support in Linux Block Layer. In *Hotstorage'17. (HotStorage '17)*. 22. https://dl.acm.org/doi/10.5555/3154601.3154623.

Kadekodi, Rohan, Se Kwon Lee, Sanidhya Kashyap, Taesoo Kim, Aasheesh Kolli, and Vijay Chidambaram. 2019. SplitFS: Reducing Software Overhead in File Systems for Persistent Memory. *Proceedings of the 27th ACM Symposium on Operating Systems Principles*, 494–508.

Kalia, A., M. Kaminsky, and D. G. Andersen. 2014. Using RDMA Efficiently for Key-Value Services. *Proceeding of SIGCOMM '14, the 2014 ACM conference on SIGCOMM (SIGCOMM '14). Association for Computing Machinery*, New York, NY, USA, 295–306. https://doi.org/10.1145/2619239.2626299.

Kalia, Anuj, Michael Kaminsky, and David Andersen. 2019. Datacenter RPCs Can Be General and Fast. *16th USENIX Symposium on Networked Systems Design and Implementation ({NSDI} 19)*, 1–16.

Kalia, Anuj, Michael Kaminsky, and David G. Andersen. 2016. Design Guidelines for High Performance RDMA Systems. *2016 USENIX Annual Technical Conference (USENIX ATC 16)*, 437–450.

Kallman, Robert, Hideaki Kimura, Jonathan Natkins, Andrew Pavlo, Alexander Rasin, Stanley Zdonik, Evan P. C. Jones, Samuel Madden, Michael Stonebraker, Yang Zhang, et al. 2008. H-Store: A High-performance, Distributed Main Memory Transaction Processing System. *Proceedings of the VLDB Endowment* 1 (2): 1496–1499.

Kang, Jeong-Uk, Jeeseok Hyun, Hyunjoo Maeng, and Sangyeun Cho. 2014. The Multi-streamed Solid-state Drive. *6th USENIX Workshop on Hot Topics in Storage and File Systems (HotStorage '14)*, 13. https://dl.acm.org/doi/10.5555/2696578.2696591

Kang, Yangwook, Rekha Pitchumani, Pratik Mishra, Yang-suk Kee, Francisco Londono, Sangyoon Oh, Jongyeol Lee, and Daniel D. G. Lee. 2019. Towards Building a High-performance, Scale-in Key-Value Storage System. *Proceedings of the 12th ACM International Conference on Systems and Storage. SYSTOR '19*, 144–154.

Kannan, Sudarsun, Andrea C. Arpaci-Dusseau, Remzi H. Arpaci-Dusseau, Yuangang Wang, Jun Xu, and Gopinath Palani. 2018. Designing a True Direct-access File System with DevFS. *16th USENIX Conference on File and Storage Technologies (FAST 18)*, 241–256.

Kanwar, Pankaj, Peter Brandt, and Zongwei Zhou. 2020. TensorFlow 2 MLPerf Submissions Demonstrate Best-in-Class Performance on Google Cloud. https://blog.tensorflow.org/2020/07/tensorflow-2-mlperf-submissions.html.

Kastuar, Vidhi, Will Ochandarena, and Tushar Saxena. 2019. Speed Up Training on Amazon SageMaker Using Amazon FSx for Lustre and Amazon EFS File Systems. https://aws.amazon.com/blogs/machine-learning/speed-up-training-on-amazon-sagemaker-using-amazon-efs-or-amazon-fsx-for-lustre-file-systems/.

Katevenis, Manolis, Stefanos Sidiropoulos, and Costas Courcoubetis. 1991. Weighted Round-Robin Cell Multi-plexing in a General-purpose ATM Switch Chip. *IEEE Journal on Selected Areas in Communications* 9 (8): 1265–1279.

Kehne, Jens, Jonathan Metter, and Frank Bellosa. GPUswap: Enabling Oversubscription of GPU Memory through Transparent Swapping. *Vee '15*, 65–77.

Kemper, Alfons, and Thomas Neumann. 2021. HyPer: Hybrid OLTP&OLAP High-Performance Database System. https://hyper-db.de/.

Kemper, Alfons, and Thomas Neumann. 2011. HyPer: A Hybrid OLTP&OLAP Main Memory Database System Based on Virtual Memory Snapshots. *2011 IEEE 27th International Conference on Data Engineering*, 195–206. https://doi.org/10.1109/ICDE.2011.5767867.

Khorasani, Farzad, Keval Vora, Rajiv Gupta, and Laxmi N. Bhuyan. 2014. CuSha: Vertex-centric Graph Process-ing on GPUs. *Proceedings of the 23rd International Symposium on High-performance Parallel and Distributed Computing. HPDC '14*, 239–252. https://doi.org/10.1145/2600212.2600227.

Kim, Changkyu, Tim Kaldewey, Victor W. Lee, Eric Sedlar, Anthony D. Nguyen, Nadathur Satish, Jatin Chhugani, Andrea Di Blas, and Pradeep Dubey. 2009. Sort vs. Hash Revisited: Fast Join Implementation on Modern Multi-Core CPUs. *Proceedings of the VLDB Endowment* 2 (2): 1378–1389.

Kim, Hyeong-Jun, Young-Sik Lee, and Jin-Soo Kim. 2016. NVMeDirect: A User-space I/O Framework for Application-specific Optimization on NVMe SSDs. *8th USENIX Workshop on Hot Topics in Storage and File Systems (HotStorage '16)*. https://www.usenix.org/conference/hotstorage16/workshop-program/presentation/kim.

Kim, Wook-Hee, Jinwoong Kim, Woongki Baek, Beomseok Nam, and Youjip Won. 2016. NVWAL: Exploiting NVRAM in Write-ahead Logging. SIGPLAN Not. 51, 4 (April 2016), 385–398. https://doi.org/10.1145/2954679.2872392.

Kinetica. 2021. The Database for Time and Space: Fuse, Analyze, and Act in Real Time. https://www.kinetica.com.

Kingma, Diederik P., and Jimmy Ba. 2014. Adam: A Method for Stochastic Optimization. *3rd International Conference for Learning Representations*, http://arxiv.org/abs/1412.6980.

Kingsbury, Kyle, Pierre-Yves Ritschard, and James Turnbull. 2021. Riemann Monitors Distributed Systems. https://riemann.io/.

Kissinger, Thomas, Tim Kiefer, Benjamin Schlegel, Dirk Habich, Daniel Molka, and Wolfgang Lehner. 2014. ERIS: A NUMA-aware In-memory Storage Engine for Analytical Workloads. *Proceedings of the VLDB Endowment* 7 (14): 1–12.

Klimovic, Ana, Heiner Litz, and Christos Kozyrakis. 2017. ReFlex: Remote Flash ≈ Local Flash. *ACM SIGARCH Computer Architecture News* 45 (1): 345–359.

Klimovic, Ana, Yawen Wang, Patrick Stuedi, Animesh Trivedi, Jonas Pfefferle, and Christos Kozyrakis. 2018. Pocket: Elastic Ephemeral Storage for Serverless Analytics. *13th USENIX Symposium on Operating Systems Design and Implementation (OSDI 18)*, 427–444.

Konduit. 2021. Deep Learning for Java: Open-source, Distributed, Deep Learning Library for the JVM. https://deeplearning4j.org.

Kourtis, Kornilios, Nikolas Ioannou, and Ioannis Koltsidas. 2019. Reaping the Performance of Fast {NVM} Storage with uDepot. *17th USENIX Conference on File and Storage Technologies (FAST 19)*, 1–15.

Krizhevsky, Alex. 2009. Learning Multiple Layers of Features from Tiny Images. https://www.cs.toronto.edu/~kriz/learning-features-2009-TR.pdf.

Krizhevsky, Alex. 2021. The CIFAR-10 Dataset. http://www.cs.toronto.edu/~kriz/cifar.html.

Krizhevsky, Alex, Ilya Sutskever, and Geoffrey E. Hinton. 2012. ImageNet Classification with Deep Convolutional Neural Networks. *Proceedings of the 25th International Conference on Neural Information Processing Systems—Volume 1 (NIPS '12). Curran Associates Inc.* 1097–1105.

Kubernetes Team. 2021. Kubernetes: Production-grade Container Orchestration. https://kubernetes.io/.

Kulkarni, Sanjeev, Nikunj Bhagat, Maosong Fu, Vikas Kedigehalli, Christopher Kellogg, Sailesh Mittal, Jig-nesh M. Patel, Karthik Ramasamy, and Siddarth Taneja. 2015. Twitter Heron: Stream Processing at Scale.

Proceedings of the 2015 ACM SIGMOD International Conference on Management of Data. SIGMOD '15, 239–250.

Kültürsay, Emre, Mahmut Kandemir, Anand Sivasubramaniam, and Onur Mutlu. 2013. Evaluating STT-RAM as an Energy-efficient Main Memory Alternative. 2013 IEEE International Symposium on *Performance Analysis of Systems and Software (ISPASS)*, 256–267.

Kurth, T., J. Zhang, N. Satish, I. Mitliagkas, E. Racah, M. A. Patwary, T. Malas, N. Sundaram, W. Bhimji, M. Smorkalov, et al. 2017. Deep Learning at 15PF: Supervised and Semi-supervised Classification for Scientific Data. *Proceedings of the International Conference for High Performance Computing, Networking, Storage and Analysis (SC '17)*. Association for Computing Machinery, Article 7, 1–11. DOI:https://doi.org/10.1145/3126908.3126916.

Kwak, Haewoon, Changhyun Lee, Hosung Park, and Sue Moon. 2010. What Is Twitter, a Social Network or a News Media? *Proceedings of the 19th International Conference on World Wide Web. WWW '10*, 591–600. https://doi.org/10.1145/1772690.1772751.

Kwon, Dongup, Junehyuk Boo, Dongryeong Kim, and Jangwoo Kim. 2020. FVM: FPGA-assisted Virtual Device Emulation for Fast, Scalable, and Flexible Storage Virtualization. *14th USENIX Symposium on Operating Systems Design and Implementation (OSDI 20)*, 955–971. https://www.usenix.org/conference/osdi20/presentation/kwon.

Kwon, Y., Fingler, H., Hunt, T., Peter, S., Witchel, E., & Anderson, T. (2017, October). Strata: A Cross Media File S System. *Proceedings of the 26th Symposium on Operating Systems Principles* (pp. 460–477). https://dl.acm.org/doi/10.1145/3132747.3132770

Lagrange, Veronica, Changho Choi, and Vijay Balakrishnan. 2016. Accelerating OLTP Performance with NVMe SSDs. https://www.samsung.com/us/labs/pdfs/collateral/samsung-tpcc-mysql-whitepaper-final-1.pdf.

Lavasani, Maysam, Hari Angepat, and Derek Chiou. 2014. An FPGA-based In-line Accelerator for Memcached. *IEEE Computer Architecture Letters* 13 (2): 57–60.

Lawrence Livermore National Laboratory. 2020. LLNL and HPE to Partner with AMD on El Capitan, Projected as World's Fastest Supercomputer. https://www.llnl.gov/news/llnl-and-hpe-partner-amd-el-capitan-projected-worlds-fastest-supercomputer.

Lawrence, Steve, C. Lee Giles, Ah Chung Tsoi, and Andrew D. Back. 1997. Face Recognition: A Convolutional Neural-Network Approach. *IEEE Transactions on Neural Networks* 8 (1): 98–113.

LDBC. 2021. Linked Data Benchmark Council (LDBC): The Graph and RDF Benchmark Reference. 27–31. https://doi.org/10.1145/2627692.2627697.

LeCun, Yann, Corinna Cortes, and Christopher J. C. Burges. 2021. The MNIST Database of Handwritten Digits. http://yann.lecun.com/exdb/mnist/.

Lee, Benjamin C., Engin Ipek, Onur Mutlu, and Doug Burger. 2009. Architecting Phase Change Memory as a Scalable DRAM Alternative. *Proceedings of the 36th annual international symposium on Computer architecture (ISCA '09). Association for Computing Machinery*, 2–13. https://doi.org/10.1145/1555754.1555758.

Lee, Benjamin C., Ping Zhou, Jun Yang, Youtao Zhang, Bo Zhao, Engin Ipek, Onur Mutlu, and Doug Burger. 2010. Phase-change Technology and the Future of Main Memory. *IEEE Micro* 30 (1). 143–143. https://10.1109/MM.2010.24.

Lee, Changman, Dongho Sim, Jooyoung Hwang, and Sangyeun Cho. 2015. F2FS: A New File System for Flash Storage. *13th USENIX Conference on File and Storage Technologies (FAST 15)*, 273–286.

Lee, Hyungro, and Geoffrey Fox. 2019. Big Data Benchmarks of High-performance Storage Systems on Commercial Bare Metal Clouds. *2019 IEEE 12th International Conference on Cloud Computing (CLOUD)*, 1–8. https://doi.org/10.1109/CLOUD.2019.00014.

Leibiusky, Jonathan. 2021. Jedis. https://github.com/redis/jedis.

Leis, Viktor, Peter Boncz, Alfons Kemper, and Thomas Neumann. 2014. Morsel-driven Parallelism: A NUMA-aware Query Evaluation Framework for the Many-core Age. *Proceedings of the 2014 ACM SIGMOD International Conference on Management of Data*, 743–754.

Lenharth, Andrew, and Keshav Pingali. 2015. Scaling Runtimes for Irregular Algorithms to Large-scale NUMA Systems. *Computer* 48 (8): 35–44.

Lepak, Kevin, Gerry Talbot, Sean White, Noah Beck, and Sam Naffziger. 2017. The Next Generation amd Enterprise Server Product Architecture. *IEEE Hot Chips* 29. https://old.hotchips.org/wp-content/uploads/hc_archives/hc29/HC29.22-Tuesday-Pub/HC29.22.90-Server-Pub/HC29.22.921-EPYC-Lepak-AMD-v2.pdf.

Lepers, Baptiste, Oana Balmau, Karan Gupta, and Willy Zwaenepoel. 2019. KVell: The Design and Implementation of a Fast Persistent Key-Value Store. SOSP '19: *Proceedings of the 27th ACM Symposium on Operating Systems Principles*, 447–461. https://doi.org/10.1145/3341301.3359628.

Li, Feng, Sudipto Das, Manoj Syamala, and Vivek R. Narasayya. 2016. Accelerating Relational Databases by Leveraging Remote Memory and RDMA. *Proceedings of the 2016 International Conference on Management of Data*, 355–370.

Li, Haoyuan, Ali Ghodsi, Matei Zaharia, Scott Shenker, and Ion Stoica. 2014. Tachyon: Reliable, Memory Speed Storage for Cluster Computing Frameworks. SoCC '14: *Proceedings of the ACM Symposium on Cloud Computing*, 1–15.

Li, Min, Jian Tan, Yandong Wang, Li Zhang, and Valentina Salapura. 2015. SparkBench: A Comprehensive Benchmarking Suite for in Memory Data Analytic Platform Spark. *Proceedings of the 12th ACM International Conference on Computing Frontiers (CF '15)*, 53–1538.

Li, Mingzhe, Xiaoyi Lu, Khaled Hamidouche, Jie Zhang, and D. K. Panda. 2016. Mizan-RMA: Accelerating Mizan Graph Processing Framework with MPI RMA. *2016 IEEE 23rd International Conference on High Performance Computing (HPC)*, 42–51. https://doi.org/10.1109/HiPC.2016.015.

Li, Mu, David G. Andersen, Jun Woo Park, Alexander J. Smola, Amr Ahmed, Vanja Josifovski, James Long, Eugene J. Shekita, and Bor-Yiing Su. 2014. Scaling Distributed Machine Learning with the Parameter Server. *Proceedings of the 11th USENIX Conference on Operating Systems Design and Implementation (OSDI '14)*, 583–598.

Li, M., X. Lu, S. Potluri, K. Hamidouche, J. Jose, K. Tomko, and D. K. Panda. 2014. Scalable Graph500 Design with MPI-3 RMA. *2014 IEEE International Conference on Cluster Computing (CLUSTER)*, 230–238.

Li, Peilong, Yan Luo, Ning Zhang, and Yu Cao. 2015. Heterospark: A Heterogeneous CPU/GPU Spark Platform for Machine Learning Algorithms. *2015 IEEE International Conference on Networking, Architecture and Storage (NAS)*, 347–348.

Li, Tianxi, Dipti Shankar, Shashank Gugnani, and Xiaoyi Lu. 2020. RDMP-KV: Designing Remote Direct Memory Persistence Based Key-Value Stores with PMEM. *Proceedings of the International Conference for High Performance Computing, Networking, Storage and Analysis (SC '20)*.

Li, Yixing, Zichuan Liu, Kai Xu, Hao Yu, and Fengbo Ren. 2018. A GPU-outperforming FPGA Accelerator Architecture for Binary Convolutional Neural Networks. *Journal of Emerging Technologies in Computing Systems* 14 (2). doi:10.1145/3154839. https://doi.org/10.1145/3154839.

LightNVM. 2018. Open-Channel SSD. http://lightnvm.io/.

Lim, H., D. Han, D. G. Andersen, and M. Kaminsky. 2014. MICA: A Holistic Approach to Fast In-memory Key-Value Storage. *Proceedings of the 11th USENIX Conference on Networked Systems Design and Implementation (NSDI '14)*.

Lin, Jimmy, and Alek Kolcz. 2012. Large-scale Machine Learning at Twitter. *Proceedings of the 2012 ACM SIGMOD International Conference on Management of Data (SIGMOD '12)*, 793–804. https://doi.org/10.1145/2213836.2213958.

Lin, Tsung-Yi, Michael Maire, Serge Belongie, James Hays, Pietro Perona, Deva Ramanan, Piotr Dollár, and C. Lawrence Zitnick. 2014. Microsoft COCO: Common Objects in Context. *European Conference on Computer Vision*, 740–755.

Linux. 2021a. AIO—POSIX Asynchronous I/O Overview. https://man7.org/linux/man-pages/man7/aio.7.html.

Linux. 2021b. lseek—Linux Manual Page. https://man7.org/linux/man-pages/man2/lseek.2.html.

Linux RDMA. 2021. RDMA Core Userspace Libraries and Daemons. https://github.com/linux-rdma/rdma-core.

Liu, Feilong, Lingyan Yin, and Spyros Blanas. 2019. Design and Evaluation of an RDMA-Aware Data Shuffling Operator for Parallel Database Systems. *ACM Transactions on Database Systems* 44 (4):1–45. https://doi.org/10.1145/3360900.

Liu, Jiuxing. 2010. Evaluating Standard-based Self-virtualizing Devices: A Performance Study on 10 GbE NICs with SR-IOV Support. *Proceedings of IEEE International Parallel and Distributed Processing Symposium (IPDPS)*, 1–12.

Liu, Xin, Yu-tong Lu, Jie Yu, Peng-fei Wang, Jie-ting Wu, and Ying Lu. 2017. ONFS: A Hierarchical Hybrid File System Based on Memory, SSD, and HDD for High Performance Computers. *Frontiers of Information Technology and Electronic Engineering* 18 (12): 1940–1971.

Lockwood, Glenn. 2017. What's So Bad about POSIX I/O? https://www.nextplatform.com/2017/09/11/whats-bad-posix-io/.

Low, Yucheng, Danny Bickson, Joseph Gonzalez, Carlos Guestrin, Aapo Kyrola, and Joseph M. Hellerstein. 2012. Distributed GraphLab: A Framework for Machine Learning and Data Mining in the Cloud. *Proceedings of the VLDB Endowment* 5 (8): 716–727.

Low, Yucheng, Joseph E. Gonzalez, Aapo Kyrola, Danny Bickson, Carlos E. Guestrin, and Joseph Hellerstein. 2014. GraphLab: A New Framework For Parallel Machine Learning. *arXiv* preprint. arXiv:1408.2041.

Lu, J., Y. Wan, Y. Li, C. Zhang, H. Dai, Y. Wang, G. Zhang, and B. Liu. 2019. Ultra-fast Bloom Filters using SIMD Techniques. *IEEE Transactions on Parallel and Distributed Systems* 30 (4): 953–964.

Lu, Lanyue, Thanumalayan Sankaranarayana Pillai, Hariharan Gopalakrishnan, Andrea C. Arpaci-Dusseau, and Remzi H. Arpaci-Dusseau. 2017. WiscKey: Separating Keys from Values in SSD-Conscious Storage. *ACM Transactions on Storage (TOS)* 13 (1): 1–28.

Lu, Ruirui, Gang Wu, Bin Xie, and Jingtong Hu. 2014. Stream Bench: Towards Benchmarking Modern Distributed Stream Computing Frameworks. *Proceedings of the 2014 IEEE/ACM 7th International Conference on Utility and Cloud Computing. (UCC '14)*, 69–78.

Lu, X., H. Shi, H. Javed, R. Biswas, and D. K. Panda. 2017. Characterizing Deep Learning over Big Data (DLoBD) Stacks on RDMA-capable Networks. *The 25th Annual Symposium on High-Performance Interconnects (HOTI)*.

Lu, X., H. Shi, R. Biswas, M. H. Javed, and D. K. Panda. 2018. Dlobd: A comprehensive study of deep learning over big data stacks on hpc clusters. *IEEE Transactions on Multi-Scale Computing Systems* 4 (4): 635–648. doi:10.1109/TMSCS.2018.2845886.

Lu, Xiaoyi, Bin Wang, Li Zha, and Zhiwei Xu. 2011. Can MPI Benefit Hadoop and MapReduce Applications? *2011 40th International Conference on Parallel Processing Workshops*, 371–379.

Lu, Xiaoyi, Dipti Shankar, Shashank Gugnani, and Dhabaleswar K. Panda. 2016. High-performance Design of Apache Spark with RDMA and Its Benefits on Various Workloads. *Proceedings of the 2016 IEEE International Conference on Big Data (Big Data)*. 253–262.

Lu, Xiaoyi, Fan Liang, Bin Wang, Li Zha, and Zhiwei Xu. 2014. DataMPI: Extending MPI to Hadoop-like Big Data Computing. *2014 IEEE 28th International Parallel and Distributed Processing Symposium*, 829–838.

Lu, Xiaoyi, Haiyang Shi, Dipti Shankar, and Dhabaleswar K. Panda. 2017. Performance Characterization and Acceleration of Big Data Workloads on OpenPOWER System. *2017 IEEE International Conference on Big Data (Big Data)*. 213–222. doi: 10.1109/BigData.2017.8257929.

Lu, Xiaoyi, Md. Wasi-ur Rahman, Nusrat Sharmin Islam, and Dhabaleswar K. (DK) Panda. 2014. A Micro-benchmark Suite for Evaluating Hadoop RPC on High-Performance Networks. In *Advancing Big Data Benchmarks*, 32–42. Lecture Notes in Computer Science 8585. New York: Springer.

Lu, Xiaoyi, M. W. U. Rahman, N. Islam, D. Shankar, and D. K. Panda. 2014. Accelerating Spark with RDMA for Big Data Processing: Early Experiences. 2014 IEEE 22nd Annual Symposium on *High-Performance Interconnects (HOTI)*, 9–16.

Lu, Xiaoyi, Nusrat S. Islam, Md. Wasi. Rahman, Jithin Jose, Hari Subramoni, Hao Wang, and Dhabaleswar K. Panda. 2013. High-performance Design of Hadoop RPC with RDMA over InfiniBand. *Proceedings of IEEE 42nd International Conference on Parallel Processing (ICPP)*. 641–650. https://doi.org/10.1109/ICPP.2013.78.

Lu, X., N. Islam, W. Rahman, and D. Panda. 2017. NRCIO: NVM-aware RDMA-based Communication and I/O Schemes for Big Data Analytics. Eighth Annual Non-Volatile Memories Workshop (NVMW '17) [Presentation]. http://nvmw.eng.ucsd.edu/2017/assets/slides/71.

Lu, Youyou, Jiwu Shu, and Wei Wang. 2014. ReconFS: A Reconstructable File System on Flash Storage. *12th USENIX Conference on File and Storage Technologies (FAST 14)*, 75–88.

Lu, Youyou, Jiwu Shu, Youmin Chen, and Tao Li. 2017. Octopus: An RDMA-enabled Distributed Persistent Memory File System. *2017 USENIX Annual Technical Conference (USENIX ATC 17)*, 773–785. https://www.usenix.org/conference/atc17/technical-sessions/presentation/lu.

Ma, Lingxiao, Zhi Yang, Han Chen, Jilong Xue, and Yafei Dai. 2017. Garaph: Efficient GPU-accelerated Graph Processing on a Single Machine with Balanced Replication. *2017 USENIX Annual Technical Conference (USENIX ATC 17)*, 195–207. https://www.usenix.org/conference/atc17/technical-sessions/presentation/ma.

Malewicz, Grzegorz, Matthew H. Austern, Aart J. C. Bik, James C. Dehnert, Ilan Horn, Naty Leiser, and Grzegorz Czajkowski. 2010. Pregel: A System for Large-Scale Graph Processing. *Proceedings of the 2010 ACM SIGMOD International Conference on Management of Data*, 135–146.

Markham, A., and Y. Jia. 2017. Caffe2: Portable High-performance Deep Learning Framework from Facebook. NVIDIA Developer (the blog): https://developer.nvidia.com/blog/caffe2-deep-learning-framework-facebook/? If so, change to URL.

Markthub, Pak, Mehmet E. Belviranli, Seyong Lee, Jeffrey S. Vetter, and Satoshi Matsuoka. Dragon: Breaking GPU Memory Capacity Limits with Direct NVM Access. In *(SC '18)*, 32, 1–13.

Marmol, Leonardo, Swaminathan Sundararaman, Nisha Talagala, Raju Rangaswami, Sushma Devendrappa, Bharath Ramsundar, and Sriram Ganesan. 2014. NVMKV: A Scalable and Lightweight Flash Aware Key-Value Store. *Proceedings of the 2015 USENIX Conference on Usenix Annual Technical Conference (USENIX ATC '15). USENIX Association, USA, 207–219.*

Massie, Matt. 2018. Ganglia Monitoring System. http://ganglia.info/.

Mattson, Peter, Christine Cheng, Gregory Diamos, Cody Coleman, Paulius Micikevicius, David Patterson, Hanlin Tang, Gu-Yeon Wei, Peter Bailis, Victor Bittorf, et al. 2020. MLPerf Training Benchmark. *Proceedings of Machine Learning and Systems* 2: 336–349. https://proceedings.mlsys.org/paper/2020/file/02522a2b2726fb0a03bb19f2d8d9524d-Paper.pdf.

Mellanox. 2010. CORE-Direct: The Most Advanced Technology for MPI/SHMEM Collectives Offloads. Technology brief. https://www.mellanox.com/related-docs/whitepapers/TB_CORE-Direct.pdf.

Mellanox. 2016. Understanding Erasure Coding Offload. https://community.mellanox.com/docs/DOC-2414.

Mellanox. NVIDIA. 2021. NVIDIA Bluefield Data Processing Units. https://www.nvidia.com/en-us/networking/products/data-processing-unit/.

Mellanox. 2018. Introducing 200G HDR InfiniBand Solutions. White paper. https://www.mellanox.com/related-docs/whitepapers/WP_Introducing_200G_HDR_InfiniBand_Solutions.pdf.

Mellanox, NVIDIA. 2011. Mellanox Announces Availability of UDA 2.0 for Big Data Analytic Acceleration. https://www.mellanox.com/news/press_release/mellanox-announces-availability-uda-20-big-data-analytic-acceleration.

Mellanox, NVIDIA. 2018. SparkRDMA ShuffleManager Plugin. https://github.com/Mellanox/SparkRDMA/.

Mellanox, NVIDIA. 2021. End-to-End High-Speed Ethernet and InfiniBand Interconnect Solutions. https://www.nvidia.com/en-us/networking/" and title " Accelerated Networks for Modern Workloads.

Meng, Xiangrui, Joseph Bradley, Burak Yavuz, Evan Sparks, Shivaram Venkataraman, Davies Liu, Jeremy Freeman, D. B. Tsai, Manish Amde, Sean Owen, et al. 2016. MLlib: Machine Learning in Apache Spark. *Journal of Machine Learning Research* 17 (1): 1235–1241.

Mickens, James, Edmund B. Nightingale, Jeremy Elson, Darren Gehring, Bin Fan, Asim Kadav, Vijay Chidambaram, Osama Khan, and Krishna Nareddy. 2014. Blizzard: Fast, Cloud-scale Block Storage for Cloud-oblivious Applications. *11th USENIX Symposium on Networked Systems Design and Implementation (NSDI 14)*, 257–273.

Micron. 2019. 3D XPoint technology.

Microsoft. 2021a. Create an Azure VM with Accelerated Networking using Azure CLI—Microsoft Docs. https://docs.microsoft.com/en-us/azure/virtual-network/create-vm-accelerated-networking-cli.

Microsoft. 2021b. GitHub - MicrosoftResearch/Dryad: This Is a Research Prototype of the Dryad and DryadLINQ Data-parallel Processing Frameworks Running on Hadoop YARN. https://github.com/MicrosoftResearch/Dryad.

Min, Changwoo, Sanidhya Kashyap, Steffen Maass, and Taesoo Kim. 2016. Understanding Manycore Scalability of File Systems. *USENIX Annual Technical Conference (USENIX ATC '16)*, 71–85.

Mitchell, C., Y. Geng, and J. Li. 2013. Using One-sided RDMA Reads to Build a Fast, CPU-efficient Key-Value Store. *Proceedings of USENIX Annual Technical Conference (USENIX ATC '13)*.

MLBench Team. 2021. MLBench: Distributed Machine Learning Benchmark. https://mlbench.github.io/.

MLCommons. 2021. MLCommons Aims to Accelerate Machine Learning Innovation to Benefit Everyone. https://mlcommons.org/en/.

Monroe, Don. 2020. Fugaku takes the lead. *Communications of the ACM* 64 (1): 16–18.

Moody, Adam, Danielle Sikich, Ned Bass, Michael J. Brim, Cameron Stanavige, Hyogi Sim, Joseph Moore, Tony Hutter, Swen Boehm, Kathryn Mohror, et al.; USDOE National Nuclear Security Administration. 2017. UnifyFS: A Distributed Burst Buffer File System 0.1.0. https://doi.org/10.11578/dc.20200519.19. https://www.osti.gov//servlets/purl/1408515.

Moor Insights and Strategy. 2020. The Graphcore Second Generation IPU. https://www.graphcore.ai/hubfs/MK2-%20The%20Graphcore%202nd%20Generation%20IPU%20Final%20v7.14.2020.pdf?h.

Morgan, Timothy Prickett. 2019. Doing the Math on Future Exascale Supercomputers. https://www.nextplatform.com/2019/11/20/doing-the-mathon-future-exascale-supercomputers/.

Moritz, Philipp, Robert Nishihara, Ion Stoica, and Michael I. Jordan. 2015. SparkNet: Training Deep Networks in Spark. *CoRR, abs/1511.06051.*

Mouzakitis, Evan. 2016. How to monitor Hadoop with Datadog. https://www.datadoghq.com/blog/monitor-hadoop-metrics-datadog/.

Mouzakitis, Evan, and David Lentz. 2018. Monitor Redis using Datadog. https://www.datadoghq.com/blog/monitor-redis-using-datadog/.

MPI, Forum. 1993. MPI: a message passing interface. In *Proceedings of the 1993 ACM/IEEE conference on Supercomputing (Supercomputing '93)*. Association for Computing Machinery, New York, NY, USA, 878–883. DOI:https://doi.org/10.1145/169627.169855.

Murphy, Barbara. 2018. How to Shorten Deep Learning Training Times. https://www.weka.io/blog/shorten-deep-learning-training-times-for-ai/.

MySQL. 2020. MySQL Database. http://www.mysql.com.

National Energy Research Scientific Computing Center (NERSC). 2021a. Cori. https://docs.nersc.gov/systems/cori/.

National Energy Research Scientific Computing Center (NERSC). 2021b. Perlmutter. https://www.nersc.gov/systems/perlmutter/.

Netty Project, The. 2021. Netty Project. http://netty.io.

Network Based Computing Lab (NOWLAB). 2021a. High-Performance Big Data (HiBD). http://hibd.cse.ohio-state.edu/static/media/ohb/changelogs/ohb-0.9.3.txt.

Network Based Computing Lab (NOWLAB). 2021b. MVAPICH: MPI over InfiniBand, Omni-Path, Ethernet/iWARP, and RoCE. http://mvapich.cse.ohio-state.edu/.

Network Based Computing Lab (NOWLAB). 2022. High-Performance Big Data (HiBD). http://hibd.cse.ohio-state.edu.

Neumann, Thomas, and Gerhard Weikum. 2008. RDF-3X: A RISC-style Engine for RDF. *Proceedings of the VLDB Endowment* 1 (1): 647–659.

NLM, National Library of Medicine. 2020. PubChemRDF. https://pubchemdocs.ncbi.nlm.nih.gov/rdf.

Norton, Alex, Steve Conway, and Earl Joseph. 2020. Bringing HPC Expertise to Cloud Computing. White paper. Hyperion Research.

NoSQL Database. 2021. NoSQL—Your Ultimate Guide to the Non-Relational Universe!. https://hostingdata.co.uk/nosql-database/.

Nowoczynski, P., N. Stone, J. Yanovich, and J. Sommerfield. 2008. Zest Checkpoint Storage System for Large Supercomputers. *2008 3rd Petascale Data Storage Workshop*, 1–5.

NumFOCUS. 2021. Pandas. https://pandas.pydata.org/.

NumPy. 2021. NumPy. http://www.numpy.org/.

NVIDIA. 2017. NVIDIA Tesla V100 GPU. ARCHITECTURE. https://images.nvidia.com/content/volta-architecture/pdf/volta-architecture-whitepaper.pdf.

NVIDIA. 2020. cuStreamz: A Journey to Develop GPU-Accelerated Streaming Using RAPIDS. https://www.nvidia.com/en-us/on-demand/session/gtcfall20-a21437/.

NVIDIA. 2021a. CUDA Zone. https://developer.nvidia.com/cuda-zone.

NVIDIA. 2021b. cuGraph. https://github.com/rapidsai/cugraph.

NVIDIA. 2021c. cuML. https://github.com/rapidsai/cuml.

NVIDIA. 2021d. Developing a Linux Kernel Module using GPUDirect RDMA. https://docs.nvidia.com/cuda/gpudirect-rdma/.

NVIDIA. 2021e. GDRCopy: A Low-latency GPU Memory Copy Library based on NVIDIA GPUDirect RDMA Technology. https://github.com/NVIDIA/gdrcopy.

NVIDIA (Mellanox Technologies). 2021f. Apache Spark RDMA plugin. https://community.mellanox.com/s/article/apache-spark-rdma-plugin.

NVIDIA. 2021g. NVIDIA Ampere Architecture: The Heart of the World's Highest-performing Elastic Data Centers. https://www.nvidia.com/en-us/data-center/ampere-architecture/.

NVIDIA. 2021h. NVIDIA DGX-1: Essential Instrument of AI Research. https://www.nvidia.com/en-us/data-center/dgx-1/.

NVIDIA. 2021i. NVIDIA DGX-2: Break through the Barriers to AI Speed and Scale. https://www.nvidia.com/en-us/data-center/dgx-2/.

NVIDIA. 2021j. NVIDIA DGX Systems: Purpose-built for the Unique Demands of AI. https://www.nvidia.com/en-us/data-center/dgx-systems/.

NVIDIA. 2021k. NVIDIA Pascal Architecture: Infinite Compute for Infinite Opportunities. https://www.nvidia.com/en-us/data-center/pascal-gpu-architecture/.

NVIDIA. 2021l. NVIDIA Turing GPU Architecture: Graphics Reinvented. White paper. https://www.nvidia.com/content/dam/en-zz/Solutions/design-visualization/technologies/turing-architecture/NVIDIA-Turing-Architecture-Whitepaper.pdf.

NVIDIA. 2021m. About Us. https://www.nvidia.com/en-us/about-nvidia/.

NVIDIA. 2021n. RAPIDS—Open GPU Data Science. https://github.com/rapidsai.

NVIDIA. 2021o. Virtual GPU Software User Guide. https://docs.nvidia.com/grid/latest/grid-vgpu-user-guide/.

NVMe Express. 2016. NVMe over Fabrics. http://www.nvmexpress.org/wp-content/uploads/NVMe_Over_Fabrics.pdf.

NVM Express. 2021. NVM Express. https://nvmexpress.org/.

Oak Ridge National Laboratory. 2018. Summit: America's Newest and Smartest Supercomputer. https://www.olcf.ornl.gov/summit/.

Oak Ridge National Laboratory. 2021a. Frontier: Direction of Discovery: ORNL"s Exascale Supercomputer Designed to Deliver World-Leading Performance in 2021. https://www.olcf.ornl.gov/frontier/.

Oak Ridge National Laboratory. 2021b. Summit: Oak Ridge National Laboratory's 200 Petaflop Supercomputer. http://www.olcf.ornl.gov/olcf-resources/compute-systems/summit/.

OpenFabrics Alliance. 2021. OpenFabrics Alliance—Innovation in High Speed Fabrics. https://www.openfabrics.org/.

Open Group, The. 2011. POSIX^{TM} 1003.1 Frequently Asked Questions (FAQ Version 1.18). http://www.opengroup.org/austin/papers/posix_faq.html.

OpenMP. 2018. OpenMP API Specification Version 5.0 November 2018: 2.9.3 SIMD Directives. https://www.openmp.org/spec-html/5.0/openmpsu42.html.

OpenSFS and EOFS. 2021. Lustre® Filesystem. http://lustre.org/.

OpenStack. 2021a. Cinder. https://wiki.openstack.org/wiki/Cinder.

OpenStack. 2021b. OpenStack Object Storage (Swift). https://wiki.openstack.org/wiki/Swift.

OpenVINO. 2021. OpenVINO Toolkit. https://github.com/openvinotoolkit/openvino.

Oracle. 2021. MySQL. https://www.mysql.com/.

OrangeFS. 2021. The OrangeFS Project. http://www.orangefs.org/.

Ott, David. 2011. Optimizing Applications for NUMA. https://software.intel.com/en-us/articles/optimizing-applications-for-numa.

Ould-Ahmed-Vall, Elmoustapha, Mahmoud Abuzaina, Md. Faijul Amin, Jayaram Bobba, Roman S. Dubtsov, Evarist M. Fomenko, Mukesh Gangadhar, Niranjan Hasabnis, Jing Huang, Deepthi Karkada, et al. 2017. TensorFlow Optimizations on Modern Intel Architecture. https://software.intel.com/content/www/us/en/develop/articles/tensorflowoptimizations-on-modern-intel-architecture.html.

Ousterhout, John, Arjun Gopalan, Ashish Gupta, Ankita Kejriwal, Collin Lee, Behnam Montazeri, Diego Ongaro, Seo Jin Park, Henry Qin, Mendel Rosenblum, et al. 2015. The RAMCloud Storage System. *ACM Transactions on Computer Systems (TOCS)* 33 (3): 7.

Ozery, Itay. 2018. Mellanox Accelerates Apache Spark Performance with RDMA and RoCE Technologies. https://blog.mellanox.com/2018/12/mellanox-accelerates-apache-spark-rdma-and-roce-technologies/.

Padua, David, ed. 2011. Partitioned Global Address Space (PGAS) Languages. In *Encyclopedia of Parallel Computing*, 1465. Boston, MA: Springer.

Pagh, Rasmus, and Flemming Friche Rodler. 2004. Cuckoo Hashing. *Journal of Algorithms* 51 (2): 122–144.

Palit, Tapti, Yongming Shen, and Michael Ferdman. 2016. Demystifying Cloud Benchmarking. *2016 IEEE International Symposium on Performance Analysis of Systems and Software (ISPASS)*, 122–132. https://doi.org/10.1109/ISPASS.2016.7482080.

Panda, Biswanath, Joshua S. Herbach, Sugato Basu, and Roberto J. Bayardo. 2011. MapReduce and Its Application to Massively Parallel Learning of Decision Tree Ensembles. In *Scaling Up Machine Learning*, eds. Ron Bekkerman, Mikhail Bilenko, and John Langford, 23–48. Cambridge: Cambridge University Press. http://dx.doi.org/10.1017/CBO9781139042918.003.

Panda, Dhabaleswar K., Xiaoyi Lu, and Hari Subramoni. 2018. Networking and Communication Challenges for Post-exascale Systems. *Frontiers of Information Technology and Electronic Engineering* 19: 1230–1235.

Panigrahy, Rina. 2004. Efficient Hashing with Lookups in Two Memory Accesses. *arXiv/CoRR* cs.DS/0407023. http://arxiv.org/abs/cs.DS/0407023.

Paszke, Adam, Sam Gross, Francisco Massa, Adam Lerer, James Bradbury, Gregory Chanan, Trevor Killeen, Zeming Lin, Natalia Gimelshein, Luca Antiga, et al. 2019. Pytorch: An Imperative Style, High-performance Deep Learning Library. *arXiv* preprint. arXiv:1912.01703.

PCI-SIG. 2019. Pioneering the Interconnect Industry: PCI-SIG® Announces Upcoming PCIe® 6.0 Specification. https://pcisig.com/pioneering-interconnect-industry-pci-sig%C2%AE-announces-upcoming-pcie%C2%AE-60-specification.

PCI-SIG. 2021. Single Root I/O Virtualization and Sharing Specification Revision 1.1. https://pcisig.com/single-root-io-virtualization-and-sharing-specification-revision-11

Pelley, Steven, Thomas F. Wenisch, Brian T. Gold, and Bill Bridge. 2013. Storage Management in the NVRAM Era. *Proceedings of the VLDB Endowment* 7 (2): 121–132.

PlatformLab. 2021. RAMCloud. https://github.com/PlatformLab/RAMCloud.

Plotly. 2021. Plotly. https://github.com/plotly/plotly.py.

Pmemkv. 2018. pmemkv. https://github.com/pmem/pmemkv.

Poke, Marius, and Torsten Hoefler. 2015. DARE: High-performance State Machine Replication on RDMA Networks. *Proceedings of the 24th International Symposium on High-performance Parallel and Distributed Computing*, 107–118.

Polychroniou, Orestis, and Kenneth A. Ross. 2014. Vectorized Bloom Filters for Advanced SIMD Processors. *Proceedings of the Tenth International Workshop on Data Management on New Hardware*, 6.

Polychroniou, Orestis, Arun Raghavan, and Kenneth A. Ross. 2015. Rethinking SIMD Vectorization for In-memory Databases. *Proceedings of the 2015 ACM SIGMOD International Conference on Management of Data*, 1493–1508.

Porobic, Danica, Erietta Liarou, Pınar Tözün, and Anastasia Ailamaki. 2014. Atrapos: Adaptive Transaction Processing on Hardware Islands. *2014 IEEE 30th International Conference on Data Engineering*, 688–699.

Powell, Brett. 2017. *Microsoft Power BI Cookbook: Creating Business Intelligence Solutions of Analytical Data Models, Reports, and Dashboards*. Birmingham, UK: Packt Publishing Ltd.

Project Jupyter. 2021. Jupyter. https://jupyter.org/.

Prometheus. 2021. Prometheus—From Metrics to Insight. https://prometheus.io/.

Psaroudakis, Iraklis, Tobias Scheuer, Norman May, Abdelkader Sellami, and Anastasia Ailamaki. 2015. Scaling Up Concurrent Main-Memory Column-Store Scans: Towards Adaptive NUMA-aware Data and Task Placement. *Proceedings of the VLDB Endowment* 8 (CONF): 1442–1453.

Psaroudakis, Iraklis, Tobias Scheuer, Norman May, Abdelkader Sellami, and Anastasia Ailamaki. 2016. Adaptive NUMA-aware Data Placement and Task Scheduling for Analytical Workloads in Main-Memory Column-Stores. *Proceedings of the VLDB Endowment* 10 (2): 37–48.

Pumma, Sarunya, Min Si, Wu-Chun Feng, and Pavan Balaji. 2019. Scalable Deep Learning via I/O Analysis and Optimization. *ACM Transactions on Parallel Computing* 6 (2): article 6. https://doi.org/10.1145/3331526.

Python. 2021. Threading—Thread-based Parallelism. https://docs.python.org/3/library/threading.html.

Qureshi, Moinuddin K., Vijayalakshmi Srinivasan, and Jude A. Rivers. 2009. Scalable High Performance Main Memory System Using Phase-change Memory Technology. *ACM SIGARCH Computer Architecture News* 37 (3): 24–33.

Rahman, Md. Wasi-ur, Nusrat Sharmin Islam, Xiaoyi Lu, Jithin Jose, Hari Subramoni, Hao Wang, and Dhabaleswar K. Panda. 2013. High-performance RDMA-based Design of Hadoop MapReduce over InfiniBand. *2013 IEEE 27th International Parallel and Distributed Processing Symposium Workshops and PHD Forum (IPDPSW)*, 1908–1917.

Rahman, Md. Wasi ur, Nusrat Sharmin Islam, Xiaoyi Lu, and Dhabaleswar K. Panda. 2017. NVMD: Non-Volatile Memory Assisted Design for Accelerating MapReduce and DAG Execution Frameworks on HPC Systems. *Proceedings of IEEE International Conference on Big Data, BigData '17*, 369–374.

Rahman, M. W., Xiaoyi Lu, Nusrat S. Islam, and D. K. Panda. 2014. HOMR: A Hybrid Approach to Exploit Maximum Overlapping in MapReduce over High Performance Interconnects. *Proceedings of the 28th ACM international conference on Supercomputing (ICS '14)*. Association for Computing Machinery, 33–42. https://doi.org/10.1145/2597652.2597684.

Rahman, M. W., Xiaoyi Lu, Nusrat S. Islam, Raghunath Rajachadrasekar, and D. K. Panda. 2015. High-performance Design of YARN MapReduce on Modern HPC Clusters with Lustre and RDMA. *2015 IEEE International Parallel and Distributed Processing Symposium*, 2015, pp. 291–300, doi: 10.1109/IPDPS.2015.83.

Raja, Raghu. 2019. Amazon Elastic Fabric Adapter: Anatomy, Capabilities, and the Road Ahead. *15th Annual OpenFabrics Alliance Workshop*. https://www.openfabrics.org/wp-content/uploads/2019-workshop-presentations/205_RRaja.pdf.

Rajpurkar, Pranav, Jian Zhang, Konstantin Lopyrev, and Percy Liang. 2016. SQuAD: 100,000+ Questions for Machine Comprehension of Text. *arXiv* preprint. arXiv:1606.05250.

Raju, Pandian, Rohan Kadekodi, Vijay Chidambaram, and Ittai Abraham. 2017. PebblesDB: Building Key-Value Stores Using Fragmented Log-Structured Merge Trees. *Proceedings of the 26th symposium on Operating Systems Principles (SOSP '17)*. Association for Computing Machinery, New York, NY, USA, 497–514. https://doi.org/10.1145/3132747.3132765.

RapidLoop. 2021. OpsDash. https://www.opsdash.com/integrations.

RAPIDS. 2021. cuDF—GPU DataFrames Library. https://github.com/rapidsai/cudf.

RDMA Consortium. 2016. Architectural Specifications for RDMA over TCP/IP. http://www.rdmaconsortium.org/.

Reddi, Vijay Janapa, Christine Cheng, David Kanter, Peter Mattson, Guenther Schmuelling, Carole-Jean Wu, Brian Anderson, Maximilien Breughe, Mark Charlebois, William Chou, et al. 2020. MLPerf Inference Benchmark. *2020 ACM/IEEE 47th Annual International Symposium on Computer Architecture (ISCA)*, 446–459. https://doi.org/10.1109/ISCA45697.2020.00045.

Red Hat, Inc. 2021. Gluster Is a Free and Open Source Software Scalable Network Filesystem. https://www.gluster.org/.

Redis Labs. 2021a. RedisGraph: A Graph Database Module for Redis. https://oss.redislabs.com/redisgraph/.

Redis Labs. 2021b. Redis Cluster Specification. https://redis.io/topics/cluster-spec.

Redis Labs. 2021c. Redis Sentinel Documentation. https://redis.io/topics/sentinel.

Redis Labs. 2021d. Redis. https://redis.io.

Ren, Kun, Alexander Thomson, and Daniel J. Abadi. 2014. An Evaluation of the Advantages and Disadvantages of Deterministic Database Systems. *Proceedings of the VLDB Endowment* 7 (10): 821–832.

Reynolds, Douglas A., and Richard C. Rose. 1995. Robust Text-independent Speaker Identification Using Gaussian Mixture Speaker Models. *IEEE Transactions on Speech and Audio Processing* 3 (1): 72–83.

Rho, Eunhee, Kanchan Joshi, Seung-Uk Shin, Nitesh Jagadeesh Shetty, Jooyoung Hwang, Sangyeun Cho, Daniel D. G. Lee, and Jaeheon Jeong. 2018. FStream: Managing Flash Streams in the File System. *16th USENIX Conference on File and Storage Technologies (FAST 18)*, 257–264.

RIKEN Center for Computational Science. 2020. Fugaku (supercomputer). https://en.wikipedia.org/wiki/Fugaku _(supercomputer).

Rödiger, Wolf, Tobias Mühlbauer, Alfons Kemper, and Thomas Neumann. 2015. High-speed Query Processing over High-speed Networks. *Proceedings of the VLDB Endowment* 9 (4): 228–239.

Rohloff, Kurt, and Richard E. Schantz. 2010. High-performance, Massively Scalable Distributed Systems Using the MapReduce Software Framework: The SHARD Triple-Store. *Programming Support Innovations for Emerging Distributed Applications (PSI ETA '10)*, 4–145.

Ronan Clément, Koray, and Soumith. 2021. Torch - Scientific computing for LuaJIT. http://torch.ch/.

Ross, Kenneth A. 2007. Efficient Hash Probes on Modern Processors. *IEEE 23rd International Conference on Data Engineering (ICDE '07)*, 1297–1301.

Rudoff, Andy. 2013. Programming Models for Emerging Non-volatile Memory Technologies. *;login:* 38 (3): 40–45.

Rudoff, Andy. 2017. Persistent Memory Programming. *;login:* 42: 34–40.

Ruprecht, Adam, Danny Jones, Dmitry Shiraev, Greg Harmon, Maya Spivak, Michael Krebs, Miche Baker-Harvey, and Tyler Sanderson. 2018. VM Live Migration at Scale. *VEE '18: Proceedings of the 14th ACM SIGPLAN/SIGOPS International Conference on Virtual Execution Environments*, 45–56. https://doi.org/10.1145 /3186411.3186415.

Sadasivam, Satish Kumar, Brian W. Thompto, Ron Kalla, and William J. Starke. 2017. IBM Power9 Processor Architecture. *IEEE Micro* 37 (2): 40–51.

SambaNova Systems. 2021a. (Compute) Power to the People: Democratizing AI: A Conversation with AI Visionaries from SambaNova Systems. https://sambanova.ai/blog/compute-power-to-the-people-democratizing-ai -a-conversation-with-ai-visionaries-from-sambanova-systems/.

SambaNova Systems. 2021b. Accelerated Computing with a Reconfigurable Dataflow Architecture. https:// sambanova.ai/wp-content/uploads/2021/06/SambaNova_RDA_Whitepaper_English.pdf.

Saxena, Mohit, Michael M. Swift, and Yiying Zhang. 2012. FlashTier: A Lightweight, Consistent and Durable Storage Cache. *Proceedings of the 7th ACM European Conference on Computer Systems*, 267–280.

SchedMD. 2020a. Slurm Workload Manager—Documentation. https://slurm.schedmd.com/.

SchedMD. 2020b. Slurm Workload Manager—Overview. https://slurm.schedmd.com/overview.html.

Schmuck, Frank B., and Roger L Haskin. 2002. GPFS: A Shared-Disk File System for Large Computing Clusters. *Proceedings of the Conference on File and Storage Technologies (FAST '02)*, 231–244.

Scouarnec, Nicolas Le. 2018. Cuckoo++ Hash Tables: High-performance Hash Tables for Networking Applications. *Proceedings of the 2018 Symposium on Architectures for Networking and Communications Systems*, 41–54.

Scylla. 2021. The Real-time Big Data Database. www.scylladb.com.

SDSC, San Diego Supercomputer Center. 2021a. SDSC Comet User Guide. https://www.sdsc.edu/support/user _guides/comet.html.

SDSC, San Diego Supercomputer Center. 2021b. SDSC Gordon User Guide. https://www.sdsc.edu/support/user _guides/gordon.html.

Segal, Oren, and Martin Margala. 2016. Exploring the Performance Benefits of Heterogeneity and Reconfigurable Architectures in a Commodity Cloud. *2016 International Conference on High Performance Computing and Simulation (HPCS)*, 132–139.

Segal, Oren, Martin Margala, Sai Rahul Chalamalasetti, and Mitch Wright. 2014. High Level Programming Framework for FPGAs in the Data Center. *2014 24th International Conference on Field Programmable Logic and Applications (FPL)*, 1–4. https://doi.org/10.1109/FPL.2014.6927442.

Seide, Frank, and Amit Agarwal. 2016. CNTK: Microsoft's Open-Source Deep-Learning Toolkit. *Proceedings of the 22nd ACM SIGKDD International Conference on Knowledge Discovery and Data Mining*, 2135–2135. *(KDD '16)*. Association for Computing Machinery, 2135. https://doi.org/10.1145/2939672.2945397.

Semiodesk. 2021. Trinity RDF: Entity Framework for Graph Databases. https://trinity-rdf.net/.

Shankar, Dipti, Xiaoyi Lu, and Dhabaleswar K. Panda. 2019a. SCOR-KV: SIMD-aware Client-centric and Optimistic RDMA-based Key-Value Store for Emerging CPU Architectures. *2019 IEEE 26th International Conference on High Performance Computing, Data, and Analytics (HIPC)*, 257–266.

Shankar, Dipti, Xiaoyi Lu, and Dhabaleswar K. Panda. 2019b. SimdHT-Bench: Characterizing SIMD-aware Hash Table Designs on Emerging CPU Architectures. *2019 IEEE International Symposium on Workload Characterization (IISWC)*, 178–188.

Shankar, Dipti, Xiaoyi Lu, Md. W. Rahman, Nusrat Islam, and D. K. Panda. 2015. Benchmarking Key-Value Stores on High-performance Storage and Interconnects for Web-scale Workloads. *BIG DATA '15: Proceedings of the 2015 IEEE International Conference on Big Data*, 539–544

Shankar, Dipti, Xiaoyi Lu, M. W. Rahman, Nusrat Islam, and Dhabaleswar K. Panda. 2014. A Micro-benchmark Suite for Evaluating Hadoop MapReduce on High-performance Networks. *Proceedings of the Fifth Workshop on Big Data Benchmarks, Performance Optimization, and Emerging Hardware (BE-5)*, 19–33. Lecture Notes in Computer Science 8807. Hangzhou, China: Springer.

Shankar, D., X. Lu, and D. K. Panda. 2016. Boldio: A Hybrid and Resilient Burst-Buffer over Lustre for Accelerating Big Data I/O. *2016 IEEE International Conference on Big Data (BIG DATA)*, 404–409.

Shankar, D., X. Lu, N. Islam, M. Wasi-Ur Rahman, and D. K. Panda. 2016. High-performance Hybrid Key-Value Store on Modern Clusters with RDMA Interconnects and SSDs: Non-blocking Extensions, Designs, and Benefits. *2016 IEEE International Parallel and Distributed Processing Symposium (IPDPS)*, 393–402.

Sharma, Upendra, Prashant Shenoy, Sambit Sahu, and Anees Shaikh. 2011. A Cost-aware Elasticity Provisioning System for the Cloud. *2011 31st International Conference on Distributed Computing Systems*, 559–570.

Shen, Zhaoyan, Feng Chen, Yichen Jia, and Zili Shao. 2018. DIDACache: An Integration of Device and Application for Flash-based Key-Value Caching. *ACM Transactions on Storage* 14 (3): article 26.

Shi, Haiyang, and Xiaoyi Lu. 2019. TriEC: Tripartite Graph Based Erasure Coding NIC Offload. *Proceedings of the International Conference for High Performance Computing, Networking, Storage and Analysis (SC '19)*. https://doi.org/10.1145/3295500.3356178.

Shi, Haiyang, and Xiaoyi Lu. 2020. INEC: Fast and Coherent In-network Erasure Coding. *Proceedings of the International Conference for High Performance Computing, Networking, Storage and Analysis (SC '20) IEEE Press*, Article 66, 1–17. https://dl.acm.org/doi/abs/10.5555/3433701.3433788.

Shi, Haiyang, Xiaoyi Lu, Dipti Shankar, and Dhabaleswar K. Panda. UMR-EC: A Unified and Multi-rail Erasure Coding Library for High-performance Distributed Storage Systems. *Proceedings of the 28th International Symposium on High-performance Parallel and Distributed Computing (HPDC '19)*, 219–230.

Shi, Jiaxin, Youyang Yao, Rong Chen, Haibo Chen, and Feifei Li. 2016. Fast and Concurrent {RDF} Queries with RDMA-based Distributed Graph Exploration. *12th USENIX Symposium on Operating Systems Design and Implementation (OSDI 16)*, 317–332.

Shi, Lin, Hao Chen, and Jianhua Sun. 2009. vCUDA: GPU Accelerated High Performance Computing in Virtual Machines. *2009 IEEE International Symposium on Parallel Distributed Processing*, 1–11.

Shreedhar, Madhavapeddi, and George Varghese. 1996. Efficient Fair Queuing Using Deficit Round-Robin. *IEEE/ACM Transactions on Networking* 4 (3): 375–385.

Shue, David, and Michael J. Freedman. From Application Requests to Virtual IOPs: Provisioned Key-Value Storage with Libra. *(EuroSys '14)*. Association for Computing Machinery, New York, NY, USA, Article 17, 1–14. DOI:https://doi.org/10.1145/2592798.2592823.

Shun, Julian, and Guy E. Blelloch. 2013. Ligra: A Lightweight Graph Processing Framework for Shared Memory. *Proceedings of the 18th ACM SIGPLAN Symposium on Principles and Practice of Parallel Programming*, 135–146.

Shvachko, Konstantin, Hairong Kuang, Sanjay Radia, and Robert Chansler. 2010. The Hadoop Distributed File System. *Proceedings of the 26th Symposium on Mass Storage Systems and Technologies (MSST)*, 1–10.

Simonyan, Karen, and Andrew Zisserman. 2014. Very Deep Convolutional Networks for Large-scale Image Recognition. *Third International Conference on Learning Representations (ICLR 2015)*. https://arxiv.org/abs/1409.1556.

Singh, Teja, Sundar Rangarajan, Deepesh John, Russell Schreiber, Spence Oliver, Rajit Seahra, and Alex Schaefer. 2020. 2.1 Zen 2: The AMD 7nm Energy-Efficient High-Performance x86-64 Microprocessor Core. *2020 IEEE International Solid-state Circuits Conference (ISSCC)*, 42–44.

Singhal, Amit. 2012. Introducing the Knowledge Graph: Things, Not Strings. https://blog.google/products/search/introducing-knowledge-graph-things-not/.

SingleStore Inc. 2021. SingleStore: The Single Database for All Data-Intensive Applications. https://www.singlestore.com/.

Sivathanu, Muthian, Tapan Chugh, Sanjay S. Singapuram, and Lidong Zhou. 2019. Astra: Exploiting Predictability to Optimize Deep Learning. *Proceedings of the 24th International Conference on Architectural Support for Programming Languages and Operating Systems (ASPLOS '19)*, 909–923. https://doi.org/10.1145/3297858.3304072.

Smola, Alexander, and Shravan Narayanamurthy. 2010. An Architecture for Parallel Topic Models. *Proceedings of the VLDB Endowment* 3 (1-2): 703–710. https://doi.org/10.14778/1920841.1920931.

Song, Xiang, Jian Yang, and Haibo Chen. 2014. Architecting Flash-based Solid-state Drive for High-performance I/O Virtualization. *IEEE Computer Architecture Letters* 13 (2): 61–64.

spdk. io. 2021. SPDK Hello World. https://github.com/spdk/spdk/blob/master/examples/nvme/hello_world/hello_world.c.

SQream. 2021. Bringing the Power of the GPU to the Era of Massive Data. https://sqream.com/product/data-acceleration-platform/sql-gpu-database/.

Stanford DAWN Team. 2021. DAWNBench: An End-to-End Deep Learning Benchmark and Competition. https://dawn.cs.stanford.edu/benchmark/.

Stanford Vision Lab. 2021. ImageNet. https://image-net.org/.

Sterling, Thomas, Ewing Lusk, and William Gropp. 2003. *Beowulf Cluster Computing with Linux*. Cambridge, MA: MIT Press.

Sterling, Thomas, Ewing Lusk, and William Gropp. 2003b. Beowulf Cluster Computing with Linux. In *Mit press*. Cambridge, MA.

Streamz. 2021. Real-time Stream Processing for Python. https://github.com/python-streamz/streamz.

Strukov, Dmitri B., Gregory S. Snider, Duncan R. Stewart, and R. Stanley Williams. 2008. The Missing Memristor Found. *Nature* 453 (7191): 80.

Stuart, Jeff A., and John D. Owens. 2011. Multi-GPU MapReduce on GPU Clusters. *2011 IEEE International Parallel and Distributed Processing Symposium*, 1068–1079.

Stuedi, Patrick, Animesh Trivedi, Jonas Pfefferle, Radu Stoica, Bernard Metzler, Nikolas Ioannou, and Ioannis Koltsidas. 2017. Crail: A High-performance I/O Architecture for Distributed Data Processing. *IEEE Data Engineering Bulletin* 40 (1): 38–49.

Su, Maomeng, Mingxing Zhang, Kang Chen, Zhenyu Guo, and Yongwei Wu. 2017. RFP: When RPC Is Faster than Server-bypass with RDMA. *Proceedings of the 12th European Conference on Computer Systems. Eurosys '17*, 1–15.

Suzuki, Yusuke, Shinpei Kato, Hiroshi Yamada, and Kenji Kono. 2014. GPUvm: Why Not Virtualizing GPUs at the Hypervisor? *Usenix Annual Technical Conference*, 109–120.

Szegedy, Christian, Wei Liu, Yangqing Jia, Pierre Sermanet, Scott Reed, Dragomir Anguelov, Dumitru Erhan, Vincent Vanhoucke, and Andrew Rabinovich. 2015. Going Deeper with Convolutions. *Proceedings of the IEEE Conference on Computer Vision and Pattern Recognition*, 1–9.

Szegedy, Christian, Sergey Ioffe, Vincent Vanhoucke, and Alex Alemi. 2016. Inception-v4, Inception-ResNet and the Impact of Residual Connections on Learning. *Proceedings of the Thirty-First AAAI Conference on Artificial Intelligence (AAAI '17)*. 4278–4284.

Tai, Kai Xin. 2020. Monitor Apache Flink with Datadog. https://www.datadoghq.com/blog/monitor-apache-flink-with-datadog/.

Taleb, Yacine, Ryan Stutsman, Gabriel Antoniu, and Toni Cortes. 2018. Tailwind: Fast and Atomic RDMA-based Replication. *2018 USENIX Annual Technical Conference (USENIX ATC 18)*, 851–863.

Talpey, Tom. 2015. Remote Access to Ultra-low-latency Storage. https://www.snia.org/sites/default/files/SDC15_presentations/persistant_mem/Talpey-Remote_Access_Storage.pdf.

Talpey, Tom. 2016. RDMA Extensions for Remote Persistent Memory Access. *12th Annual Open Fabrics Alliance Workshop*. https://www.openfabrics.org/images/eventpresos/2016presentations/215RDMAforRemPerMem.pdf.

Talpey, Tom. 2019. RDMA Persistent Memory Extensions. *15th Annual Open Fabrics Alliance Workshop*. https://www.openfabrics.org/wp-content/uploads/209_TTalpey.pdf.

Talpey, Tom, and Jim Pinkerton. 2016. Rdma Durable Write Commit. *Internet Engineering Task Force* (IETF) Internet-Draft, https://datatracker.ietf.org/doc/html/draft-talpey-rdma-commit-00.

Tang, Haodong, Jian Zhang, and Fred Zhang. 2018. Accelerating Ceph with RDMA and NVMe-oF. https://www.openfabrics.org/images/2018workshop/presentations/206_HTang_AcceleratingCephRDMANVMe-oF.pdf.

TensorFlow. 2021. TensorBoard: TensorFlow's Visualization Toolkit. https://www.tensorflow.org/tensorboard.

Thakur, Rajeev, William Gropp, and Ewing Lusk. 1998. A Case for Using MPI's Derived Datatypes to Improve I/O Performance. *Proceedings of the 1998 ACM/IEEE conference on Supercomputing (SC '98)*. IEEE Computer Society, 1–10. https://dl.acm.org/doi/fullHtml/10.5555/509058.509059.

Thaler, David, and Chinya V. Ravishankar. 1996. A Name-based Mapping Scheme for Rendezvous. Technical report CSE-TR-316-96, University of Michigan.

Thomson, Alexander, Thaddeus Diamond, Shu-Chun Weng, Kun Ren, Philip Shao, and Daniel J. Abadi. 2012. Calvin: Fast Distributed Transactions for Partitioned Database Systems. *Proceedings of the 2012 ACM SIGMOD International Conference on Management of Data (SIGMOD '12)*, 1–12.

Tian, Kun, Yaozu Dong, and David Cowperthwaite. 2014. A Full GPU Virtualization Solution with Mediated Pass-Through. *2014 USENIX Annual Technical Conference (USENIX ATC 14)*. 121–132. https://dl.acm.org/doi/10.5555/2643634.2643647.

Toon, Nigel. 2020. Introducing 2nd Generation IPU Systems for AI at Scale. https://www.graphcore.ai/posts/introducing-second-generation-ipusystems-for-ai-at-scale.

TOP500.org. TOP500 Supercomputing Sites. http://www.top500.org/.

TOP500.org. 2020. Highlights—November 2020. https://www.top500.org/lists/top500/2020/11/highs/.

Toshniwal, Ankit, Siddarth Taneja, Amit Shukla, Karthik Ramasamy, Jignesh M. Patel, Sanjeev Kulkarni, Jason Jackson, Krishna Gade, Maosong Fu, Jake Donham, et al. 2014. Storm@ Twitter. *Proceedings of the 2014 ACM SIGMOD International Conference on Management of Data, (SIGMOD '14). Association for Computing Machinery*, 147–156. DOI:https://doi.org/10.1145/2588555.2595641.

TPC-H Version 2 and Version 3. 2021. TPC-H Benchmark. http://www.tpc.org/tpch/.

Transaction Processing Performance Council. TPC—Homepage. http://www.tpc.org.

Tu, Stephen, Wenting Zheng, Eddie Kohler, Barbara Liskov, and Samuel Madden. 2013. Speedy Transactions in Multicore In-memory Databases. *Proceedings of the 24th ACM Symposium on Operating Systems Principles*, 18–32.

Tulapurkar, A. A., Y. Suzuki, A. Fukushima, H. Kubota, H. Maehara, K. Tsunekawa, D. D. Djayaprawira, N. Watanabe, and S. Yuasa. 2005. Spin-Torque Diode Effect in Magnetic Tunnel Junctions. *Nature* 438 (7066): 339.

Twitter. 2017. Fatcache: Memcache on SSD. https://github.com/twitter/fatcache.

Twitter. 2019. Twemcache: Twitter Memcached. https://github.com/twitter/twemcache.

Valiant, Leslie G. 1990. A Bridging Model for Parallel Computation. *Communications of the ACM* 33 (8): 103–111.

Vavilapalli, Vinod Kumar, Arun C. Murthy, Chris Douglas, Sharad Agarwal, Mahadev Konar, Robert Evans, Thomas Graves, Jason Lowe, Hitesh Shah, Siddharth Seth, et al. 2013. Apache Hadoop YARN: Yet Another Resource Negotiator. *Proceedings of the 4th Annual Symposium on Cloud Computing (SoCC)*, 5.

Vazhkudai, Sudharshan S., Bronis R. de Supinski, Arthur S. Bland, Al Geist, James Sexton, Jim Kahle, Christopher J. Zimmer, Scott Atchley, Sarp Oral, Don E. Maxwell, et al. 2018. The Design, Deployment, and Evaluation of the CORAL Pre-exascale Systems. *Proceedings of the International Conference for High Performance Computing, Networking, Storage, and Analysis. (SC '18)*, 52.

Verma, Abhishek, Luis Pedrosa, Madhukar R. Korupolu, David Oppenheimer, Eric Tune, and John Wilkes. 2015. Large-scale Cluster Management at Google with Borg. *Proceedings of the European Conference on Computer Systems (EuroSys '15)*. Association for Computing Machinery, New York, NY, USA, Article 18, 1–17. DOI:https://doi.org/10.1145/2741948.2741964.

Volos, Haris, Sanketh Nalli, Sankarlingam Panneerselvam, Venkatanathan Varadarajan, Prashant Saxena, and Michael M. Swift. 2014. Aerie: Flexible File-system Interfaces to Storage-class Memory. *Proceedings of the 9th European Conference on Computer Systems*, 1–14.

Wang, Chao, Lei Gong, Qi Yu, Xi Li, Yuan Xie, and Xuehai Zhou. 2016. DLAU: A Scalable Deep Learning Accelerator Unit on FPGA. *IEEE Transactions on Computer-Aided Design of Integrated Circuits and Systems* 36 (3): 513–517.

Wang, Cheng, Jianyu Jiang, Xusheng Chen, Ning Yi, and Heming Cui. 2017. APUS: Fast and Scalable Paxos on RDMA. SoCC '17: *Proceedings of the 2017 Symposium on Cloud Computing*, 94–107. https://doi.org/10.1145/3127479.3128609.

Wang, Guanhua, Shivaram Venkataraman, Amar Phanishayee, Nikhil Devanur, Jorgen Thelin, and Ion Stoica. 2020. Blink: Fast and Generic Collectives for Distributed ML. *Proceedings of Machine Learning and Systems*. 2: 172–186. https://proceedings.mlsys.org/paper/2020/file/43ec517d68b6edd3015b3edc9a11367b-Paper.pdf.

Wang, Kuang-Ching, James Griffioen, Ronald Hutchins, and Zongming Fei. 2020. Large Scale Networking (LSN) Workshop on Huge Data: A Computing, Networking and Distributed Systems Perspective. https://protocols.netlab.uky.edu/~hugedata2020/.

Wang, Lei, Jianfeng Zhan, Chunjie Luo, Yuqing Zhu, Qiang Yang, Yongqiang He, Wanling Gao, Zhen Jia, Yingjie Shi, Shujie Zhang, et al. 2014. BigDataBench: A Big Data Benchmark Suite from Internet Services. *2014 IEEE 20th International Symposium on High Performance Computer Architecture (HPCA)*, 488–499. https://doi.org/10.1109/HPCA.2014.6835958.

Wang, Peng, Guangyu Sun, Song Jiang, Jian Ouyang, Shiding Lin, Chen Zhang, and Jason Cong. 2014. An Efficient Design and Implementation of LSM-Tree Based Key-Value Store on Open-Channel SSD. *Proceedings of the 9th European Conference on Computer Systems*, 1–14.

Wang, Teng, Kathryn Mohror, Adam Moody, Weikuan Yu, and Kento Sato. 2015. BurstFS: A Distributed Burst Buffer File System for Scientific Applications. *Proceedings of the International Conference for High Performance Computing, Networking, Storage and Analysis, SC '16*. 807–818, doi: 10.1109/SC.2016.68.

Wang, Teng, S. Oral, Yandong Wang, B. Settlemyer, S. Atchley, and Weikuan Yu. 2014. BurstMem: A High-performance Burst Buffer System for Scientific Applications. *2014 IEEE International Conference on Big Data. (Big Data)*. 71–79. IEEE.

Wang, Yandong, Li Zhang, Jian Tan, Min Li, Yuqing Gao, Xavier Guerin, Xiaoqiao Meng, and Shicong Meng. 2015. HydraDB: A Resilient RDMA-driven Key-Value Middleware for In-memory Cluster Computing. *Proceedings of the International Conference for High Performance Computing, Networking, Storage and Analysis. (SC '15)*. *Association for Computing Machinery, New York, NY, USA, Article 22*, 1–11. DOI:https://doi.org/10.1145/2807591.2807614.

Wang, Yandong, Xiaoqiao Meng, Li Zhang, and Jian Tan. 2014. C-Hint: An Effective and Reliable Cache Management for RDMA-accelerated Key-Value Stores. *Proceedings of the ACM Symposium on Cloud Computing*, 1–13.

Wang, Yandong, Xinyu Que, Weikuan Yu, Dror Goldenberg, and Dhiraj Sehgal. 2011. Hadoop Acceleration through Network Levitated Merge. SC '11: *Proceedings of International Conference for High Performance Computing, Networking, Storage and Analysis*, 1–10.

Wang, Yiheng, Xin Qiu, Ding Ding, Yao Zhang, Yanzhang Wang, Xianyan Jia, Yan Wan, Zhichao Li, Jiao Wang, Shengsheng Huang, et al. 2018. BigDL: A Distributed Deep Learning Framework for Big Data. *arXiv* preprint. arXiv:1804.05839.

Wei, Q., M. Xue, J. Yang, C. Wang, and C. Cheng. 2015. Accelerating Cloud Storage System with Byte-addressable Non-volatile Memory. *2015 IEEE 21st International Conference on Parallel and Distributed Systems (ICPADS)*, 354–361.

Wei, Xingda, Jiaxin Shi, Yanzhe Chen, Rong Chen, and Haibo Chen. 2015. Fast In-memory Transaction Processing Using RDMA and HTM. *Proceedings of the 25th Symposium on Operating Systems Principles*, 87–104. https://doi.org/10.1145/2815400.2815419.

Welsh, Matt, David Culler, and Eric Brewer. 2001. SEDA: An Architecture for Well-conditioned, Scalable Internet Services. *Proceedings of the 18th ACM Symposium on Operating Systems Principles (SOSP '01). Association for Computing Machinery, New York, NY, USA*, 230–243. DOI:https://doi.org/10.1145/502034.502057.

Wong, H. S., S. Raoux, S. Kim, J. Liang, J. P. Reifenberg, B. Rajendran, M. Asheghi, and K. E. Goodson. 2010. Phase Change Memory. *Proceedings of the IEEE* 98 (12): 2201–2227. https://doi.org/10.1109/JPROC.2010.2070050.

Wu, Xiaojian, and A. L. Reddy. SCMFS: A File System for Storage Class Memory. In *Sc'11*, 39.

Wu, Xiaojian, and A. L. Narasimha Reddy. 2011. SCMFS: A File System for Storage Class Memory *and its Extensions. ACM Trans. Storage 9, 3, Article 7 (2013)*, 23 pages. https://doi.org/10.1145/2501620.2501621.

Xia, Fei, Dejun Jiang, Jin Xiong, and Ninghui Sun. 2017. HiKV: A Hybrid Index Key-Value Store for DRAM-NVM Memory Systems. *2017 USENIX Annual Technical Conference (USENIX ATC 17)*, 349–362.

Xilinx. 2021. FPGA Leadership across Multiple Process Nodes. https://www.xilinx.com/products/silicon-devices /fpga.html.

Xilinx, AMD. 2021. Field Programmable Gate Array: What Is an FPGA?. https://www.xilinx.com/products /silicon-devices/fpga/what-is-an-fpga.html.

X IO. 2021. Axellio Edge Computing Systems, from XIO Technologies. https://nvmexpress.org/portfolio-items /axellio-super-io-platform-from-xio-technologies/.

XLA. 2021. XLA: Optimizing Compiler for Machine Learning. https://www.tensorflow.org/xla.

Xu, Jian, and Steven Swanson. 2016. NOVA: A Log-structured File System for Hybrid Volatile/Non-volatile Main Memories. *14th USENIX Conference on File and Storage Technologies (FAST 16)*, 323–338.

Xu, Qiumin, Huzefa Siyamwala, Mrinmoy Ghosh, Tameesh Suri, Manu Awasthi, Zvika Guz, Anahita Shayesteh, and Vijay Balakrishnan. 2015. Performance Analysis of NVMe SSDs and Their Implication on Real World Databases. *Proceedings of the 8th ACM International Systems and Storage Conference*, 6.

Xu, Yuehai, Eitan Frachtenberg, Song Jiang, and Mike Paleczny. 2014. Characterizing Facebook's Memcached Workload. *IEEE Internet Computing* 18 (2): 41–49.

Yahoo. 2018. CaffeOnSpark: Distributed Deep Learning on Hadoop and Spark Clusters. https://github.com/yahoo /CaffeOnSpark.

Yahoo. 2021. TensorFlowOnSpark Brings Scalable Deep Learning to Apache Hadoop and Apache Spark Clusters. https://github.com/yahoo/TensorFlowOnSpark.

Yahoo. 2021. Webscope Datasets. https://webscope.sandbox.yahoo.com/catalog.php.

Yang, Carl, Aydin Buluç, and John D. Owens. 2019. GraphBLAST: A High-performance Linear Algebra-based Graph Framework on the GPU. *arXiv/CoRR* abs/1908.01407. http://arxiv.org/abs/1908.01407.

Yang, Jian, Joseph Izraelevitz, and Steven Swanson. 2019. Orion: A Distributed File System for Non-volatile Main Memory and RDMA-capable Networks. *17th USENIX Conference on File and Storage Technologies (FAST 19)*, 221–234. https://dl.acm.org/doi/10.5555/3323298.3323319.

Yang, Ziye, Luse E. Paul, James R. Harris, Benjamin Walker, Daniel Verkamp, Changpeng Liu, Cunyin Chang, Gang Cao, Jonathan Stern, and Vishal Verma. 2017. SPDK: A Development Kit to Build High Performance Storage Applications. *2017 IEEE International Conference on Cloud Computing Technology and Science (CloudCom)*, 154–161.

Yoshimura, Takeshi, Tatsuhiro Chiba, and Hiroshi Horii. 2019. EvFS: User-level, Event-driven File System for Non-volatile Memory. *11th USENIX Workshop on Hot Topics in Storage and File Systems (HotStorage 19)*. https://dl.acm.org/doi/10.5555/3357062.3357083.

Yuan, Yuan, Rubao Lee, and Xiaodong Zhang. 2013. The Yin and Yang of Processing Data Warehousing Queries on GPU Devices. *Proceedings of the VLDB Endowment* 6 (10): 817–828. https://doi.org/10.14778/2536206.2536210.

Yuan, Yuan, Meisam Fathi Salmi, Yin Huai, Kaibo Wang, Rubao Lee, and Xiaodong Zhang. 2016. Spark-GPU: An Accelerated In-memory Data Processing Engine on Clusters. *2016 IEEE International Conference on Big Data (Big Data)*, 273–283.

Zadok, Erez, Dean Hildebrand, Geoff Kuenning, and Keith A. Smith. 2017. POSIX Is Dead! Long Live ... errr ... What Exactly? Proceedings of the 9th USENIX Workshop on Hot Topics in Storage and File Systems *(HotStorage '17)*, 12–12.

Zaharia, Matei, Mosharaf Chowdhury, Michael J. Franklin, Scott Shenker, and Ion Stoica. 2010. Spark: Cluster Computing with Working Sets. *Proceedings of the 2nd USENIX Conference on Hot Topics in Cloud Computing. (HotStorage '17).* USENIX Association. https://dl.acm.org/doi/10.5555/1863103.1863113. Boston, MA.

Zaharia, Matei, Mosharaf Chowdhury, Tathagata Das, Ankur Dave, Justin Ma, Murphy McCauly, Michael J. Franklin, Scott Shenker, and Ion Stoica. 2012. Resilient Distributed Datasets: A Fault-Tolerant Abstraction for In-memory Cluster Computing. *9th USENIX Symposium on Networked Systems Design and Implementation (NSDI 12)*, 15–28.

Zaharia, Matei, Reynold S. Xin, Patrick Wendell, Tathagata Das, Michael Armbrust, Ankur Dave, Xiangrui Meng, Josh Rosen, Shivaram Venkataraman, Michael J. Franklin, et al. 2016. Apache Spark: A Unified Engine for Big Data Processing. *Communications of the ACM* 59 (11): 56–65.

Zaharia, Matei, Tathagata Das, Haoyuan Li, Timothy Hunter, Scott Shenker, and Ion Stoica. 2013. Discretized Streams: Fault-tolerant Streaming Computation at Scale. *Proceedings of the 24th ACM Symposium on Operating Systems Principles*, 423–438.

Zamanian, Erfan, Xiangyao Yu, Michael Stonebraker, and Tim Kraska. 2019. Rethinking Database High Availability with RDMA Networks. *Proceedings of the VLDB Endowment* 12 (11): 1637–1650. https://doi.org/10.14778/3342263.3342639.

Zhang, Chen, Peng Li, Guangyu Sun, Yijin Guan, Bingjun Xiao, and Jason Cong. 2015. Optimizing FPGA-based Accelerator Design for Deep Convolutional Neural Networks. *Proceedings of the 2015 ACM/SIGDA International Symposium on Field-programmable Gate Arrays*, 161–170.

Zhang, Jie, David Donofrio, John Shalf, Mahmut T. Kandemir, and Myoungsoo Jung. 2015. NVMMU: A Non-volatile Memory Management Unit for Heterogeneous GPU-SSD Architectures. *2015 International Conference on Parallel Architecture and Compilation (PACT '15)*, 13–24.

Zhang, Jie, Xiaoyi Lu, and Dhabaleswar K. Panda. 2017. High-performance Virtual Machine Migration Framework for MPI Applications on SR-IOV Enabled InfiniBand Clusters. *2017 IEEE International Parallel and Distributed Processing Symposium (IPDPS)*, 143–152.

Zhang, Jie, Xiaoyi Lu, Ching-Hsiang Chu, and Dhabaleswar K. Panda. 2019. C-GDR: High-performance Container-aware GPUDirect MPI Communication Schemes on RDMA Networks. *2019 IEEE International Parallel and Distributed Processing Symposium (IPDPS)*, 242–251.

Zhang, Jie, Xiaoyi Lu, Jithin Jose, Mingzhe Li, Rong Shi, and Dhabaleswar K. (DK) Panda. 2014. High Performance MPI Library over SR-IOV Enabled InfiniBand Clusters. *2014 21st International Conference on High Performance Computing (HIPC)*, 1–10.

Zhang, Kai, Kaibo Wang, Yuan Yuan, Lei Guo, Rubao Lee, and Xiaodong Zhang. 2015. Mega-KV: A Case for GPUs to Maximize the Throughput of In-memory Key-Value Stores. *Proceedings of the VLDB Endowment* 8 (11): 1226–1237. https://doi.org/10.14778/2809974.2809984.

Zhang, Kaiyuan, Rong Chen, and Haibo Chen. 2015. NUMA-aware Graph-structured Analytics. *Proceedings of the 20th ACM SIGPLAN Symposium on Principles and Practice of Parallel Programming (PPOPP 2015)*, 183–193. https://doi.org/10.1145/2688500.2688507.

Zhang, K., J. Hu, B. He, and B. Hua. 2017. DIDO: Dynamic Pipelines for In-memory Key-Value Stores on Coupled CPU-GPU Architectures. 2017 IEEE 33rd International Conference on Data Engineering (ICDE), 671–682.

Zhang, X., X. Zhou, M. Lin, and J. Sun. 2018. ShuffleNet: An Extremely Efficient Convolutional Neural Network for Mobile Devices. *2018 IEEE/CVF Conference on Computer Vision and Pattern Recognition*, 6848–6856. https://doi.org/10.1109/CVPR.2018.00716.

Zhang, Yiying, Jian Yang, Amirsaman Memaripour, and Steven Swanson. 2015. Mojim: A Reliable and Highly-available Non-volatile Memory System. *ACM SIGARCH Computer Architecture News* 43: 3–18.

Zhang, Yunming, Mengjiao Yang, Riyadh Baghdadi, Shoaib Kamil, Julian Shun, and Saman Amarasinghe. 2018. GraphIt: A High-performance DSL for Graph Analytics. *arXiv* preprint. arXiv:1805.00923.

Zhang, Yunming, Vladimir Kiriansky, Charith Mendis, Saman Amarasinghe, and Matei Zaharia. 2017. Making Caches Work for Graph Analytics. *2017 IEEE International Conference on Big Data (Big Data)*, 293–302.

Zheng, Chao, Lukas Rupprecht, Vasily Tarasov, Douglas Thain, Mohamed Mohamed, Dimitrios Skourtis, Amit S. Warke, and Dean Hildebrand. 2018. Wharf: Sharing Docker Images in a Distributed File System. *Proceedings of the ACM Symposium on Cloud Computing (SOCC '18)*, 174–185. https://doi.org/10.1145/3267809.3267836.

Zheng, Shengan, Linpeng Huang, Hao Liu, Linzhu Wu, and Jin Zha. 2016. HMVFS: A Hybrid Memory Versioning File System. *2016 32nd Symposium on Mass Storage Systems and Technologies (MSST)*, 1–14.

Zheng, Shengan, Morteza Hoseinzadeh, and Steven Swanson. 2019. Ziggurat: A Tiered File System for Non-Volatile Main Memories and Disks. *17th USENIX Conference on File and Storage Technologies (FAST 19)*, 207–219.

Zhou, Jingren, and Kenneth A. Ross. 2002. Implementing Database Operations Using SIMD Instructions. *Proceedings of the 2002 ACM SIGMOD International Conference on Management of Data*, 145–156.

Zhou, Shijie, Rajgopal Kannan, Viktor K. Prasanna, Guna Seetharaman, and Qing Wu. 2019. HitGraph: High-throughput Graph Processing Framework on FPGA. *IEEE Transactions on Parallel and Distributed Systems* 30 (10): 2249–2264. 10.1109/TPDS.2019.2910068.

Zhu, Xiaowei, Wentao Han, and Wenguang Chen. 2015. GridGraph: Large-scale Graph Processing on a Single Machine Using 2-Level Hierarchical Partitioning. *2015 USENIX Annual Technical Conference (USENIX ATC 15)*, 375–386. https://www.usenix.org/conference/atc15/technical-session/presentation/zhu.

Zinkevich, Martin, Markus Weimer, Lihong Li, and Alex Smola. 2010. Parallelized Stochastic Gradient Descent. *Advances in Neural Information Processing Systems (nips)*, eds. J. Lafferty, C. Williams, J. Shawe-Taylor, R. Zemel, and A. Culotta, *vol. 2 (NIPS '10)*. Curran Associates Inc., 2595–2603. https://dl.acm.org/doi/10.5555/2997046.2997185.

Index